Microbiology: An Introduction

Microbiology: An Introduction

Edited by
Andre Douglas

Larsen & Keller
www.larsen-keller.com

Microbiology: An Introduction
Edited by Andre Douglas
ISBN: 978-1-63549-184-5 (Hardback)

目 Larsen & Keller

Published by Larsen and Keller Education,
5 Penn Plaza,
19th Floor,
New York, NY 10001, USA

Cataloging-in-Publication Data

Microbiology : an introduction / edited by Andre Douglas.
      p. cm.
   Includes bibliographical references and index.
   ISBN 978-1-63549-184-5
   1. Microbiology. 2. Microorganisms. 3. Biology.
I. Douglas, Andre.
QR41.2 .M53 2017
579--dc23

The publisher's policy is to use permanent paper from mills that operate a sustainable forestry policy. Furthermore, the publisher ensures that the text paper and cover boards used have met acceptable environmental accreditation standards.

Printed and bound in the United States of America.

For more information regarding Larsen and Keller Education and its products, please visit the publisher's website www.larsen-keller.com

Table of Contents

Preface

Microbiology is a part of biology that studies the structure, functions and development of microscopic organisms such as acellular, multicellular and unicellular organisms. The field is further divided into two groups, pure and applied microbiology. This book contains chapters on both of these parts and their sub-divisions as well as their applications and essential theories. This book outlines the processes and applications of the subject in detail. It is a compilation of chapters that discuss the most vital concepts of this field. This book, with its detailed analyses and data, will prove immensely beneficial to students involved in this area at various levels. For all those who are interested in microbiology, this textbook can prove to be an essential guide.

A foreword of all Chapters of the book is provided below:

Chapter 1 - Microbiology studies minute organisms; they can only be studied under the microscope. These organisms are multicellular, unicellular or acellular. Unicellular organisms have a single cell, multicellular organisms have a cell colony and acellular organisms are the ones that lack cells. This chapter will provide an integrated understanding of microbiology; **Chapter 2** - Microorganisms are single-celled or multicellular organisms. The study of these organisms is called microbiology. The organisms discussed within this text are prokaryote, bacteria, eukaryote, protist and fungus. The section is an overview of the subject matter incorporating all the major aspects of microorganisms; **Chapter 3** - Some of the essential topics of microbiology are sterilization, endosymbiont, microbiota, microbial cyst, colony –forming unit and viral eukaryogenesis. Sterilization refers to the process that eliminates biological agents and viruses. Sterilization is majorly used to sanitize and pasteurize, and is usually achieved by using chemicals, irradiation and filtration. The section serves as a source to understand the major concepts related to microbiology; **Chapter 4** - The theories of microbiology elucidated in this section are germ theory of disease, fermentation theory, hologenome theory of evolution and germ theory denialism. The germ theory of disease claims that some microorganisms cause diseases in organisms. The growth of these microorganisms in their hosts causes diseases. The following chapter will not only provide an overview, it will also delve into the topics related to it; **Chapter 5** - Methods and techniques are important components of any field of study. Bacteriological water analysis is a method that is usually used to analyze water in order to estimate the number of bacteria present in the water. Microbiological culture is the method applied on organisms to multiply them and then they are used as a tool to determine the cause of an infectious disease. This text helps the reader in developing an understanding of the methods and techniques that are used in microbiology; **Chapter 6** - The subdisciplines explained in this text are immunology, virology, microbial ecology, mycology, microbial genetics etc. The branch of biology that covers the analyses of immune systems is known as immunology whereas virology is the study of viruses. This section is a compilation of the various subdisciplines of microbiology that forms an integral part of the broader subject matter; **Chapter 7** - Humans use industrial fermentation to make products useful. Algae fuel is used as a substitute to liquid fossil fuels. A number of companies and governments have been trying to reduce the operating cost to make algae fuel production commercially feasible. The aspects elucidated in this chapter are of vital importance, and provides a better understanding of microbiology.

I would like to thank the entire editorial team who made sincere efforts for this book and my family who supported me in my efforts of working on this book. I take this opportunity to thank all those who have been a guiding force throughout my life.

Editor

Introduction to Microbiology

Microbiology studies minute organisms; they can only be studied under the microscope. These organisms are multicellular, unicellular or acellular. Unicellular organisms have a single cell, multicellular organisms have a cell colony and acellular organisms are the ones that lack cells. This chapter will provide an integrated understanding of microbiology.

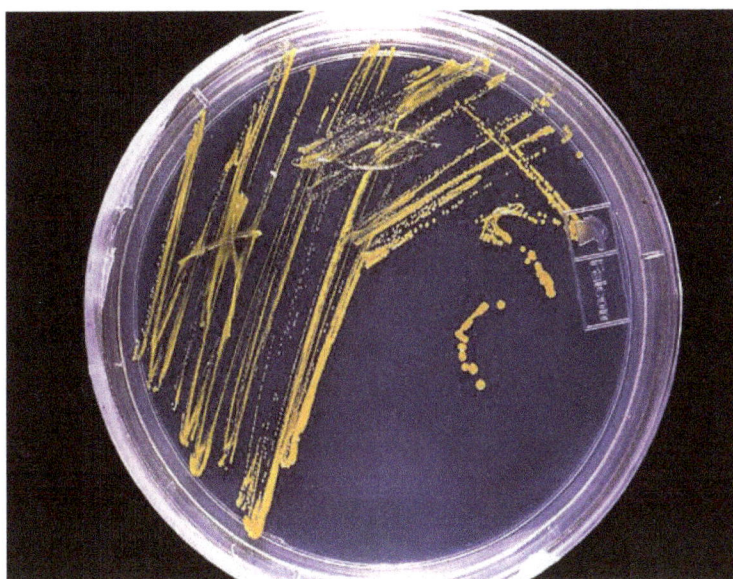

An agar plate streaked with microorganisms

Microbiology is the study of microscopic organisms, those being unicellular (single cell), multicellular (cell colony), or acellular (lacking cells). Microbiology encompasses numerous sub-disciplines including virology, mycology, parasitology, and bacteriology.

Eukaryotic micro-organisms possess membrane-bound cell organelles and include fungi and protists, whereas prokaryotic organisms—all of which are microorganisms—are conventionally classified as lacking membrane-bound organelles and include eubacteria and archaebacteria. Microbiologists traditionally relied on culture, staining, and microscopy. However, less than 1% of the microorganisms present in common environments can be cultured in isolation using current means. Microbiologists often rely on extraction or detection of nucleic acid, either DNA or RNA sequences.

Viruses have been variably classified as organisms, as they have been considered either as very simple microorganisms or very complex molecules. Prions, never considered microorganisms, have been investigated by virologists, however, as the clinical effects traced to them were originally presumed due to chronic viral infections, and virologists took search—discovering "infectious proteins".

As an application of microbiology, medical microbiology is often introduced with medical principles of immunology as *microbiology and immunology*. Otherwise, microbiology, virology, and immunology as basic sciences have greatly exceeded the medical variants, applied sciences.

Branches

The branches of microbiology can be classified into pure and applied sciences. Microbiology can be also classified based on taxonomy, in the cases of bacteriology, mycology, protozoology, and phycology. There is considerable overlap between the specific branches of microbiology with each other and with other disciplines, and certain aspects of these branches can extend beyond the traditional scope of microbiology.

Pure Microbiology

- Bacteriology: The study of bacteria.

- Mycology: The study of fungi.

- Protozoology: The study of protozoa.

- Phycology/algology: The study of algae.

- Parasitology: The study of parasites.

- Immunology: The study of the immune system.

- Virology: The study of viruses.

- Nematology: The study of nematodes.

- Microbial cytology: The study of microscopic and submicroscopic details of microorganisms.

- Microbial physiology: The study of how the microbial cell functions biochemically. Includes the study of microbial growth, microbial metabolism and microbial cell structure.

- Microbial ecology: The relationship between microorganisms and their environment.

- Microbial genetics: The study of how genes are organized and regulated in microbes in relation to their cellular functions. Closely related to the field of molecular biology.

- Cellular microbiology: A discipline bridging microbiology and cell biology.

- Evolutionary microbiology: The study of the evolution of microbes. This field can be subdivided into:

  o Microbial taxonomy: The naming and classification of microorganisms.

  o Microbial systematic: The study of the diversity and genetic relationship of microorganisms.

- Generation microbiology: The study of those microorganisms that have the same characters as their parents.

- Systems microbiology: A discipline bridging systems biology and microbiology.

- Molecular microbiology: The study of the molecular principles of the physiological processes in microorganisms.

Other

- Nano microbiology: The study of those organisms on nano level.

- Exo microbiology (or Astro microbiology): The study of microorganisms in outer space

- Biological agent: The study of those microorganisms which are being used in weapon industries.

- Predictive microbiology: The quantification of relations between controlling factors in foods and responses of pathogenic and spoilage microorganisms using mathematical modelling

Applied Microbiology

- Medical microbiology: The study of the pathogenic microbes and the role of microbes in human illness. Includes the study of microbial pathogenesis and epidemiology and is related to the study of disease pathology and immunology. This area of microbiology also covers the study of human microbiota, cancer, and the tumor microenvironment.

- Pharmaceutical microbiology: The study of microorganisms that are related to the production of antibiotics, enzymes, vitamins,vaccines, and other pharmaceutical products and that cause pharmaceutical contamination and spoil.

- Industrial microbiology: The exploitation of microbes for use in industrial processes. Examples include industrial fermentation and wastewater treatment. Closely linked to the biotechnology industry. This field also includes brewing, an important application of microbiology.

- Microbial biotechnology: The manipulation of microorganisms at the genetic and molecular level to generate useful products.

- Food microbiology: The study of microorganisms causing food spoilage and foodborne illness. Using microorganisms to produce foods, for example by fermentation.

- Agricultural microbiology: The study of agriculturally relevant microorganisms. This field can be further classified into the following:

  o Plant microbiology and Plant pathology: The study of the interactions between microorganisms and plants and plant pathogens.

  o Soil microbiology: The study of those microorganisms that are found in soil.

- Veterinary microbiology: The study of the role of microbes in veterinary medicine or animal taxonomy.

Food microbiology laboratory at the Faculty of Food Technology, Latvia University of Agriculture.

- Environmental microbiology: The study of the function and diversity of microbes in their natural environments. This involves the characterization of key bacterial habitats such as the rhizosphere and phyllosphere, soil and groundwater ecosystems, open oceans or extreme environments (extremophiles). This field includes other branches of microbiology such as:

    o Microbial ecology

    o Microbially mediated nutrient cycling

    o Geomicrobiology

    o Microbial diversity

    o Bioremediation

- Water microbiology (or Aquatic microbiology): The study of those microorganisms that are found in water.

- Aeromicrobiology (or Air microbiology): The study of airborne microorganisms.

Benefits

Fermenting tanks with yeast being used to brew beer

While some fear microbes due to the association of some microbes with various human illnesses, many microbes are also responsible for numerous beneficial processes such as industrial fermentation (e.g. the production of alcohol, vinegar and dairy products), antibiotic production and as vehicles for cloning in more complex organisms such as plants. Scientists have also exploited their knowledge of microbes to produce biotechnologically important enzymes such as Taq polymerase, reporter genes for use in other genetic systems and novel molecular biology techniques such as the yeast two-hybrid system.

Bacteria can be used for the industrial production of amino acids. *Corynebacterium glutamicum* is one of the most important bacterial species with an annual production of more than two million tons of amino acids, mainly L-glutamate and L-lysine. Since some bacteria have the ability to synthesize antibiotics, they are used for medicinal purposes, such as Streptomyces to make aminoglycoside antibiotics.

A variety of biopolymers, such as polysaccharides, polyesters, and polyamides, are produced by microorganisms. Microorganisms are used for the biotechnological production of biopolymers with tailored properties suitable for high-value medical application such as tissue engineering and drug delivery. Microorganisms are used for the biosynthesis of xanthan, alginate, cellulose, cyanophycin, poly(gamma-glutamic acid), levan, hyaluronic acid, organic acids, oligosaccharides and polysaccharide, and polyhydroxyalkanoates.

Microorganisms are beneficial for microbial biodegradation or bioremediation of domestic, agricultural and industrial wastes and subsurface pollution in soils, sediments and marine environments. The ability of each microorganism to degrade toxic waste depends on the nature of each contaminant. Since sites typically have multiple pollutant types, the most effective approach to microbial biodegradation is to use a mixture of bacterial and fungal species and strains, each specific to the biodegradation of one or more types of contaminants.

Symbiotic microbial communities are known to confer various benefits to their human and animal hosts health including aiding digestion, production of beneficial vitamins and amino acids, and suppression of pathogenic microbes. Some benefit may be conferred by consuming fermented foods, probiotics (bacteria potentially beneficial to the digestive system) and/or prebiotics (substances consumed to promote the growth of probiotic microorganisms). The ways the microbiome influences human and animal health, as well as methods to influence the microbiome are active areas of research.

Research has suggested that microorganisms could be useful in the treatment of cancer. Various strains of non-pathogenic clostridia can infiltrate and replicate within solid tumors. Clostridial vectors can be safely administered and their potential to deliver therapeutic proteins has been demonstrated in a variety of preclinical models.

History

Ancient Times

The existence of microorganisms was hypothesized for many centuries before their actual discovery. The existence of unseen microbiological life was postulated by Jainism which is based on Mahavira's teachings as early as 6th century BCE. Paul Dundas notes that Mahavira asserted existence of unseen microbiological creatures living in earth, water, air and fire. Jain scriptures also describe nigodas which are sub-microscopic creatures living in large clusters and having a very short life and are said to pervade each and every part of the universe, even in tissues of plants and flesh of animals. The Roman Marcus Terentius Varro made references to microbes when he warned against locating a homestead in the vicinity of swamps "because there are bred certain minute creatures which cannot be seen by the eyes, which float in the air and enter the body through the mouth and nose and there by cause serious diseases."

Medieval Islamic World

Avicenna "ibn Sina"

At the golden age of Islamic civilization, some scientists had knowledge about microorganisms, such as Ibn Sina in his book *The Canon of Medicine*, Ibn Zuhr (also known as Avenzoar) who discovered scabies mites, and Al-Razi who gave the earliest known description of smallpox in his book *The Virtuous Life* (al-Hawi).

In 1546, Girolamo Fracastoro proposed that epidemic diseases were caused by transferable seedlike entities that could transmit infection by direct or indirect contact, or vehicle transmission.

However, early claims about the existence of microorganisms were speculative, and not based on microscopic observation. Actual observation and discovery of microbes had to await the invention of the microscope in the 17th century.

Modern Times

In 1676, Anton van Leeuwenhoek, who lived most of his life in Delft, Holland, observed bacteria and other microorganisms using a single-lens microscope of his own design. While Van Leeuwenhoek is often cited as the first to observe microbes, Robert Hooke made the first recorded microscopic observation, of the fruiting bodies of moulds, in 1665. It has, however, been suggested that a Jesuit priest called Athanasius Kircher was the first to observe micro-organisms.

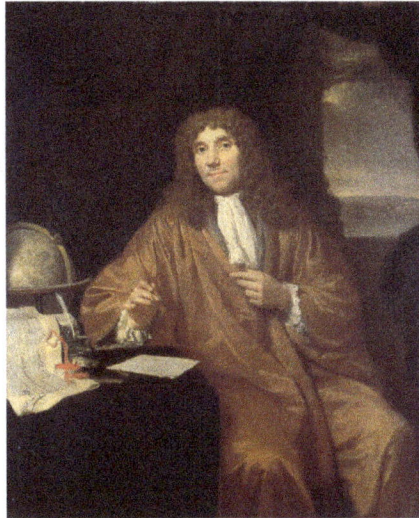

Antonie van Leeuwenhoek, is considered to be the one of the first to observe microorganisms using a microscope.

He was among the first to design magic lanterns for projection purposes, so he must have been well acquainted with the properties of lenses. One of his books contains a chapter in Latin, which reads in translation – "Concerning the wonderful structure of things in nature, investigated by Microscope." Here, he wrote "who would believe that vinegar and milk abound with an innumerable multitude of worms." He also noted that putrid material is full of innumerable creeping animalcule. These observations antedate Robert Hooke's *Micrographia* by nearly 20 years and were published some 29 years before van Leeuwenhoek saw protozoa and 37 years before he described having seen bacteria. Joseph Lister was the first person who said infectious diseases are caused by micro-organism and was first person who used phenol as disinfectant on the open wounds of patients.

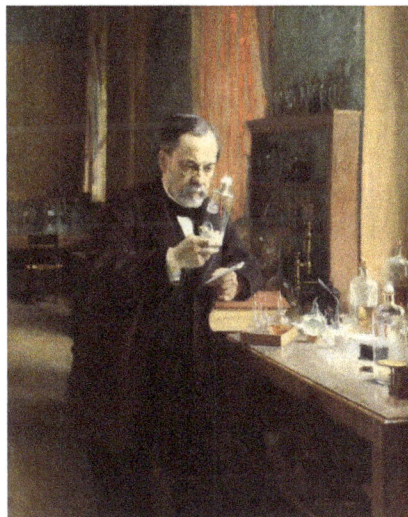

Innovative laboratory glassware and experimental methods developed by Louis Pasteur and other biologists contributed to the young field of bacteriology in the late 19th century.

The field of bacteriology (later a subdiscipline of microbiology) was founded in the 19th century by Ferdinand Cohn, a botanist whose studies on algae and photosynthetic bacteria led him to describe several bacteria including *Bacillus* and *Beggiatoa*. Cohn was also the first to formulate a scheme for the taxonomic classification of bacteria and discover spores. Louis Pasteur and Robert Koch were contemporaries of Cohn's and are often considered to be the father of microbiology and medical microbiology, respectively. Pasteur is most famous for his series of experiments designed to disprove the then widely held theory of spontaneous generation, thereby solidifying microbiology's identity as a biological science. Pasteur also designed methods for food preservation (pasteurization) and vaccines against several diseases such as anthrax, fowl cholera and rabies. Koch is best known for his contributions to the germ theory of disease, proving that specific diseases were caused by specific pathogenic micro-organisms. He developed a series of criteria that have become known as the Koch's postulates. Koch was one of the first scientists to focus on the isolation of bacteria in pure culture resulting in his description of several novel bacteria including *Mycobacterium tuberculosis*, the causative agent of tuberculosis.

While Pasteur and Koch are often considered the founders of microbiology, their work did not accurately reflect the true diversity of the microbial world because of their exclusive focus on micro-organisms having direct medical relevance. It was not until the late 19th century and the work of Martinus Beijerinck and Sergei Winogradsky, the founders of general microbiology (an older term encompassing aspects of microbial physiology, diversity and ecology), that the true breadth of microbiology was revealed. Beijerinck made two major contributions to microbiology: the discovery of viruses and the development of enrichment culture techniques. While his work on the Tobacco Mosaic Virus established the basic principles of virology, it was his development of enrichment culturing that had the most immediate impact on microbiology by allowing for the cultivation of a wide range of microbes with wildly different physiologies. Winogradsky was the first to develop the concept of chemolithotrophy and to thereby reveal the essential role played by micro-organisms in geochemical processes. He was responsible for the first isolation and description of both nitrifying and nitrogen-fixing bacteria. French-Canadian microbiologist Felix d'Herelle co-discovered bacteriophages and was one of the earliest applied microbiologists.

References

- Madigan M, Martinko J (editors) (2006). Brock Biology of Microorganisms (13th ed.). Pearson Education. p. 1096. ISBN 0-321-73551-X.

- Symbiosis : An Introduction to Biological Associations: An Introduction to Biological Associations (2nd ed.). New York: Oxford University Press. 2000. p. 10. ISBN 9780198027881. Retrieved 2016-03-25.

- Pharmaceutical Microbiology Principles and Applications. Nirali Prakashan. pp. 1.1–1.2. ISBN 978-81-85790-61-9. Retrieved 18 June 2011.

- Burkovski A (editor). (2008). Corynebacteria: Genomics and Molecular Biology. Caister Academic Press. ISBN 1-904455-30-1. Retrieved 2016-03-25.

- Rehm BHA (editor). (2008). Microbial Production of Biopolymers and Polymer Precursors: Applications and Perspectives. Caister Academic Press. ISBN 978-1-904455-36-3. Retrieved 2016-03-25.

- Diaz E (editor). (2008). Microbial Biodegradation: Genomics and Molecular Biology (1st ed.). Caister Academic Press. ISBN 1-904455-17-4. Retrieved 2016-03-25.

- Tannock GW (editor). (2005). Probiotics and Prebiotics: Scientific Aspects. Caister Academic Press. ISBN 978-1-904455-01-1. Retrieved 2016-03-25.

- Mengesha; et al. (2009). "Clostridia in Anti-tumor Therapy". Clostridia: Molecular Biology in the Post-genomic Era. Caister Academic Press. ISBN 978-1-904455-38-7.

- Mahavira is dated 599 BC - 527 BC. See. Dundas, Paul; John Hinnels ed. (2002). The Jain. London: Routledge. ISBN 0-415-26606-8. p. 24

- Jaini, Padmanabh (1998). The Jaina Path of Purification. New Delhi: Motilal Banarsidass. p. 109. ISBN 81-208-1578-5.

- Wainwright, Milton (2003). "An Alternative View of the Early History of Microbiology". Advances in Applied Microbiology. Advances in Applied Microbiology. 52: 333–55. doi:10.1016/S0065-2164(03)01013-X. ISBN 978-0-12-002654-8. PMID 12964250.

- Johnson J (2001) [1998]. "Martinus Willem Beijerinck". APSnet. American Phytopathological Society. Archived from the original on 2010-06-20. Retrieved May 2, 2010. Retrieved from Internet Archive January 12, 2014.

2

Microorganism: An Overview

Microorganisms are single-celled or multicellular organisms. The study of these organisms is called microbiology. The organisms discussed within this text are prokaryote, bacteria, eukaryote, protist and fungus. The section is an overview of the subject matter incorporating all the major aspects of microorganisms.

Microorganism

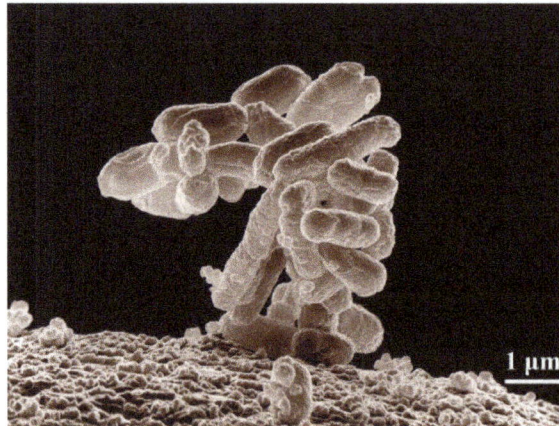

A cluster of *Escherichia coli* bacteria magnified 10,000 times

A microorganism or microbe is a microscopic living organism, which may be single-celled or multicellular. The study of microorganisms is called microbiology, a subject that began with the discovery of microorganisms in 1674 by Antonie van Leeuwenhoek, using a microscope of his own design.

Microorganisms are very diverse and include all bacteria, archaea and most protozoa. This group also contains some species of fungi, algae, and certain microscopic animals, such as rotifers. Many macroscopic animals and plants have microscopic juvenile stages. Some microbiologists also classify viruses (and viroids) as microorganisms, but others consider these as nonliving. In July 2016, scientists reported identifying a set of 355 genes from the last universal common ancestor of all life, including microorganisms, living on Earth.

Microorganisms live in every part of the biosphere, including soil, hot springs, "seven miles deep" in the ocean, "40 miles high" in the atmosphere and inside rocks far down within the Earth's crust. Microorganisms, under certain test conditions, have been observed to thrive in the vacuum of outer space. According to some estimates, microorganisms outweigh "all other living things combined thousands of times over". The mass of prokaryote microorganisms — which includes bacteria and archaea, but not the nucleated eukaryote microorganisms — may be as much as 0.8 trillion tons of carbon (of the total biosphere mass, estimated at between 1 and 4 trillion tons). On 17 March

2013, researchers reported data that suggested microbial life forms thrive in the Mariana Trench. the deepest spot in the Earth's oceans. Other researchers reported related studies that microorganisms thrive inside rocks up to 580 m (1,900 ft; 0.36 mi) below the sea floor under 2,590 m (8,500 ft; 1.61 mi) of ocean off the coast of the northwestern United States, as well as 2,400 m (7,900 ft; 1.5 mi) beneath the seabed off Japan. On 20 August 2014, scientists confirmed the existence of microorganisms living 800 m (2,600 ft; 0.50 mi) below the ice of Antarctica. According to one researcher, "You can find microbes everywhere — they're extremely adaptable to conditions, and survive wherever they are."

Microorganisms are crucial to nutrient recycling in ecosystems as they act as decomposers. As some microorganisms can fix nitrogen, they are a vital part of the nitrogen cycle, and recent studies indicate that airborne microorganisms may play a role in precipitation and weather. Microorganisms are also exploited in biotechnology, both in traditional food and beverage preparation, and in modern technologies based on genetic engineering. A small proportion of microorganisms are pathogenic, causing disease and even death in plants and animals.

Evolution

Single-celled microorganisms were the first forms of life to develop on Earth, approximately 3–4 billion years ago. Further evolution was slow, and for about 3 billion years in the Precambrian eon, all organisms were microscopic. So, for most of the history of life on Earth, the only forms of life were microorganisms. Bacteria, algae and fungi have been identified in amber that is 220 million years old, which shows that the morphology of microorganisms has changed little since the Triassic period. The newly discovered biological role played by nickel, however — especially that engendered by volcanic eruptions from the Siberian Traps (site of the modern city of Norilsk) — is thought to have accelerated the evolution of methanogens towards the end of the Permian–Triassic extinction event.

Microorganisms tend to have a relatively fast rate of evolution. Most microorganisms can reproduce rapidly, and bacteria are also able to freely exchange genes through conjugation, transformation and transduction, even between widely divergent species. This horizontal gene transfer, coupled with a high mutation rate and many other means of genetic variation, allows microorganisms to swiftly evolve (via natural selection) to survive in new environments and respond to environmental stresses. This rapid evolution is important in medicine, as it has led to the recent development of "super-bugs", pathogenic bacteria that are resistant to modern antibiotics.

Pre-microbiology

The possibility that microorganisms exist was discussed for many centuries before their discovery in the 17th century. The existence of unseen microbiological life was postulated by Jainism, which is based on Mahavira's teachings as early as 6th century BCE. Paul Dundas notes that Mahavira asserted the existence of unseen microbiological creatures living in earth, water, air and fire. The Jain scriptures also describe nigodas, which are sub-microscopic creatures living in large clusters and having a very short life, which are said to pervade every part of the universe, even the tissues of plants and animals. The earliest known idea to indicate the possibility of diseases spreading by yet unseen organisms was that of the Roman scholar Marcus Terentius Varro in a 1st-century BC book titled *On Agriculture* in which he warns against locating a homestead near swamps:

... and because there are bred certain minute creatures that cannot be seen by the eyes, which float in the air and enter the body through the mouth and nose and they cause serious diseases.

In *The Canon of Medicine* (1020), Abū Alī ibn Sīnā (Avicenna) hypothesized that tuberculosis and other diseases might be contagious.

In 1546, Girolamo Fracastoro proposed that epidemic diseases were caused by transferable seed-like entities that could transmit infection by direct or indirect contact, or even without contact over long distances.

All these early claims about the existence of microorganisms were speculative and while grounded on indirect observations, they were not based on direct observation of microorganisms or systematized empirical investigation, e.g. experimentation. Microorganisms were neither proven, observed, nor accurately described until the 17th century. The reason for this was that all these early studies lacked the microscope.

History of Discovery

Lazzaro Spallanzani showed that boiling a broth stopped it from decaying.

Robert Koch showed that microorganisms caused disease.

Antonie Van Leeuwenhoek (1632–1723) was one of the first people to observe microorganisms, using microscopes of his own design. Robert Hooke, a contemporary of Leeuwenhoek, also used microscopes to observe microbial life; his 1665 book *Micrographia* describes these observations and coined the term *cell*.

Before Leeuwenhoek's discovery of microorganisms in 1675, it had been a mystery why grapes could be turned into wine, milk into cheese, or why food would spoil. Leeuwenhoek did not make the connection between these processes and microorganisms, but using a microscope, he did establish that there were signs of life that were not visible to the naked eye. Leeuwenhoek's discovery, along with subsequent observations by Spallanzani and Pasteur, ended the long-held belief that life spontaneously appeared from non-living substances during the process of spoilage.

Lazzaro Spallanzani (1729–1799) found that boiling broth would sterilise it, killing any microorganisms in it. He also found that new microorganisms could only settle in a broth if the broth was exposed to air.

Louis Pasteur (1822–1895) expanded upon Spallanzani's findings by exposing boiled broths to the air, in vessels that contained a filter to prevent all particles from passing through to the growth medium, and also in vessels with no filter at all, with air being admitted via a curved tube that would not allow dust particles to come in contact with the broth. By boiling the broth beforehand, Pasteur ensured that no microorganisms survived within the broths at the beginning of his experiment. Nothing grew in the broths in the course of Pasteur's experiment. This meant that the living organisms that grew in such broths came from outside, as spores on dust, rather than spontaneously generated within the broth. Thus, Pasteur dealt the death blow to the theory of spontaneous generation and supported germ theory.

In 1876, Robert Koch (1843–1910) established that microorganisms can cause disease. He found that the blood of cattle which were infected with anthrax always had large numbers of *Bacillus anthracis*. Koch found that he could transmit anthrax from one animal to another by taking a small sample of blood from the infected animal and injecting it into a healthy one, and this caused the healthy animal to become sick. He also found that he could grow the bacteria in a nutrient broth, then inject it into a healthy animal, and cause illness. Based on these experiments, he devised criteria for establishing a causal link between a microorganism and a disease and these are now known as Koch's postulates. Although these postulates cannot be applied in all cases, they do retain historical importance to the development of scientific thought and are still being used today.

On 8 November 2013, scientists reported the discovery of what may be the earliest signs of life on Earth—the oldest complete fossils of a microbial mat (associated with sandstone in Western Australia) estimated to be 3.48 billion years old.

Classification and Structure

Microorganisms can be found almost anywhere in the taxonomic organization of life on the planet. Bacteria and archaea are almost always microscopic, while a number of eukaryotes are also microscopic, including most protists, some fungi, as well as some animals and plants. Viruses are generally regarded as not living and therefore not considered as microorganisms, although the field of microbiology also encompasses the study of viruses.

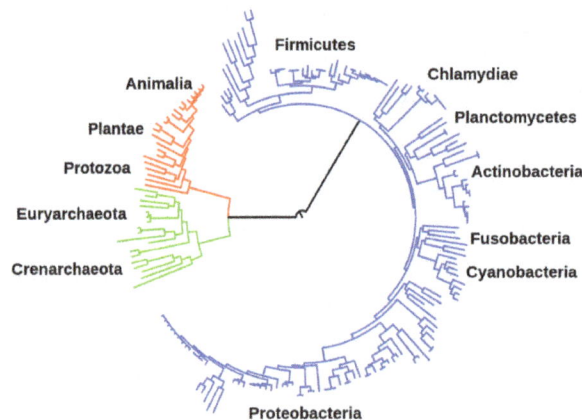

Evolutionary tree showing the common ancestry of all three domains of life. Bacteria are colored blue, eukaryotes red, and archaea green. Relative positions of some phyla are shown around the tree.

Prokaryotes

Prokaryotes are organisms that lack a cell nucleus and the other membrane bound organelles. They are almost always unicellular, although some species such as myxobacteria can aggregate into complex structures as part of their life cycle.

Consisting of two domains, bacteria and archaea, the prokaryotes are the most diverse and abundant group of organisms on Earth and inhabit practically all environments where the temperature is below +140 °C. They are found in water, soil, air, animals' gastrointestinal tracts, hot springs and even deep beneath the Earth's crust in rocks. Practically all surfaces that have not been specially sterilized are covered by prokaryotes. The number of prokaryotes on Earth is estimated to be around five million trillion trillion, or 5×10^{30}, accounting for at least half the biomass on Earth.

The biodiversity of the prokaryotes is unknown, but may be very large. A May 2016 estimate, based on laws of scaling from known numbers of species against the size of organism, gives an estimate of perhaps 1 trillion species on the planet, of which most would be microorganisms. Currently, only one-thousandth of one percent of that total have been described.

Bacteria

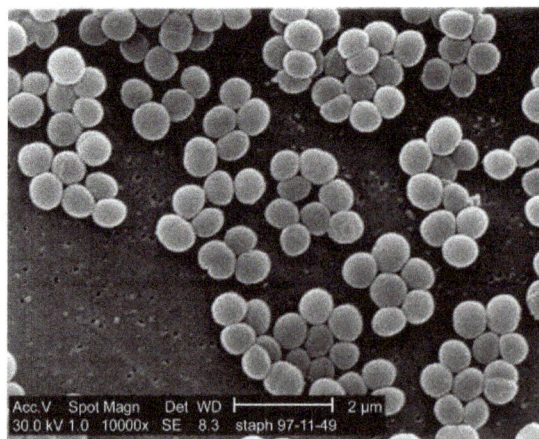

Staphylococcus aureus bacteria magnified about 10,000x

Almost all bacteria are invisible to the naked eye, with a few extremely rare exceptions, such as *Thiomargarita namibiensis*. They lack a nucleus and other membrane-bound organelles, and can function and reproduce as individual cells, but often aggregate in multicellular colonies. Their genome is usually a single loop of DNA, although they can also harbor small pieces of DNA called plasmids. These plasmids can be transferred between cells through bacterial conjugation. Bacteria are surrounded by a cell wall, which provides strength and rigidity to their cells. They reproduce by binary fission or sometimes by budding, but do not undergo meiotic sexual reproduction. However, many bacterial species can transfer DNA between individual cells by a process referred to as natural transformation. Some species form extraordinarily resilient spores, but for bacteria this is a mechanism for survival, not reproduction. Under optimal conditions bacteria can grow extremely rapidly and can double as quickly as every 20 minutes.

Archaea

Archaea are also single-celled organisms that lack nuclei. In the past, the differences between bacteria and archaea were not recognised and archaea were classified with bacteria as part of the kingdom Monera. However, in 1990 the microbiologist Carl Woese proposed the three-domain system that divided living things into bacteria, archaea and eukaryotes. Archaea differ from bacteria in both their genetics and biochemistry. For example, while bacterial cell membranes are made from phosphoglycerides with ester bonds, archaean membranes are made of ether lipids.

Archaea were originally described in extreme environments, such as hot springs, but have since been found in all types of habitats. Only now are scientists beginning to realize how common archaea are in the environment, with Crenarchaeota being the most common form of life in the ocean, dominating ecosystems below 150 m in depth. These organisms are also common in soil and play a vital role in ammonia oxidation.

Eukaryotes

Most living things that are visible to the naked eye in their adult form are eukaryotes, including humans. However, a large number of eukaryotes are also microorganisms. Unlike bacteria and archaea, eukaryotes contain organelles such as the cell nucleus, the Golgi apparatus and mitochondria in their cells. The nucleus is an organelle that houses the DNA that makes up a cell's genome. DNA (Deoxyribonuclaic acid) itself is arranged in complex chromosomes. Mitochondria are organelles vital in metabolism as they are the site of the citric acid cycle and oxidative phosphorylation. They evolved from symbiotic bacteria and retain a remnant genome. Like bacteria, plant cells have cell walls, and contain organelles such as chloroplasts in addition to the organelles in other eukaryotes. Chloroplasts produce energy from light by photosynthesis, and were also originally symbiotic bacteria.

Unicellular eukaryotes consist of a single cell throughout their life cycle. This qualification is significant since most multicellular eukaryotes consist of a single cell called a zygote only at the beginning of their life cycles. Microbial eukaryotes can be either haploid or diploid, and some organisms have multiple cell nuclei.

Unicellular eukaryotes usually reproduce asexually by mitosis under favorable conditions. However, under stressful conditions such as nutrient limitations and other conditions associated with DNA damage, they tend to reproduce sexually by meiosis and syngamy.

Protists

Of eukaryotic groups, the protists are most commonly unicellular and microscopic. This is a highly diverse group of organisms that are not easy to classify. Several algae species are multicellular protists, and slime molds have unique life cycles that involve switching between unicellular, colonial, and multicellular forms. The number of species of protists is unknown since we may have identified only a small portion. Studies from 2001-2004 have shown that a high degree of protist diversity exists in oceans, deep sea-vents, river sediment and an acidic river which suggests that a large number of eukaryotic microbial communities have yet to be discovered.

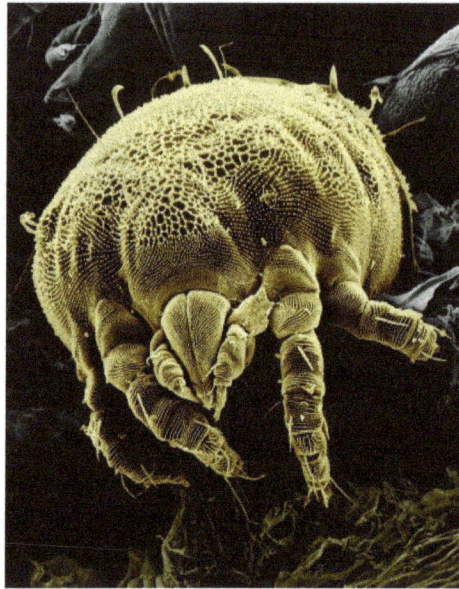

A microscopic mite *Lorryia formosa*

Animals

Some micro-animals are multicellular but at least one animal group, Myxozoa, is unicellular in its adult form. Microscopic arthropods include dust mites and spider mites. Microscopic crustaceans include copepods, some cladocera and water bears. Many nematodes are also too small to be seen with the naked eye. A common group of microscopic animals are the rotifers, which are filter feeders that are usually found in fresh water. Some micro-animals reproduce both sexually and asexually and may reach new habitats by producing eggs which can survive harsh environments that would kill the adult animal. However, some simple animals, such as rotifers, tardigrades and nematodes, can dry out completely and remain dormant for long periods of time.

Fungi

The fungi have several unicellular species, such as baker's yeast (*Saccharomyces cerevisiae*) and fission yeast (*Schizosaccharomyces pombe*). Some fungi, such as the pathogenic yeast *Candida albicans*, can undergo phenotypic switching and grow as single cells in some environments, and filamentous hyphae in others. Fungi reproduce both asexually, by budding or binary fission, as well by producing spores, which are called conidia when produced asexually, or basidiospores when produced sexually.

Plants

The green algae are a large group of photosynthetic eukaryotes that include many microscopic organisms. Although some green algae are classified as protists, others such as charophyta are classified with embryophyte plants, which are the most familiar group of land plants. Algae can grow as single cells, or in long chains of cells. The green algae include unicellular and colonial flagellates, usually but not always with two flagella per cell, as well as various colonial, coccoid, and filamentous forms. In the Charales, which are the algae most closely related to higher plants, cells differentiate into several distinct tissues within the organism. There are about 6000 species of green algae.

Habitats and Ecology

Microorganisms are found in almost every habitat present in nature. Even in hostile environments such as the poles, deserts, geysers, rocks, and the deep sea. Some types of microorganisms have adapted to the extreme conditions and sustained colonies; these organisms are known as extremophiles. Extremophiles have been isolated from rocks as much as 7 kilometres below the Earth's surface, and it has been suggested that the amount of living organisms below the Earth's surface is comparable with the amount of life on or above the surface. Extremophiles have been known to survive for a prolonged time in a vacuum, and can be highly resistant to radiation, which may even allow them to survive in space. Many types of microorganisms have intimate symbiotic relationships with other larger organisms; some of which are mutually beneficial (mutualism), while others can be damaging to the host organism (parasitism). If microorganisms can cause disease in a host they are known as pathogens and then they are sometimes referred to as microbes.

Extremophiles

Extremophiles are microorganisms that have adapted so that they can survive and even thrive in conditions that are normally fatal to most life-forms. For example, some species have been found in the following extreme environments:

- Temperature: as high as 130 °C (266 °F), as low as −17 °C (1 °F)

- Acidity/alkalinity: less than pH 0, up to pH 11.5

- Salinity: up to saturation

- Pressure: up to 1,000-2,000 atm, down to 0 atm (e.g. vacuum of space)

- Radiation: up to 5kGy

Extremophiles are significant in different ways. They extend terrestrial life into much of the Earth's hydrosphere, crust and atmosphere, their specific evolutionary adaptation mechanisms to their extreme environment can be exploited in bio-technology, and their very existence under such extreme conditions increases the potential for extraterrestrial life.

Soil Microorganisms

The nitrogen cycle in soils depends on the fixation of atmospheric nitrogen. One way this can occur is in the nodules in the roots of legumes that contain symbiotic bacteria of the genera *Rhizobium*, *Mesorhizobium*, *Sinorhizobium*, *Bradyrhizobium*, and *Azorhizobium*.

Symbiotic Microorganisms

Symbiotic microorganisms such as fungi and algae form an association in lichen. Certain fungi form mycorrhizal symbioses with trees that increase the supply of nutrients to the tree.

Applications

Microorganisms are vital to humans and the environment, as they participate in the carbon and nitrogen cycles, as well as fulfilling other vital roles in virtually all ecosystems, such as recycling other organisms' dead remains and waste products through decomposition. Microorganisms also have an important place in most higher-order multicellular organisms as symbionts. Many blame the failure of Biosphere 2 on an improper balance of microorganisms.

Food Production

Microorganisms are used to make yoghurt, cheese and curd. They are used to leaven bread, and to convert sugars to alcohol in wine and beer.

Soil

Microbes can make nutrients and minerals in the soil available to plants, produce hormones that spur growth, stimulate the plant immune system and trigger or dampen stress responses. In general a more diverse soil microbiome results in fewer plant diseases and higher yield.

Fermentation

Microorganisms are used in brewing, wine making, baking, pickling and other food-making processes.

They are also used to control the fermentation process in the production of cultured dairy products such as yogurt and cheese. The cultures also provide flavor and aroma, and inhibit undesirable organisms.

Hygiene

Hygiene is the avoidance of infection or food spoiling by eliminating microorganisms from the surroundings. As microorganisms, in particular bacteria, are found virtually everywhere, the levels of harmful microorganisms can be reduced to acceptable levels. However, in some cases, it is required that an object or substance be completely sterile, i.e. devoid of all living entities and viruses. A good example of this is a hypodermic needle.

In food preparation microorganisms are reduced by preservation methods (such as the addition of vinegar), clean utensils used in preparation, short storage periods, or by cool temperatures. If complete sterility is needed, the two most common methods are irradiation and the use of an autoclave, which resembles a pressure cooker.

There are several methods for investigating the level of hygiene in a sample of food, drinking water, equipment, etc. Water samples can be filtrated through an extremely fine filter. This filter is then placed in a nutrient medium. Microorganisms on the filter then grow to form a visible colony. Harmful microorganisms can be detected in food by placing a sample in a nutrient broth designed to enrich the organisms in question. Various methods, such as selective media or polymerase chain reaction, can then be used for detection. The hygiene of hard surfaces, such as cooking pots, can be tested by touching them with a solid piece of nutrient medium and then allowing the microorganisms to grow on it.

There are no conditions where all microorganisms would grow, and therefore often several methods are needed. For example, a food sample might be analyzed on three different nutrient mediums designed to indicate the presence of "total" bacteria (conditions where many, but not all, bacteria grow), molds (conditions where the growth of bacteria is prevented by, e.g., antibiotics) and coliform bacteria (these indicate a sewage contamination).

Water Treatment

The majority of all oxidative sewage treatment processes rely on a large range of microorganisms to oxidise organic constituents which are not amenable to sedimentation or flotation. Anaerobic microorganisms are also used to reduce sludge solids producing methane gas (amongst other gases) and a sterile mineralised residue. In potable water treatment, one method, the slow sand filter, employs a complex gelatinous layer composed of a wide range of microorganisms to remove both dissolved and particulate material from raw water.

Energy

Microorganisms are used in fermentation to produce ethanol, and in biogas reactors to produce methane. Scientists are researching the use of algae to produce liquid fuels, and bacteria to convert various forms of agricultural and urban waste into usable fuels.

Chemicals, Enzymes

Microorganisms are used for many commercial and industrial production of chemicals, enzymes and other bioactive molecules.
Examples of organic acid produced include

- Acetic acid: Produced by the bacterium *Acetobacter aceti* and other acetic acid bacteria (AAB)

- Butyric acid (butanoic acid): Produced by the bacterium *Clostridium butyricum*

- Lactic acid: *Lactobacillus* and others commonly called as lactic acid bacteria (LAB)

- Citric acid: Produced by the fungus *Aspergillus niger*

Microorganisms are used for preparation of bioactive molecules and enzymes.

- Streptokinase produced by the bacterium *Streptococcus* and modified by genetic engineering is used as a clot buster for removing clots from the blood vessels of patients who have undergone myocardial infarctions leading to heart attack.

- Cyclosporin A is a bioactive molecule used as an immunosuppressive agent in organ transplantation

- Statins produced by the yeast *Monascus purpureus* are commercialised as blood cholesterol lowering agents which act by competitively inhibiting the enzyme responsible for synthesis of cholesterol.

Science

Microorganisms are essential tools in biotechnology, biochemistry, genetics, and molecular biology. The yeasts (*Saccharomyces cerevisiae*) and fission yeast (*Schizosaccharomyces pombe*) are important model organisms in science, since they are simple eukaryotes that can be grown rapidly in large numbers and are easily manipulated. They are particularly valuable in genetics, genomics and proteomics. Microorganisms can be harnessed for uses such as creating steroids and treating skin diseases. Scientists are also considering using microorganisms for living fuel cells, and as a solution for pollution.

Warfare

In the Middle Ages, diseased corpses were thrown into castles during sieges using catapults or other siege engines. Individuals near the corpses were exposed to the pathogen and were likely to spread that pathogen to others.

Human Health

Human Digestion

Microorganisms can form an endosymbiotic relationship with other, larger organisms. For example, the bacteria that live within the human digestive system contribute to gut immunity, synthesise vitamins such as folic acid and biotin, and ferment complex indigestible carbohydrates.

Disease

Microorganisms are the causative agents (pathogens) in many infectious diseases. The organisms involved include pathogenic bacteria, causing diseases such as plague, tuberculosis and anthrax; protozoa, causing diseases such as malaria, sleeping sickness, dysentery and toxoplasmosis; and also fungi causing diseases such as ringworm, candidiasis or histoplasmosis. However, other diseases such as influenza, yellow fever or AIDS are caused by pathogenic viruses, which are not usually classified as living organisms and are not, therefore, microorganisms by the strict definition. No clear examples of archaean pathogens are known, although a relationship has been proposed between the presence of some archaean methanogens and human periodontal disease.

Ecology

Microorganisms play critical roles in Earth's biogeochemical cycles as they are responsible for decomposition and play vital parts in carbon and nitrogen fixation, as well as oxygen production.

Prokaryote

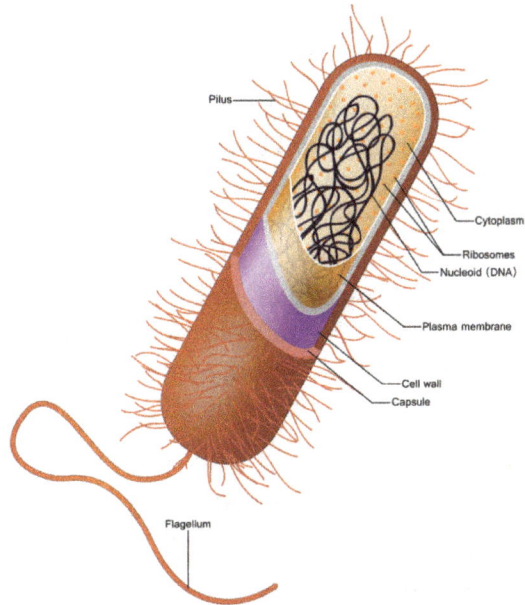

A diagram of a typical prokaryotic bacteria cell.

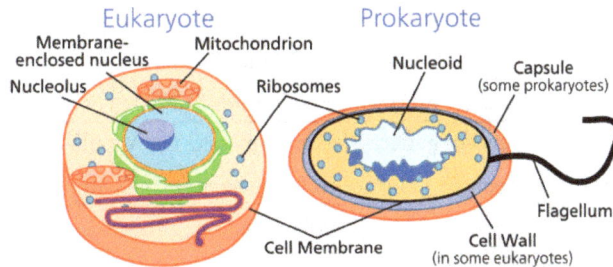

Comparison of eukaryotes vs. prokaryotes

A prokaryote is a single-celled organism that lacks a membrane-bound nucleus (karyon), mitochondria, or any other membrane-bound organelle. Prokaryotes can be divided into two domains, Archaea and Bacteria. In contrast, species with nuclei and organelles are placed in the domain Eukaryota.

In the prokaryotes, all the intracellular water-soluble components (proteins, DNA and metabolites) are located together in the cytoplasm enclosed by the cell membrane, rather than in separate cellular compartments. Bacteria, however, do possess protein-based bacterial microcompartments, which are thought to act as primitive organelles enclosed in protein shells. Some prokaryotes, such as cyanobacteria may form large colonies. Others, such as myxobacteria, have multicellular stages in their life cycles.

Molecular studies have provided insight into the evolution and interrelationships of the three domains of biological species. Eukaryotes are organisms, including humans, whose cells have a well defined membrane-bound nucleus (containing chromosomal DNA) and organelles. The division between prokaryotes and eukaryotes reflects the existence of two very different levels of cellular organization. Distinctive types of prokaryotes include extremophiles and methanogens; these are common in some extreme environments.

Structure

Prokaryotes have a prokaryotic cytoskeleton, albeit more primitive than that of the eukaryotes. Besides homologues of actin and tubulin (MreB and FtsZ), the helically arranged building-block of the flagellum, flagellin, is one of the most significant cytoskeletal proteins of bacteria, as it provides structural backgrounds of chemotaxis, the basic cell physiological response of bacteria. At least some prokaryotes also contain intracellular structures that can be seen as primitive organelles. Membranous organelles (or intracellular membranes) are known in some groups of prokaryotes, such as vacuoles or membrane systems devoted to special metabolic properties, such as photosynthesis or chemolithotrophy. In addition, some species also contain carbohydrate-enclosed microcompartments, which have distinct physiological roles (e.g. carboxysomes or gas vacuoles).

Most prokaryotes are between 1 μm and 10 μm, but they can vary in size from 0.2 μm (*Mycoplasma genitalium*) to 750 μm (*Thiomargarita namibiensis*).

Prokaryotic cell structure
Flagellum (only in some types of prokaryotes)
Long, whip-like protrusion that aids cellular locomotion.
Cell membrane
Surrounds the cell's cytoplasm and regulates the flow of substances in and out of the cell.
Cell wall (except genera Mycoplasma and Thermoplasma)
Outer covering of most cells that protects the bacterial cell and gives it shape.
Cytoplasm
A gel-like substance composed mainly of water that also contains enzymes, salts, cell components, and various organic molecules.
Ribosome
Cell structures responsible for protein production.
Nucleoid
Area of the cytoplasm that contains the single bacterial DNA molecule.
Glycocalyx (only in some types of prokaryotes)
A glycoprotein-polysaccharide covering that surrounds the cell membranes.
Inclusions

Morphology

Prokaryotic cells have various shapes; the four basic shapes of bacteria are:

- Cocci – spherical

- Bacilli – rod-shaped

- Spirochaete – spiral-shaped

- Vibrio – comma-shaped

The archaeon Haloquadratum has flat square-shaped cells.

Reproduction

Bacteria and archaea reproduce through asexual reproduction, usually by binary fission. Genetic exchange and recombination still occur, but this is a form of horizontal gene transfer and is not a replicative process, simply involving the transference of DNA between two cells, as in bacterial conjugation.

DNA Transfer

DNA transfer between prokaryotic cells occurs in bacteria and archaea, although it has been mainly studied in bacteria. In bacteria, gene transfer occurs by three processes. These are (1) bacterial virus (bacteriophage)-mediated transduction, (2) plasmid-mediated conjugation, and (3) natural transformation. Transduction of bacterial genes by bacteriophage appears to reflect an occasional error during intracellular assembly of virus particles, rather than an adaptation of the host bacteria. The transfer of bacterial DNA is under the control of the bacteriophage's genes rather than bacterial genes. Conjugation in the well-studied *E. coli* system is controlled by plasmid genes, and is an adaptation for distributing copies of a plasmid from one bacterial host to another. Infrequently during this process, a plasmid may integrate into the host bacterial chromosome, and subsequently transfer part of the host bacterial DNA to another bacterium. Plasmid mediated transfer of host bacterial DNA (conjugation) also appears to be an accidental process rather than a bacterial adaptation.

Natural bacterial transformation involves the transfer of DNA from one bacterium to another through the intervening medium. Unlike transduction and conjugation, transformation is clearly a bacterial adaptation for DNA transfer, because it depends on numerous bacterial gene products that specifically interact to perform this complex process. For a bacterium to bind, take up and recombine donor DNA into its own chromosome, it must first enter a special physiological state called competence. About 40 genes are required in *Bacillus subtilis* for the development of competence. The length of DNA transferred during *B. subtilis* transformation can be as much as a third to the whole chromosome. Transformation is a common mode of DNA transfer, and 67 prokaryotic species are thus far known to be naturally competent for transformation.

Among archaea, *Halobacterium volcanii* forms cytoplasmic bridges between cells that appear to be used for transfer of DNA from one cell to another. Another archaeon, *Sulfolobus solfataricus*, transfers DNA between cells by direct contact. Frols et al. found that exposure of *S. solfataricus* to DNA damaging agents induces cellular aggregation, and suggested that cellular aggregation may enhance DNA transfer among cells to provide increased repair of damaged DNA via homologous recombination.

Sociality

While prokaryotes are considered strictly unicellular, most can form stable aggregate communities. When such communities are encased in a stabilizing polymer matrix ("slime"), they may be called "biofilms". Cells in biofilms often show distinct patterns of gene expression (phenotypic differentiation) in time and space. Also, as with multicellular eukaryotes, these changes in expression often appear to result from cell-to-cell signaling, a phenomenon known as quorum sensing.

Biofilms may be highly heterogeneous and structurally complex and may attach to solid surfaces, or exist at liquid-air interfaces, or potentially even liquid-liquid interfaces. Bacterial biofilms are often made up of microcolonies (approximately dome-shaped masses of bacteria and matrix) separated by "voids" through which the medium (e.g., water) may flow easily. The microcolonies may join together above the substratum to form a continuous layer, closing the network of channels separating microcolonies. This structural complexity—combined with observations that oxygen limitation (a ubiquitous challenge for anything growing in size beyond the scale of diffusion) is at least partially eased by movement of medium throughout the biofilm—has led some to speculate that this may constitute a circulatory system and many researchers have started calling prokaryotic communities multicellular (for example). Differential cell expression, collective behavior, signaling, programmed cell death, and (in some cases) discrete biological dispersal events all seem to point in this direction. However, these colonies are seldom if ever founded by a single founder (in the way that animals and plants are founded by single cells), which presents a number of theoretical issues. Most explanations of co-operation and the evolution of multicellularity have focused on high relatedness between members of a group (or colony, or whole organism). If a copy of a gene is present in all members of a group, behaviors that promote cooperation between members may permit those members to have (on average) greater fitness than a similar group of selfish individuals.

Should these instances of prokaryotic sociality prove to be the rule rather than the exception, it would have serious implications for the way we view prokaryotes in general, and the way we deal with them in medicine. Bacterial biofilms may be 100 times more resistant to antibiotics than free-living unicells and may be nearly impossible to remove from surfaces once they have colonized them. Other aspects of bacterial cooperation—such as bacterial conjugation and quorum-sensing-mediated pathogenicity, present additional challenges to researchers and medical professionals seeking to treat the associated diseases.

Environment

Prokaryotes have diversified greatly throughout their long existence. The metabolism of prokaryotes is far more varied than that of eukaryotes, leading to many highly distinct prokaryotic types. For example, in addition to using photosynthesis or organic compounds for energy, as eukaryotes do, prokaryotes may obtain energy from inorganic compounds such as hydrogen sulfide. This enables prokaryotes to thrive in harsh environments as cold as the snow surface of Antarctica, studied in cryobiology or as hot as undersea hydrothermal vents and land-based hot springs.

Prokaryotes live in nearly all environments on Earth. Some archaea and bacteria thrive in harsh conditions, such as high temperatures (thermophiles) or high salinity (halophiles). Organisms such as these are referred to as extremophiles. Many archaea grow as plankton in the oceans. Symbiotic prokaryotes live in or on the bodies of other organisms, including humans.

Classification

In 1977, Carl Woese proposed dividing prokaryotes into the Bacteria and Archaea (originally Eubacteria and Archaebacteria) because of the major differences in the structure and genetics between the two groups of organisms. Archaea were originally thought to be extremophiles, living only in inhospitable conditions such as extremes of temperature, pH, and radiation but have since been found in all types of habitats. The resulting arrangement of Eukaryota (also called "Eukarya"), Bacteria, and Archaea is called the three-domain system, replacing the traditional two-empire system.

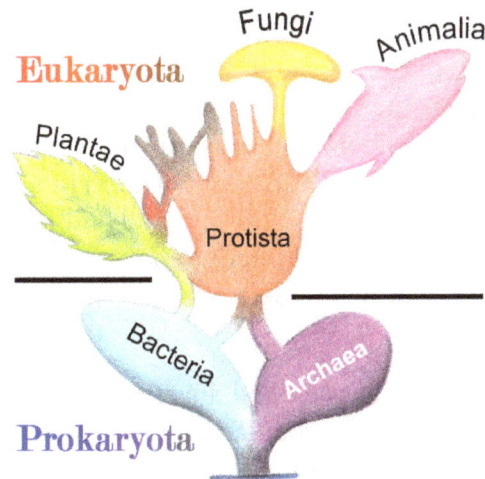

Phylogenetic and symbiogenetic tree of living organisms, showing the origins of eukaryotes and prokaryotes

One criticism of the dichotomous prokaryote-eukaryote distinction points out that the word "prokaryote" is based on what these organisms are not (they are not eukaryotic), rather than what they are (either archaea or bacteria).

Evolution of Prokaryotes

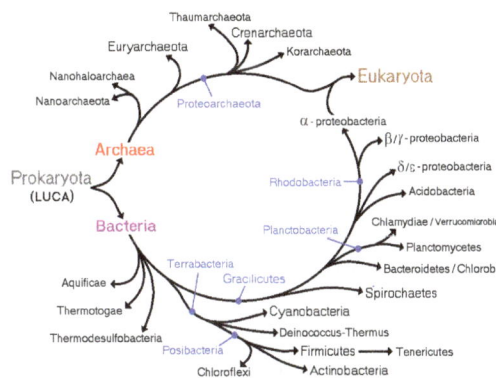

Phylogenetic ring showing the diversity of prokaryotes, and symbiogenetic origins of eukaryotes

The current model of the evolution of the first living organisms is that these were some form of prokaryotes, which may have evolved out of protocells. In general, the eukaryotes are thought to have evolved later in the history of life.

However, some authors have questioned this conclusion, arguing that the current set of prokaryotic species may have evolved from more complex eukaryotic ancestors through a process of simplification.

Others have argued that the three domains of life arose simultaneously, from a set of varied cells that formed a single gene pool.

This controversy was summarized in 2005:

There is no consensus among biologists concerning the position of the eukaryotes in the overall scheme of cell evolution. Current opinions on the origin and position of eukaryotes span a broad spectrum including the views that eukaryotes arose first in evolution and that prokaryotes descend from them, that eukaryotes arose contemporaneously with eubacteria and archeabacteria and hence represent a primary line of descent of equal age and rank as the prokaryotes, that eukaryotes arose through a symbiotic event entailing an endosymbiotic origin of the nucleus, that eukaryotes arose without endosymbiosis, and that eukaryotes arose through a symbiotic event entailing a simultaneous endosymbiotic origin of the flagellum and the nucleus, in addition to many other models, which have been reviewed and summarized elsewhere.

The oldest known fossilized prokaryotes were laid down approximately 3.5 billion years ago, only about 1 billion years after the formation of the Earth's crust. Eukaryotes only appear in the fossil record later, and may have formed from endosymbiosis of multiple prokaryote ancestors. The oldest known fossil eukaryotes are about 1.7 billion years old. However, some genetic evidence suggests eukaryotes appeared as early as 3 billion years ago.

While Earth is the only place in the universe where life is known to exist, some have suggested that there is evidence on Mars of fossil or living prokaryotes. However, this possibility remains the subject of considerable debate and skepticism.

Relationship to Eukaryotes

The division between prokaryotes and eukaryotes is usually considered the most important distinction or difference among organisms. The distinction is that eukaryotic cells have a "true" nucleus containing their DNA, whereas prokaryotic cells do not have a nucleus. Both eukaryotes and prokaryotes contain large RNA/protein structures called ribosomes, which produce protein.

Another difference is that ribosomes in prokaryotes are smaller than in eukaryotes. However, two organelles found in many eukaryotic cells, mitochondria and chloroplasts, contain ribosomes similar in size and makeup to those found in prokaryotes. This is one of many pieces of evidence that mitochondria and chloroplasts are themselves descended from free-living bacteria. This theory holds that early eukaryotic cells took in primitive prokaryotic cells by phagocytosis and adapted themselves to incorporate their structures, leading to the mitochondria we see today.

The genome in a prokaryote is held within a DNA/protein complex in the cytosol called the nucleoid, which lacks a nuclear envelope. The complex contains a single, cyclic, double-stranded molecule of stable chromosomal DNA, in contrast to the multiple linear, compact, highly organized chromosomes found in eukaryotic cells. In addition, many important genes of prokaryotes are stored in separate circular DNA structures called plasmids.

Prokaryotes lack mitochondria and chloroplasts. Instead, processes such as oxidative phosphorylation and photosynthesis take place across the prokaryotic cell membrane. However, prokaryotes do possess some internal structures, such as prokaryotic cytoskeletons. It has been suggested that the bacterial order Planctomycetes have a membrane around their nucleoid and contain other membrane-bound cellular structures. However, further investigation revealed that Planctomycetes cells are not compartmentalized or nucleated and like the other bacterial membrane systems are all interconnected.

Prokaryotic cells are usually much smaller than eukaryotic cells. Therefore, prokaryotes have a larger surface-area-to-volume ratio, giving them a higher metabolic rate, a higher growth rate, and as a consequence, a shorter generation time than eukaryotes.

Bacteria

Bacteria constitute a large domain of prokaryotic microorganisms. Typically a few micrometres in length, bacteria have a number of shapes, ranging from spheres to rods and spirals. Bacteria were among the first life forms to appear on Earth, and are present in most of its habitats. Bacteria inhabit soil, water, acidic hot springs, radioactive waste, and the deep portions of Earth's crust. Bacteria also live in symbiotic and parasitic relationships with plants and animals.

There are typically 40 million bacterial cells in a gram of soil and a million bacterial cells in a millilitre of fresh water. There are approximately 5×10^{30} bacteria on Earth, forming a biomass which exceeds that of all plants and animals. Bacteria are vital in recycling nutrients, with many of the stages in nutrient cycles dependent on these organisms, such as the fixation of nitrogen from the atmosphere and putrefaction. In the biological communities surrounding hydrothermal vents and cold seeps, bacteria provide the nutrients needed to sustain life by converting dissolved compounds, such as hydrogen sulphide and methane, to energy. On 17 March 2013, researchers reported data that suggested bacterial life forms thrive in the Mariana Trench, which with a depth of up to 11 kilometres is the deepest part of the Earth's oceans. Other researchers reported related studies that microbes thrive inside rocks up to 580 metres below the sea floor under 2.6 kilometres of ocean off the coast of the northwestern United States. According to one of the researchers, "You can find microbes everywhere — they're extremely adaptable to conditions, and survive wherever they are."

Most bacteria have not been characterised, and only about half of the bacterial phyla have species that can be grown in the laboratory. The study of bacteria is known as bacteriology, a branch of microbiology.

There are approximately ten times as many bacterial cells in the human flora as there are human cells in the body, with the largest number of the human flora being in the gut flora, and a large number on the skin. The vast majority of the bacteria in the body are rendered harmless by the protective effects of the immune system, and some are beneficial. However, several species of bacteria are pathogenic and cause infectious diseases, including cholera, syphilis, anthrax, leprosy, and bubonic plague. The most common fatal bacterial diseases are respiratory infections, with tuberculosis alone killing about 2 million people per year, mostly in sub-Saharan Africa. In developed countries, antibiotics are used to treat bacterial infections and are also used in farming, making antibiotic resistance a growing problem. In industry, bacteria are important in sewage treatment and the breakdown of oil spills, the production of cheese and yogurt through fermentation, and the recovery of gold, palladium, copper and other metals in the mining sector, as well as in biotechnology, and the manufacture of antibiotics and other chemicals.

Once regarded as plants constituting the class *Schizomycetes*, bacteria are now classified as prokaryotes. Unlike cells of animals and other eukaryotes, bacterial cells do not contain a nucleus and rarely harbour membrane-bound organelles. Although the term *bacteria* traditionally included all prokaryotes, the scientific classification changed after the discovery in the 1990s that prokaryotes consist of two very different groups of organisms that evolved from an ancient common ancestor. These evolutionary domains are called *Bacteria* and *Archaea*.

Origin and Early Evolution

The ancestors of modern bacteria were unicellular microorganisms that were the first forms of life to appear on Earth, about 4 billion years ago. For about 3 billion years, most organisms were microscopic, and bacteria and archaea were the dominant forms of life. In 2008, fossils of macroorganisms were discovered and named as the Francevillian biota. Although bacterial fossils exist, such as stromatolites, their lack of distinctive morphology prevents them from being used to examine the history of bacterial evolution, or to date the time of origin of a particular bacterial species. However, gene sequences can be used to reconstruct the bacterial phylogeny, and these studies indicate that bacteria diverged first from the archaeal/eukaryotic lineage. Bacteria were also involved in the second great evolutionary divergence, that of the archaea and eukaryotes. Here, eukaryotes resulted from the entering of ancient bacteria into endosymbiotic associations with the ancestors of eukaryotic cells, which were themselves possibly related to the Archaea. This involved the engulfment by proto-eukaryotic cells of alphaproteobacterial symbionts to form either mitochondria or hydrogenosomes, which are still found in all known Eukarya (sometimes in highly reduced form, e.g. in ancient "amitochondrial" protozoa). Later on, some eukaryotes that already contained mitochondria also engulfed cyanobacterial-like organisms. This led to the formation of chloroplasts in algae and plants. There are also some algae that originated from even later endosymbiotic events. Here, eukaryotes engulfed a eukaryotic algae that developed into a "second-generation" plastid. This is known as secondary endosymbiosis.

Morphology

Bacteria display a wide diversity of shapes and sizes, called *morphologies*. Bacterial cells are about one-tenth the size of eukaryotic cells and are typically 0.5–5.0 micrometres in length. However, a few species are visible to the unaided eye — for example, *Thiomargarita namibiensis* is up to half a millimetre long and *Epulopiscium fishelsoni* reaches 0.7 mm. Among the smallest bacteria are members of the genus *Mycoplasma*, which measure only 0.3 micrometres, as small as the largest viruses. Some bacteria may be even smaller, but these ultramicrobacteria are not well-studied.

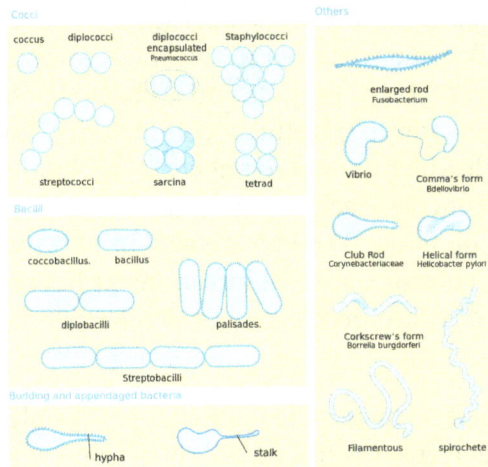

Bacteria display many cell morphologies and arrangements

Most bacterial species are either spherical, called *cocci*, or rod-shaped, called *bacilli* (*sing.* bacillus, from Latin *baculus*, stick). Elongation is associated with swimming. Some bacteria, called *vibrio*, are shaped like slightly curved rods or comma-shaped; others can be spiral-shaped, called *spirilla*, or tightly coiled, called *spirochaetes*. A small number of species even have tetrahedral or cuboidal shapes. More recently, some bacteria were discovered deep under Earth's crust that grow as branching filamentous types with a star-shaped cross-section. The large surface area to volume ratio of this morphology may give these bacteria an advantage in nutrient-poor environments. This wide variety of shapes is determined by the bacterial cell wall and cytoskeleton, and is important because it can influence the ability of bacteria to acquire nutrients, attach to surfaces, swim through liquids and escape predators.

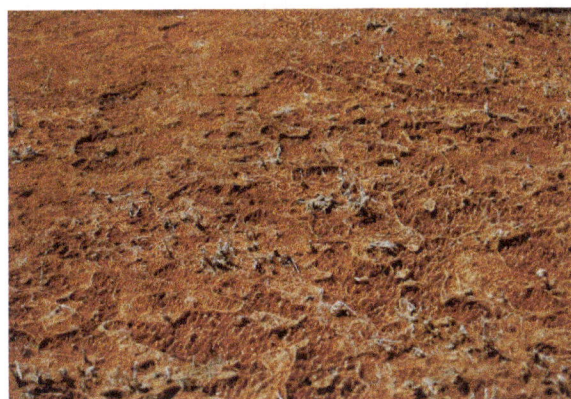

A biofilm of thermophilic bacteria in the outflow of Mickey Hot Springs, Oregon, approximately 20 mm thick.

Many bacterial species exist simply as single cells, others associate in characteristic patterns: *Neisseria* form diploids (pairs), *Streptococcus* form chains, and *Staphylococcus* group together in "bunch of grapes" clusters. Bacteria can also be elongated to form filaments, for example the Actinobacteria. Filamentous bacteria are often surrounded by a sheath that contains many individual cells. Certain types, such as species of the genus *Nocardia*, even form complex, branched filaments, similar in appearance to fungal mycelia.

Bacteria often attach to surfaces and form dense aggregations called *biofilms* or bacterial mats. These films can range from a few micrometres in thickness to up to half a metre in depth, and may contain multiple species of bacteria, protists and archaea. Bacteria living in biofilms display a complex arrangement of cells and extracellular components, forming secondary structures, such as microcolonies, through which there are networks of channels to enable better diffusion of nutrients. In natural environments, such as soil or the surfaces of plants, the majority of bacteria are bound to surfaces in biofilms. Biofilms are also important in medicine, as these structures are often present during chronic bacterial infections or in infections of implanted medical devices, and bacteria protected within biofilms are much harder to kill than individual isolated bacteria.

Even more complex morphological changes are sometimes possible. For example, when starved of amino acids, Myxobacteria detect surrounding cells in a process known as quorum sensing, migrate towards each other, and aggregate to form fruiting bodies up to 500 micrometres long and containing approximately 100,000 bacterial cells. In these fruiting bodies, the bacteria perform separate tasks; this type of cooperation is a simple type of multicellular organisation. For example, about one in 10 cells migrate to the top of these fruiting bodies and differentiate into a specialised dormant state called myxospores, which are more resistant to drying and other adverse environmental conditions than are ordinary cells.

Cellular Structure

Intracellular Structures

The bacterial cell is surrounded by a cell membrane (also known as a lipid, cytoplasmic or plasma membrane). This membrane encloses the contents of the cell and acts as a barrier to hold nutrients, proteins and other essential components of the *cytoplasm* within the cell. As they are prokaryotes, bacteria do not usually have membrane-bound organelles in their cytoplasm, and thus contain few large intracellular structures. They lack a true nucleus, mitochondria, chloroplasts and the other organelles present in eukaryotic cells. Bacteria were once seen as simple bags of cytoplasm, but structures such as the *prokaryotic cytoskeleton* and the localisation of proteins to specific locations within the cytoplasm that give bacteria some complexity have been discovered. These subcellular levels of organisation have been called "bacterial hyperstructures".

Bacterial microcompartments, such as carboxysomes, provide a further level of organisation; they are compartments within bacteria that are surrounded by polyhedral protein shells, rather than by lipid membranes. These "polyhedral organelles" localise and compartmentalise bacterial metabolism, a function performed by the membrane-bound organelles in eukaryotes.

Many important biochemical reactions, such as energy generation, use concentration gradients across membranes. The general lack of internal membranes in bacteria means reactions such as electron transport occur across the cell membrane between the cytoplasm and the *periplasmic space*. However, in many photosynthetic bacteria the plasma membrane is highly folded and fills most of the cell with layers of light-gathering membrane. These light-gathering complexes may even form lipid-enclosed structures called chlorosomes in green sulfur bacteria. Other proteins import nutrients across the cell membrane, or expel undesired molecules from the cytoplasm.

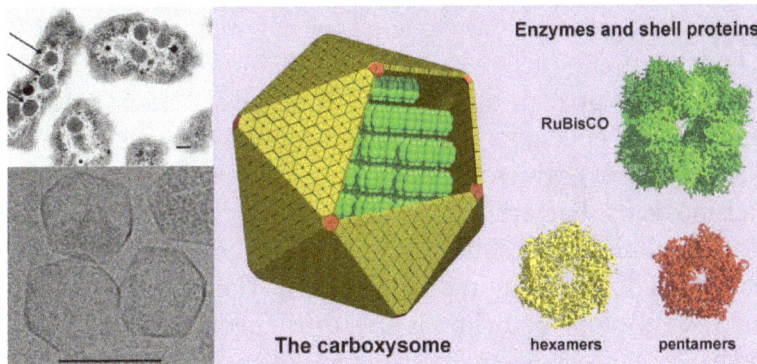

Carboxysomes are protein-enclosed bacterial organelles. Top left is an electron microscope image of carboxysomes in Halothiobacillus neapolitanus, below is an image of purified carboxysomes. On the right is a model of their structure. Scale bars are 100 nm.

Bacteria do not have a membrane-bound nucleus, and their genetic material is typically a single circular DNA chromosome located in the cytoplasm in an irregularly shaped body called the *nucleoid*. The nucleoid contains the chromosome with its associated proteins and RNA. The phylum Planctomycetes and candidate phylum Poribacteria may be exceptions to the general absence of internal membranes in bacteria, because they appear to have a double membrane around their nucleoids and contain other membrane-bound cellular structures. Like all living organisms, bacteria contain *ribosomes*, often grouped in chains called polyribosomes, for the production of proteins, but the structure of the bacterial ribosome is different from that of eukaryotes and Archaea. Bacterial ribosomes have a sedimentation rate of 70S (measured in Svedberg units): their subunits have rates of 30S and 50S. Some antibiotics bind specifically to 70S ribosomes and inhibit bacterial protein synthesis. Those antibiotics kill bacteria without affecting the larger 80S ribosomes of eukaryotic cells and without harming the host.

Some bacteria produce intracellular nutrient storage granules for later use, such as glycogen, polyphosphate, sulfur or polyhydroxyalkanoates. Certain bacterial species, such as the photosynthetic Cyanobacteria, produce internal gas vesicles, which they use to regulate their buoyancy – allowing them to move up or down into water layers with different light intensities and nutrient levels. *Intracellular membranes* called *chromatophores* are also found in membranes of phototrophic bacteria. Used primarily for photosynthesis, they contain bacteriochlorophyll pigments and carotenoids. An early idea was that bacteria might contain membrane folds termed mesosomes, but these were later shown to be artefacts produced by the chemicals used to prepare the cells for electron microscopy. *Inclusions* are considered to be nonliving components of the cell that do not possess metabolic activity and are not bounded by membranes. The most common inclusions are glycogen, lipid droplets, crystals, and pigments. *Volutin granules* are cytoplasmic inclusions of complexed inorganic polyphosphate. These granules are called *metachromatic granules* due to their displaying the metachromatic effect; they appear red or blue when stained with the blue dyes methylene blue or toluidine blue. *Gas vacuoles*, which are freely permeable to gas, are membrane-bound vesicles present in some species of *Cyanobacteria*. They allow the bacteria to control their buoyancy. *Microcompartments* are widespread, membrane-bound organelles that are made of a protein shell that surrounds and encloses various enzymes. *Carboxysomes* are bacterial microcompartments that contain enzymes involved in carbon fixation. *Magnetosomes* are bacterial microcompartments, present in magnetotactic bacteria, that contain magnetic crystals.

Extracellular Structures

In most bacteria, a *cell wall* is present on the outside of the cell membrane. The cell membrane and cell wall comprise the *cell envelope*. A common bacterial cell wall material is *peptidoglycan* (called "murein" in older sources), which is made from polysaccharide chains cross-linked by peptides containing D-amino acids. Bacterial cell walls are different from the cell walls of plants and fungi, which are made of cellulose and chitin, respectively. The cell wall of bacteria is also distinct from that of Archaea, which do not contain peptidoglycan. The cell wall is essential to the survival of many bacteria, and the antibiotic penicillin is able to kill bacteria by inhibiting a step in the synthesis of peptidoglycan.

There are broadly speaking two different types of cell wall in bacteria, a thick one in the gram-positives and a thinner one in the gram-negatives. The names originate from the reaction of cells to the Gram stain, a long-standing test for the classification of bacterial species.

Gram-positive bacteria possess a thick cell wall containing many layers of peptidoglycan and *teichoic acids*. In contrast, *gram-negative bacteria* have a relatively thin cell wall consisting of a few layers of peptidoglycan surrounded by a second lipid membrane containing *lipopolysaccharides* and lipoproteins. Lipopolysaccharides, also called *endotoxins*, are composed of polysaccharides and *lipid A* that is responsible for much of the toxicity of gram-negative bacteria. Most bacteria have the gram-negative cell wall, and only the Firmicutes and Actinobacteria have the alternative gram-positive arrangement. These two groups were previously known as the low G+C and high G+C gram-positive bacteria, respectively. These differences in structure can produce differences in antibiotic susceptibility; for instance, vancomycin can kill only gram-positive bacteria and is ineffective against gram-negative pathogens, such as *Haemophilus influenzae* or *Pseudomonas aeruginosa*. If the bacterial cell wall is entirely removed, it is called a *protoplast*, whereas if it is partially removed, it is called a *spheroplast*. β-Lactam antibiotics, such as penicillin, inhibit the formation of peptidoglycan cross-links in the bacterial cell wall. The enzyme lysozyme, found in human tears, also digests the cell wall of bacteria and is the body's main defence against eye infections.

Acid-fast bacteria, such as *Mycobacteria*, are resistant to decolorisation by acids during staining procedures. The high mycolic acid content of *Mycobacteria*, is responsible for the staining pattern of poor absorption followed by high retention. The most common staining technique used to identify acid-fast bacteria is the Ziehl-Neelsen stain or acid-fast stain, in which the acid-fast bacilli are stained bright-red and stand out clearly against a blue background. *L-form bacteria* are strains of bacteria that lack cell walls. The main pathogenic bacteria in this class is *Mycoplasma*.

In many bacteria, an *S-layer* of rigidly arrayed protein molecules covers the outside of the cell. This layer provides chemical and physical protection for the cell surface and can act as a macromolecular diffusion barrier. S-layers have diverse but mostly poorly understood functions, but are known to act as virulence factors in *Campylobacter* and contain surface enzymes in *Bacillus stearothermophilus*.

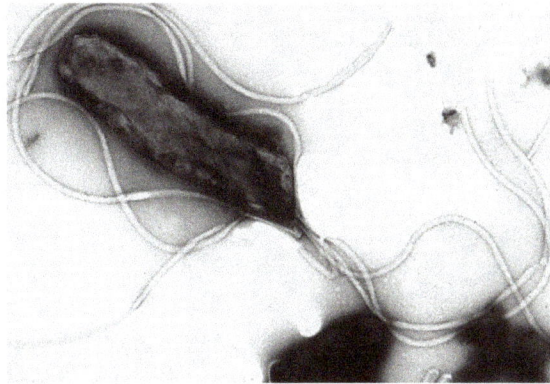

Helicobacter pylori electron micrograph, showing multiple flagella on the cell surface

Flagella are rigid protein structures, about 20 nanometres in diameter and up to 20 micrometres in length, that are used for motility. Flagella are driven by the energy released by the transfer of ions down an electrochemical gradient across the cell membrane.

Fimbriae (sometimes called "attachment pili") are fine filaments of protein, usually 2–10 nanometres in diameter and up to several micrometres in length. They are distributed over the surface of the cell, and resemble fine hairs when seen under the electron microscope. Fimbriae are believed to be involved in attachment to solid surfaces or to other cells, and are essential for the virulence of some bacterial pathogens. *Pili* (*sing.* pilus) are cellular appendages, slightly larger than fimbriae, that can transfer genetic material between bacterial cells in a process called conjugation where they are called *conjugation pili* or "sex pili". They can also generate movement where they are called *type IV pili.*

Glycocalyx are produced by many bacteria to surround their cells, and vary in structural complexity: ranging from a disorganised *slime layer* of extra-cellular polymer to a highly structured *capsule*. These structures can protect cells from engulfment by eukaryotic cells such as macrophages (part of the human immune system). They can also act as antigens and be involved in cell recognition, as well as aiding attachment to surfaces and the formation of biofilms.

The assembly of these extracellular structures is dependent on bacterial secretion systems. These transfer proteins from the cytoplasm into the periplasm or into the environment around the cell. Many types of secretion systems are known and these structures are often essential for the virulence of pathogens, so are intensively studied.

Endospores

Certain genera of gram-positive bacteria, such as *Bacillus*, *Clostridium*, *Sporohalobacter*, *Anaerobacter*, and *Heliobacterium*, can form highly resistant, dormant structures called *endospores*. In almost all cases, one endospore is formed and this is not a reproductive process, although *Anaerobacter* can make up to seven endospores in a single cell. Endospores have a central core of cytoplasm containing DNA and ribosomes surrounded by a cortex layer and protected by an impermeable and rigid coat. Dipicolinic acid is a chemical compound that composes 5% to 15% of the dry weight of bacterial spores. It is implicated as responsible for the heat resistance of the endospore.

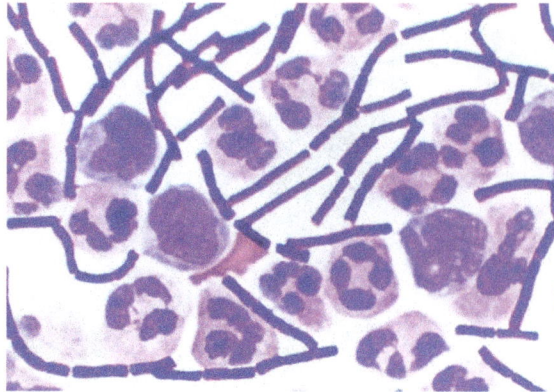

Bacillus anthracis (stained purple) growing in cerebrospinal fluid

Endospores show no detectable metabolism and can survive extreme physical and chemical stresses, such as high levels of UV light, gamma radiation, detergents, disinfectants, heat, freezing, pressure, and desiccation. In this dormant state, these organisms may remain viable for millions of years, and endospores even allow bacteria to survive exposure to the vacuum and radiation in space. According to scientist Dr. Steinn Sigurdsson, "There are viable bacterial spores that have been found that are 40 million years old on Earth — and we know they're very hardened to radiation." Endospore-forming bacteria can also cause disease: for example, anthrax can be contracted by the inhalation of *Bacillus anthracis* endospores, and contamination of deep puncture wounds with *Clostridium tetani* endospores causes tetanus.

Metabolism

Bacteria exhibit an extremely wide variety of metabolic types. The distribution of metabolic traits within a group of bacteria has traditionally been used to define their taxonomy, but these traits often do not correspond with modern genetic classifications. Bacterial metabolism is classified into nutritional groups on the basis of three major criteria: the kind of energy used for growth, the source of carbon, and the electron donors used for growth. An additional criterion of respiratory microorganisms are the electron acceptors used for aerobic or anaerobic respiration.

Nutritional types in bacterial metabolism			
Nutritional type	**Source of energy**	**Source of carbon**	**Examples**
Phototrophs	Sunlight	Organic compounds (photoheterotrophs) or carbon fixation (photoautotrophs)	Cyanobacteria, Green sulfur bacteria, Chloroflexi, or Purple bacteria
Lithotrophs	Inorganic compounds	Organic compounds (lithoheterotrophs) or carbon fixation (lithoautotrophs)	Thermodesulfobacteria, *Hydrogenophilaceae*, or Nitrospirae
Organotrophs	Organic compounds	Organic compounds (chemoheterotrophs) or carbon fixation (chemoautotrophs)	*Bacillus*, *Clostridium* or *Enterobacteriaceae*

Carbon metabolism in bacteria is either *heterotrophic*, where organic carbon compounds are used as carbon sources, or *autotrophic*, meaning that cellular carbon is obtained by fixing carbon dioxide. Heterotrophic bacteria include parasitic types. Typical autotrophic bacteria are phototrophic cyanobacteria, green sulfur-bacteria and some purple bacteria, but also many chemolithotrophic species, such as nitrifying or sulfur-oxidising bacteria. Energy metabolism of bacteria is either based on *phototrophy*, the use of light through photosynthesis, or based on *chemotrophy*, the use of chemical substances for energy, which are mostly oxidised at the expense of oxygen or alternative electron acceptors (aerobic/anaerobic respiration).

Filaments of photosynthetic cyanobacteria

Bacteria are further divided into *lithotrophs* that use inorganic electron donors and *organotrophs* that use organic compounds as electron donors. Chemotrophic organisms use the respective electron donors for energy conservation (by aerobic/anaerobic respiration or fermentation) and biosynthetic reactions (e.g., carbon dioxide fixation), whereas phototrophic organisms use them only for biosynthetic purposes. Respiratory organisms use chemical compounds as a source of energy by taking electrons from the reduced substrate and transferring them to a terminal electron acceptor in a redox reaction. This reaction releases energy that can be used to synthesise ATP and drive metabolism. In *aerobic organisms*, oxygen is used as the electron acceptor. In *anaerobic organisms* other inorganic compounds, such as nitrate, sulfate or carbon dioxide are used as electron acceptors. This leads to the ecologically important processes of denitrification, sulfate reduction, and acetogenesis, respectively.

Another way of life of chemotrophs in the absence of possible electron acceptors is fermentation, wherein the electrons taken from the reduced substrates are transferred to oxidised intermediates to generate reduced fermentation products (e.g., lactate, ethanol, hydrogen, butyric acid). Fermentation is possible, because the energy content of the substrates is higher than that of the products, which allows the organisms to synthesise ATP and drive their metabolism.

These processes are also important in biological responses to pollution; for example, sulfate-reducing bacteria are largely responsible for the production of the highly toxic forms of mercury (methyl- and dimethylmercury) in the environment. Non-respiratory anaerobes use fermentation to generate energy and reducing power, secreting metabolic by-products (such as ethanol in brewing) as waste. Facultative anaerobes can switch between fermentation and different terminal electron acceptors depending on the environmental conditions in which they find themselves.

Lithotrophic bacteria can use inorganic compounds as a source of energy. Common inorganic electron donors are hydrogen, carbon monoxide, ammonia (leading to nitrification), ferrous iron and other reduced metal ions, and several reduced sulfur compounds. In unusual circumstances, the gas methane can be used by methanotrophic bacteria as both a source of electrons and a substrate for carbon anabolism. In both aerobic phototrophy and chemolithotrophy, oxygen is used as a terminal electron acceptor, whereas under anaerobic conditions inorganic compounds are used instead. Most lithotrophic organisms are autotrophic, whereas organotrophic organisms are heterotrophic.

In addition to fixing carbon dioxide in photosynthesis, some bacteria also fix nitrogen gas (nitrogen fixation) using the enzyme nitrogenase. This environmentally important trait can be found in bacteria of nearly all the metabolic types listed above, but is not universal.

Regardless of the type of metabolic process they employ, the majority of bacteria are able to take in raw materials only in the form of relatively small molecules, which enter the cell by diffusion or through molecular channels in cell membranes. The Planctomycetes are the exception (as they are in possessing membranes around their nuclear material). It has recently been shown that *Gemmata obscuriglobus* is able to take in large molecules via a process that in some ways resembles endocytosis, the process used by eukaryotic cells to engulf external items.

Growth and Reproduction

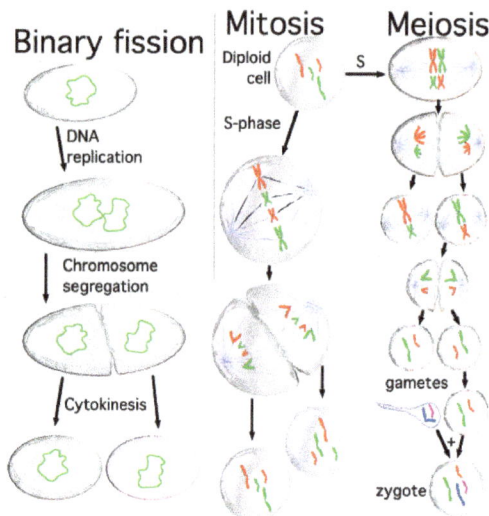

Many bacteria reproduce through binary fission, which is compared to mitosis and meiosis in this image.

Unlike in multicellular organisms, increases in cell size (cell growth) and reproduction by cell division are tightly linked in unicellular organisms. Bacteria grow to a fixed size and then reproduce through *binary fission*, a form of asexual reproduction. Under optimal conditions, bacteria can grow and divide extremely rapidly, and bacterial populations can double as quickly as every 9.8 minutes. In cell division, two identical clone daughter cells are produced. Some bacteria, while still reproducing asexually, form more complex reproductive structures that help disperse the newly formed daughter cells. Examples include fruiting body formation by *Myxobacteria* and aerial hyphae formation by *Streptomyces*, or budding. Budding involves a cell forming a protrusion that breaks away and produces a daughter cell.

In the laboratory, bacteria are usually grown using solid or liquid media. Solid *growth media*, such as agar plates, are used to isolate pure cultures of a bacterial strain. However, liquid growth media are used when measurement of growth or large volumes of cells are required. Growth in stirred liquid media occurs as an even cell suspension, making the cultures easy to divide and transfer, although isolating single bacteria from liquid media is difficult. The use of selective media (media with specific nutrients added or deficient, or with antibiotics added) can help identify specific organisms.

A colony of *Escherichia coli*

Most laboratory techniques for growing bacteria use high levels of nutrients to produce large amounts of cells cheaply and quickly. However, in natural environments, nutrients are limited, meaning that bacteria cannot continue to reproduce indefinitely. This nutrient limitation has led the evolution of different growth strategies. Some organisms can grow extremely rapidly when nutrients become available, such as the formation of algal (and cyanobacterial) blooms that often occur in lakes during the summer. Other organisms have adaptations to harsh environments, such as the production of multiple antibiotics by *Streptomyces* that inhibit the growth of competing microorganisms. In nature, many organisms live in communities (e.g., biofilms) that may allow for increased supply of nutrients and protection from environmental stresses. These relationships can be essential for growth of a particular organism or group of organisms (syntrophy).

Bacterial growth follows four phases. When a population of bacteria first enter a high-nutrient environment that allows growth, the cells need to adapt to their new environment. The first phase of growth is the *lag phase*, a period of slow growth when the cells are adapting to the high-nutrient environment and preparing for fast growth. The lag phase has high biosynthesis rates, as proteins necessary for rapid growth are produced. The second phase of growth is the *log phase*, also known as the *logarithmic or exponential phase*. The log phase is marked by rapid exponential growth. The rate at which cells grow during this phase is known as the *growth rate* (k), and the time it takes the cells to double is known as the *generation time* (g). During log phase, nutrients are metabolised at maximum speed until one of the nutrients is depleted and starts limiting growth. The third phase of growth is the *stationary phase* and is caused by depleted nutrients. The cells reduce their metabolic activity and consume non-essential cellular proteins. The stationary phase is a transition from rapid growth to a stress response state and there is increased expression of genes involved in DNA repair, antioxidant metabolism and nutrient transport. The final phase is the *death phase* where the bacteria run out of nutrients and die.

Genomes

The genomes of thousands of bacterial species have been sequenced, with at least 9,000 sequences completed and more than 42,000 left as "permanent" drafts (as of Sep 2016).

Most bacteria have a single circular chromosome that can range in size from only 160,000 base pairs in the endosymbiotic bacteria *Candidatus Carsonella ruddii*, to 12,200,000 base pairs in the soil-dwelling bacteria *Sorangium cellulosum*. The genes in bacterial genomes are usually a single continuous stretch of DNA and although several different types of introns do exist in bacteria, these are much rarer than in eukaryotes. Some bacteria, including the Spirochaetes of the genus *Borrelia* are a notable exception to this arrangement. *Borrelia burgdorferi*, the cause of Lyme disease, contains a single linear chromosome and several linear and circular plasmids.

Plasmids are small extra-chromosomal DNAs that may contain genes for antibiotic resistance or virulence factors. Plasmids replicate independently of chromosomes, so it is possible that plasmids could be lost in bacterial cell division. Against this possibility is the fact that a single bacterium can contain hundreds of copies of a single plasmid.

Genetics

Bacteria, as asexual organisms, inherit identical copies of their parent's genes (i.e., they are clonal). However, all bacteria can evolve by selection on changes to their genetic material DNA caused by genetic recombination or mutations. Mutations come from errors made during the replication of DNA or from exposure to mutagens. Mutation rates vary widely among different species of bacteria and even among different clones of a single species of bacteria. Genetic changes in bacterial genomes come from either random mutation during replication or "stress-directed mutation", where genes involved in a particular growth-limiting process have an increased mutation rate.

DNA Transfer

Some bacteria also transfer genetic material between cells. This can occur in three main ways. First, bacteria can take up exogenous DNA from their environment, in a process called *transformation*. Genes can also be transferred by the process of *transduction*, when the integration of a bacteriophage introduces foreign DNA into the chromosome. The third method of gene transfer is *conjugation*, whereby DNA is transferred through direct cell contact.

Transduction of bacterial genes by bacteriophage appears to be a consequence of infrequent errors during intracellular assembly of virus particles, rather than a bacterial adaptation. Conjugation, in the much-studied E. coli system is determined by plasmid genes, and is an adaptation for transferring copies of the plasmid from one bacterial host to another. It is seldom that a conjugative plasmid integrates into the host bacterial chromosome, and subsequently transfers part of the host bacterial DNA to another bacterium. Plasmid-mediated transfer of host bacterial DNA also appears to be an accidental process rather than a bacterial adaptation.

Transformation, unlike transduction or conjugation, depends on numerous bacterial gene products that specifically interact to perform this complex process, and thus transformation is clearly a bacterial adaptation for DNA transfer. In order for a bacterium to bind, take up and recombine donor DNA into its own chromosome, it must first enter a special physiological state termed competence. In *Bacillus subtilis*, about 40 genes are required for the development of competence. The length of DNA transferred during *B. subtilis* transformation can be between a third of a chromosome up to the whole chromosome. Transformation appears to be common among bacterial species, and thus far at least 60 species are known to have the natural ability to become competent for transformation. The development of competence in nature is usually associated with stressful environmental conditions, and seems to be an adaptation for facilitating repair of DNA damage in recipient cells.

In ordinary circumstances, transduction, conjugation, and transformation involve transfer of DNA between individual bacteria of the same species, but occasionally transfer may occur between individuals of different bacterial species and this may have significant consequences, such as the transfer of antibiotic resistance. In such cases, gene acquisition from other bacteria or the environment is called *horizontal gene transfer* and may be common under natural conditions. Gene transfer is particularly important in antibiotic resistance as it allows the rapid transfer of resistance genes between different pathogens.

Bacteriophages

Bacteriophages are viruses that infect bacteria. Many types of bacteriophage exist, some simply infect and lyse their host bacteria, while others insert into the bacterial chromosome. A bacteriophage can contain genes that contribute to its host's phenotype: for example, in the evolution of *Escherichia coli* O157:H7 and *Clostridium botulinum*, the toxin genes in an integrated phage converted a harmless ancestral bacterium into a lethal pathogen. Bacteria resist phage infection through restriction modification systems that degrade foreign DNA, and a system that uses CRISPR sequences to retain fragments of the genomes of phage that the bacteria have come into contact with in the past, which allows them to block virus replication through a form of RNA interference. This CRISPR system provides bacteria with acquired immunity to infection.

Behaviour

Secretion

Bacteria frequently secrete chemicals into their environment in order to modify it favourably. The secretions are often proteins and may act as enzymes that digest some form of food in the environment.

Bioluminescence

A few bacteria have chemical systems that generate light. This bioluminescence often occurs in bacteria that live in association with fish, and the light probably serves to attract fish or other large animals.

Multicellularity

Bacteria often function as multicellular aggregates known as biofilms, exchanging a variety of molecular signals for inter-cell communication, and engaging in coordinated multicellular behaviour.

The communal benefits of multicellular cooperation include a cellular division of labour, accessing resources that cannot effectively be used by single cells, collectively defending against antagonists, and optimising population survival by differentiating into distinct cell types. For example, bacteria in biofilms can have more than 500 times increased resistance to antibacterial agents than individual "planktonic" bacteria of the same species.

One type of inter-cellular communication by a molecular signal is called quorum sensing, which serves the purpose of determining whether there is a local population density that is sufficiently high that it is productive to invest in processes that are only successful if large numbers of similar organisms behave similarly, as in excreting digestive enzymes or emitting light.

Quorum sensing allows bacteria to coordinate gene expression, and enables them to produce, release and detect autoinducers or pheromones which accumulate with the growth in cell population.

Movement

Many bacteria can move using a variety of mechanisms: flagella are used for swimming through fluids; bacterial gliding and twitching motility move bacteria across surfaces; and changes of buoyancy allow vertical motion.

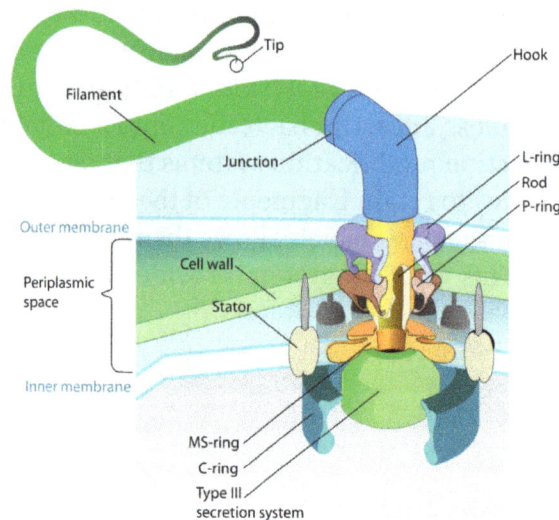

Flagellum of gram-negative bacteria. The base drives the rotation of the hook and filament.

Swimming bacteria frequently move near 10 body lengths per second and a few as fast as 100. This makes them at least as fast as fish, on a relative scale.

In bacterial gliding and twitching motility, bacteria use their *type IV pili* as a grappling hook, repeatedly extending it, anchoring it and then retracting it with remarkable force (>80 pN).

"Our observations redefine twitching motility as a rapid, highly organized mechanism of bacte-

rial translocation by which *Pseudomonas aeruginosa* can disperse itself over large areas to colonize new territories. It is also now clear, both morphologically and genetically, that twitching motility and social gliding motility, such as occurs in *Myxococcus xanthus*, are essentially the same process."

— *"A re-examination of twitching motility in Pseudomonas aeruginosa" – Semmler, Whitchurch & Mattick (1999)*

Flagella are semi-rigid cylindrical structures that are rotated and function much like the propeller on a ship. Objects as small as bacteria operate a low Reynolds number and cylindrical forms are more efficient than the flat, paddle-like, forms appropriate at human-size scale.

Bacterial species differ in the number and arrangement of flagella on their surface; some have a single flagellum (*monotrichous*), a flagellum at each end (*amphitrichous*), clusters of flagella at the poles of the cell (*lophotrichous*), while others have flagella distributed over the entire surface of the cell (*peritrichous*). The bacterial flagella is the best-understood motility structure in any organism and is made of about 20 proteins, with approximately another 30 proteins required for its regulation and assembly. The flagellum is a rotating structure driven by a reversible motor at the base that uses the electrochemical gradient across the membrane for power. This motor drives the motion of the filament, which acts as a propeller.

Many bacteria (such as *E. coli*) have two distinct modes of movement: forward movement (swimming) and tumbling. The tumbling allows them to reorient and makes their movement a three-dimensional random walk. The flagella of a unique group of bacteria, the spirochaetes, are found between two membranes in the periplasmic space. They have a distinctive helical body that twists about as it moves.

Motile bacteria are attracted or repelled by certain stimuli in behaviours called taxes: these include chemotaxis, phototaxis, energy taxis, and magnetotaxis. In one peculiar group, the myxobacteria, individual bacteria move together to form waves of cells that then differentiate to form fruiting bodies containing spores. The myxobacteria move only when on solid surfaces, unlike *E. coli*, which is motile in liquid or solid media.

Several *Listeria* and *Shigella* species move inside host cells by usurping the cytoskeleton, which is normally used to move organelles inside the cell. By promoting actin polymerisation at one pole of their cells, they can form a kind of tail that pushes them through the host cell's cytoplasm.

Classification and Identification

Classification seeks to describe the diversity of bacterial species by naming and grouping organisms based on similarities. Bacteria can be classified on the basis of cell structure, cellular metabolism or on differences in cell components, such as DNA, fatty acids, pigments, antigens and quinones. While these schemes allowed the identification and classification of bacterial strains, it was unclear whether these differences represented variation between distinct species or between strains of the same species. This uncertainty was due to the lack of distinctive structures in most bacteria, as well as lateral gene transfer between unrelated species. Due to lateral gene transfer, some closely related bacteria can have very different morphologies and metabolisms. To overcome this uncertainty, modern bacterial classification emphasises molecular systematics, using genetic techniques such as guanine cytosine ratio determination, genome-genome hybridisation, as well as sequencing genes that have not undergone extensive lateral gene transfer, such as the rRNA gene. Classification of bacteria is determined by publication in the International Journal of Systematic Bacteriology, and Bergey's Manual of Systematic Bacteriology. The International Committee on Systematic Bacteriology (ICSB) maintains international rules for the naming of bacteria and taxonomic categories and for the ranking of them in the International Code of Nomenclature of Bacteria.

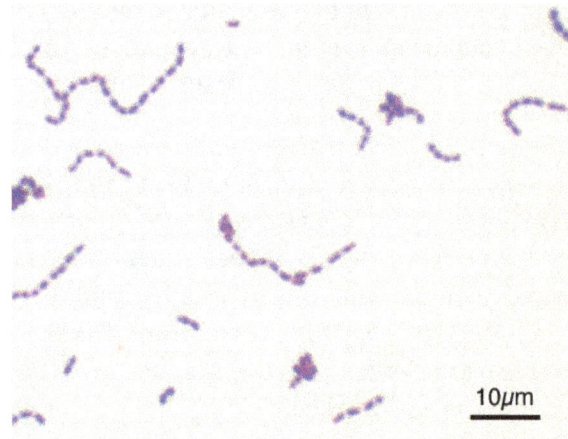

Streptococcus mutans visualised with a Gram stain

The term "bacteria" was traditionally applied to all microscopic, single-cell prokaryotes. However, molecular systematics showed prokaryotic life to consist of two separate domains, originally called *Eubacteria* and *Archaebacteria*, but now called *Bacteria* and *Archaea* that evolved independently from an ancient common ancestor. The archaea and eukaryotes are more closely related to each other than either is to the bacteria. These two domains, along with Eukarya, are the basis of the three-domain system, which is currently the most widely used classification system in microbiolology. However, due to the relatively recent introduction of molecular systematics and a rapid increase in the number of genome sequences that are available, bacterial classification remains a changing and expanding field. For example, a few biologists argue that the Archaea and Eukaryotes evolved from gram-positive bacteria.

The identification of bacteria in the laboratory is particularly relevant in medicine, where the correct treatment is determined by the bacterial species causing an infection. Consequently, the need to identify human pathogens was a major impetus for the development of techniques to identify bacteria.

The *Gram stain*, developed in 1884 by Hans Christian Gram, characterises bacteria based on the structural characteristics of their cell walls. The thick layers of peptidoglycan in the "gram-positive" cell wall stain purple, while the thin "gram-negative" cell wall appears pink. By combining morphology and Gram-staining, most bacteria can be classified as belonging to one of four groups (gram-positive cocci, gram-positive bacilli, gram-negative cocci and gram-negative bacilli). Some organisms are best identified by stains other than the Gram stain, particularly mycobacteria or *Nocardia*, which show acid-fastness on Ziehl–Neelsen or similar stains. Other organisms may need to be identified by their growth in special media, or by other techniques, such as serology.

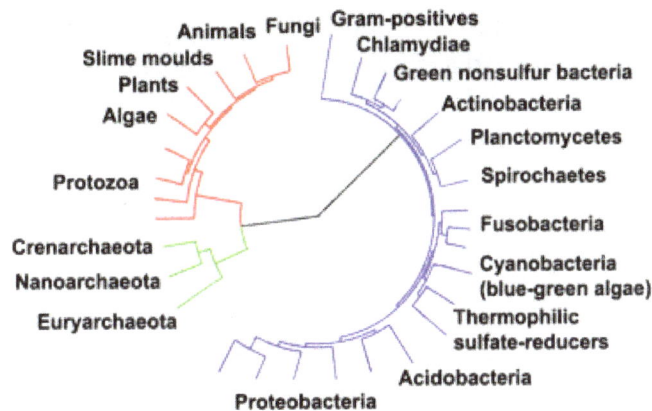

Phylogenetic tree showing the diversity of bacteria, compared to other organisms. Eukaryotes are coloured red, archaea green and bacteria blue.

Culture techniques are designed to promote the growth and identify particular bacteria, while restricting the growth of the other bacteria in the sample. Often these techniques are designed for specific specimens; for example, a sputum sample will be treated to identify organisms that cause pneumonia, while stool specimens are cultured on selective media to identify organisms that cause diarrhoea, while preventing growth of non-pathogenic bacteria. Specimens that are normally sterile, such as blood, urine or spinal fluid, are cultured under conditions designed to grow all possible organisms. Once a pathogenic organism has been isolated, it can be further characterised by its morphology, growth patterns (such as aerobic or anaerobic growth), patterns of hemolysis, and staining.

As with bacterial classification, identification of bacteria is increasingly using molecular methods. Diagnostics using DNA-based tools, such as polymerase chain reaction, are increasingly popular due to their specificity and speed, compared to culture-based methods. These methods also allow the detection and identification of "viable but nonculturable" cells that are metabolically active but non-dividing. However, even using these improved methods, the total number of bacterial species is not known and cannot even be estimated with any certainty. Following present classification, there are a little less than 9,300 known species of prokaryotes, which includes bacteria and archaea; but attempts to estimate the true number of bacterial diversity have ranged from 10^7 to 10^9 total species – and even these diverse estimates may be off by many orders of magnitude.

Interactions With Other Organisms

Despite their apparent simplicity, bacteria can form complex associations with other organisms. These symbiotic associations can be divided into parasitism, mutualism and commensalism. Due to their small size, commensal bacteria are ubiquitous and grow on animals and plants exactly as they will grow on any other surface. However, their growth can be increased by warmth and sweat, and large populations of these organisms in humans are the cause of body odour.

Predators

Some species of bacteria kill and then consume other microorganisms, these species are called *predatory bacteria.* These include organisms such as *Myxococcus xanthus*, which forms swarms of cells that kill and digest any bacteria they encounter. Other bacterial predators either attach to their prey in order to digest them and absorb nutrients, such as *Vampirovibrio chlorellavorus*, or invade another cell and multiply inside the cytosol, such as *Daptobacter*. These predatory bacteria are thought to have evolved from saprophages that consumed dead microorganisms, through adaptations that allowed them to entrap and kill other organisms.

Mutualists

Certain bacteria form close spatial associations that are essential for their survival. One such mutualistic association, called interspecies hydrogen transfer, occurs between clusters of anaerobic bacteria that consume organic acids, such as butyric acid or propionic acid, and produce hydrogen, and methanogenic Archaea that consume hydrogen. The bacteria in this association are unable to consume the organic acids as this reaction produces hydrogen that accumulates in their surroundings. Only the intimate association with the hydrogen-consuming Archaea keeps the hydrogen concentration low enough to allow the bacteria to grow.

In soil, microorganisms that reside in the rhizosphere (a zone that includes the root surface and the soil that adheres to the root after gentle shaking) carry out nitrogen fixation, converting nitrogen gas to nitrogenous compounds. This serves to provide an easily absorbable form of nitrogen for many plants, which cannot fix nitrogen themselves. Many other bacteria are found as symbionts in humans and other organisms. For example, the presence of over 1,000 bacterial species in the normal human gut flora of the intestines can contribute to gut immunity, synthesise vitamins, such as folic acid, vitamin K and biotin, convert sugars to lactic acid, as well as fermenting complex undigestible carbohydrates. The presence of this gut flora also inhibits the growth of potentially pathogenic bacteria (usually through competitive exclusion) and these beneficial bacteria are consequently sold as probiotic dietary supplements.

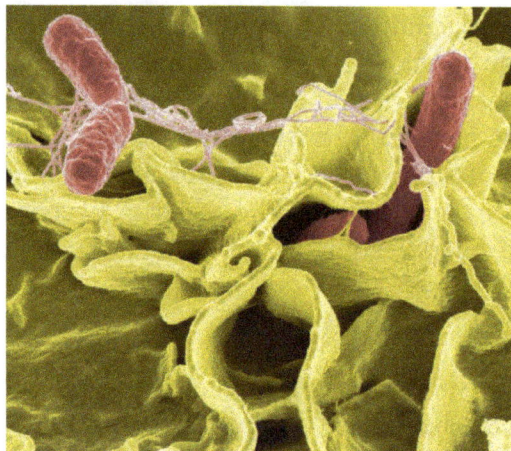

Colour-enhanced scanning electron micrograph showing Salmonella typhimurium (red) invading cultured human cells

Pathogens

If bacteria form a parasitic association with other organisms, they are classed as pathogens. Pathogenic bacteria are a major cause of human death and disease and cause infections such as tetanus, typhoid fever, diphtheria, syphilis, cholera, foodborne illness, leprosy and tuberculosis. A pathogenic cause for a known medical disease may only be discovered many years after, as was the case with *Helicobacter pylori* and peptic ulcer disease. Bacterial diseases are also important in agriculture, with bacteria causing leaf spot, fire blight and wilts in plants, as well as Johne's disease, mastitis, salmonella and anthrax in farm animals.

Each species of pathogen has a characteristic spectrum of interactions with its human hosts. Some organisms, such as *Staphylococcus* or *Streptococcus*, can cause skin infections, pneumonia, meningitis and even overwhelming sepsis, a systemic inflammatory response producing shock, massive vasodilation and death. Yet these organisms are also part of the normal human flora and usually exist on the skin or in the nose without causing any disease at all. Other organisms invariably cause disease in humans, such as the Rickettsia, which are obligate intracellular parasites able to grow and reproduce only within the cells of other organisms. One species of Rickettsia causes typhus, while another causes Rocky Mountain spotted fever. *Chlamydia*, another phylum of obligate intracellular parasites, contains species that can cause pneumonia, or urinary tract infection and may be involved in coronary heart disease. Finally, some species, such as *Pseudomonas aeruginosa*, *Burkholderia cenocepacia*, and *Mycobacterium avium*, are opportunistic pathogens and cause disease mainly in people suffering from immunosuppression or cystic fibrosis.

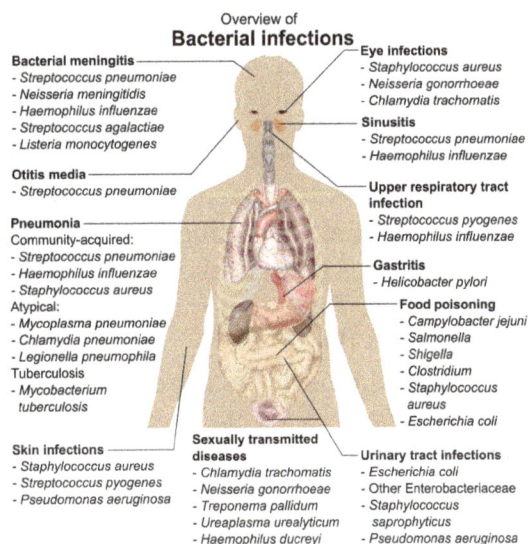

Overview of bacterial infections and main species involved.

Bacterial infections may be treated with antibiotics, which are classified as bacteriocidal if they kill bacteria, or bacteriostatic if they just prevent bacterial growth. There are many types of antibiotics and each class inhibits a process that is different in the pathogen from that found in the host. An example of how antibiotics produce selective toxicity are chloramphenicol and puromycin, which inhibit the bacterial ribosome, but not the structurally different eukaryotic ribosome. Antibiotics are used both in treating human disease and in intensive farming to promote animal growth, where they may be contributing to the rapid development of antibiotic resistance in bacterial populations. Infections can be prevented by antiseptic measures such as sterilising the skin prior to piercing it with the needle of a syringe, and by proper care of indwelling catheters. Surgical and dental instruments are also sterilised to prevent contamination by bacteria. Disinfectants such as bleach are used to kill bacteria or other pathogens on surfaces to prevent contamination and further reduce the risk of infection.

Significance in Technology and Industry

Bacteria, often lactic acid bacteria, such as *Lactobacillus* and *Lactococcus*, in combination with yeasts and moulds, have been used for thousands of years in the preparation of fermented foods, such as cheese, pickles, soy sauce, sauerkraut, vinegar, wine and yogurt.

The ability of bacteria to degrade a variety of organic compounds is remarkable and has been used in waste processing and bioremediation. Bacteria capable of digesting the hydrocarbons in petroleum are often used to clean up oil spills. Fertiliser was added to some of the beaches in Prince William Sound in an attempt to promote the growth of these naturally occurring bacteria after the 1989 *Exxon Valdez* oil spill. These efforts were effective on beaches that were not too thickly covered in oil. Bacteria are also used for the bioremediation of industrial toxic wastes. In the chemical industry, bacteria are most important in the production of enantiomerically pure chemicals for use as pharmaceuticals or agrichemicals.

Bacteria can also be used in the place of pesticides in the biological pest control. This commonly involves *Bacillus thuringiensis* (also called BT), a gram-positive, soil dwelling bacterium. Subspecies of this bacteria are used as a Lepidopteran-specific insecticides under trade names such as Dipel and Thuricide. Because of their specificity, these pesticides are regarded as environmentally friendly, with little or no effect on humans, wildlife, pollinators and most other beneficial insects.

Because of their ability to quickly grow and the relative ease with which they can be manipulated, bacteria are the workhorses for the fields of molecular biology, genetics and biochemistry. By making mutations in bacterial DNA and examining the resulting phenotypes, scientists can determine the function of genes, enzymes and metabolic pathways in bacteria, then apply this knowledge to more complex organisms. This aim of understanding the biochemistry of a cell reaches its most complex expression in the synthesis of huge amounts of enzyme kinetic and gene expression data into mathematical models of entire organisms. This is achievable in some well-studied bacteria, with models of *Escherichia coli* metabolism now being produced and tested. This understanding of bacterial metabolism and genetics allows the use of biotechnology to bioengineer bacteria for the production of therapeutic proteins, such as insulin, growth factors, or antibodies.

Because of their importance for research in general, samples of bacterial strains are isolated and preserved in Biological Resource Centers. This ensures the availability of the strain to scientists worldwide.

History of Bacteriology

Bacteria were first observed by the Dutch microscopist Antonie van Leeuwenhoek in 1676, using a single-lens microscope of his own design. He then published his observations in a series of letters to the Royal Society of London. Bacteria were Leeuwenhoek's most remarkable microscopic discovery. They were just at the limit of what his simple lenses could make out and, in one of the most striking hiatuses in the history of science, no one else would see them again for over a century.

Only then were his by-then-largely-forgotten observations of bacteria — as opposed to his famous "animalcules" (spermatozoa) — taken seriously.

Christian Gottfried Ehrenberg introduced the word "bacterium" in 1828. In fact, his *Bacterium* was a genus that contained non-spore-forming rod-shaped bacteria, as opposed to *Bacillus*, a genus of spore-forming rod-shaped bacteria defined by Ehrenberg in 1835.

Louis Pasteur demonstrated in 1859 that the growth of microorganisms causes the fermentation process, and that this growth is not due to spontaneous generation. (Yeasts and moulds, commonly associated with fermentation, are not bacteria, but rather fungi.) Along with his contemporary Robert Koch, Pasteur was an early advocate of the germ theory of disease.

Robert Koch, a pioneer in medical microbiology, worked on cholera, anthrax and tuberculosis. In his research into tuberculosis Koch finally proved the germ theory, for which he received a Nobel Prize in 1905. In *Koch's postulates*, he set out criteria to test if an organism is the cause of a disease, and these postulates are still used today.

Though it was known in the nineteenth century that bacteria are the cause of many diseases, no effective antibacterial treatments were available. In 1910, Paul Ehrlich developed the first antibiotic, by changing dyes that selectively stained *Treponema pallidum* — the spirochaete that causes syphilis — into compounds that selectively killed the pathogen. Ehrlich had been awarded a 1908 Nobel Prize for his work on immunology, and pioneered the use of stains to detect and identify bacteria, with his work being the basis of the Gram stain and the Ziehl–Neelsen stain.

A major step forward in the study of bacteria came in 1977 when Carl Woese recognised that archaea have a separate line of evolutionary descent from bacteria. This new phylogenetic taxonomy depended on the sequencing of 16S ribosomal RNA, and divided prokaryotes into two evolutionary domains, as part of the three-domain system.

Archaea

The Archaea (ⁱ/ɑːraraɪalkiːə/ or /ɑːrˈkeɪə/ *ar-KEE-ə* or *ar-KAY-ə*) constitute a domain and kingdom of single-celled microorganisms. These microbes (archaea; singular archaeon) are prokaryotes, meaning that they have no cell nucleus or any other membrane-bound organelles in their cells.

Archaea were initially classified as bacteria, receiving the name archaebacteria (in the Archaebacteria kingdom), but this classification is outdated. Archaeal cells have unique properties separating them from the other two domains of life, Bacteria and Eukaryota. The Archaea are further divided into multiple recognized phyla. Classification is difficult because the majority have not been isolated in the laboratory and have only been detected by analysis of their nucleic acids in samples from their environment.

Archaea and bacteria are generally similar in size and shape, although a few archaea have very strange shapes, such as the flat and square-shaped cells of *Haloquadratum walsbyi*. Despite this morphological similarity to bacteria, archaea possess genes and several metabolic pathways that are more closely related to those of eukaryotes, notably the enzymes involved in transcription and translation. Other aspects of archaeal biochemistry are unique, such as their reliance on ether lipids in their cell membranes, including archaeols. Archaea use more energy sources than eukaryotes: these range from organic compounds, such as sugars, to ammonia, metal ions or even hydrogen gas. Salt-tolerant archaea (the Haloarchaea) use sunlight as an energy source, and other species of archaea fix carbon; however, unlike plants and cyanobacteria, no known species of archaea does both. Archaea reproduce asexually by binary fission, fragmentation, or budding; unlike bacteria and eukaryotes, no known species forms spores.

Archaea were initially viewed as extremophiles living in harsh environments, such as hot springs and salt lakes, but they have since been found in a broad range of habitats, including soils, oceans, marshlands and the human colon, oral cavity, and skin. Archaea are particularly numerous in the oceans, and the archaea in plankton may be one of the most abundant groups of organisms on the planet. Archaea are a major part of Earth's life and may play roles in both the carbon cycle and the nitrogen cycle. No clear examples of archaeal pathogens or parasites are known, but they are often mutualists or commensals. One example is the methanogens that inhabit human and ruminant guts, where their vast numbers aid digestion. Methanogens are also used in biogas production and sewage treatment, and enzymes from extremophile archaea that can endure high temperatures and organic solvents are exploited in biotechnology.

Classification

New Domain

For much of the 20th century, prokaryotes were regarded as a single group of organisms and classified based on their biochemistry, morphology and metabolism. For example, microbiologists tried to classify microorganisms based on the structures of their cell walls, their shapes, and the substances they consume. In 1965, Emile Zuckerkandl and Linus Pauling proposed instead using the sequences of the genes in different prokaryotes to work out how they are related to each other. This approach, known as phylogenetics, is the main method used today.

Archaea were first classified as a separate group of prokaryotes in 1977 by Carl Woese and George E. Fox in phylogenetic trees based on the sequences of ribosomal RNA (rRNA) genes. These two groups were originally named the Archaebacteria and Eubacteria and treated as kingdoms or sub-kingdoms, which Woese and Fox termed *Urkingdoms*. Woese argued that this group of prokaryotes is a fundamentally different sort of life. To emphasize this difference, Woese later proposed a new natural system of organisms with three separate Domains: the Eukarya, the Bacteria and the Archaea, in what is now known as "The Woesian Revolution".

The word *archaea* comes from the Ancient meaning "ancient things", as the first representatives of the domain Archaea were methanogens and it was assumed that their metabolism reflected Earth's primitive atmosphere and the organism's antiquity. For a long time, archaea were seen as extremophiles that only exist in extreme habitats such as hot springs and salt lakes. However, as new habitats were studied, more organisms were discovered. Extreme halophilic and hyperther-mophilic microbes were also included in the Archaea. By the end of the 20th century, archaea had been identified in non-extreme environments as well. Today, they are known to be a large and diverse group of organisms that are widely distributed in nature and are common in all habitats.

This new appreciation of the importance and ubiquity of archaea came from using polymerase chain reaction (PCR) to detect prokaryotes from environmental samples (such as water or soil) by multiplying their ribosomal genes. This allows the detection and identification of organisms that have not been cultured in the laboratory.

Current Classification

The classification of archaea, and of prokaryotes in general, is a rapidly moving and contentious field. Current classification systems aim to organize archaea into groups of organisms that share structural features and common ancestors. These classifications rely heavily on the use of the sequence of ribosomal RNA genes to reveal relationships between organisms (molecular phylogenetics). Most of the culturable and well-investigated species of archaea are members of two main phyla, the Euryarchaeota and Crenarchaeota. Other groups have been tentatively created. For example, the peculiar species *Nanoarchaeum equitans*, which was discovered in 2003, has been given its own phylum, the Nanoarchaeota. A new phylum Korarchaeota has also been proposed. It contains a small group of unusual thermophilic species that shares features of both of the main phyla, but is most closely related to the Crenarchaeota. Other recently detected species of archaea are only distantly related to any of these groups, such as the Archaeal Richmond Mine acidophilic nanoorganisms (ARMAN), which were discovered in 2006 and are some of the smallest organisms known.

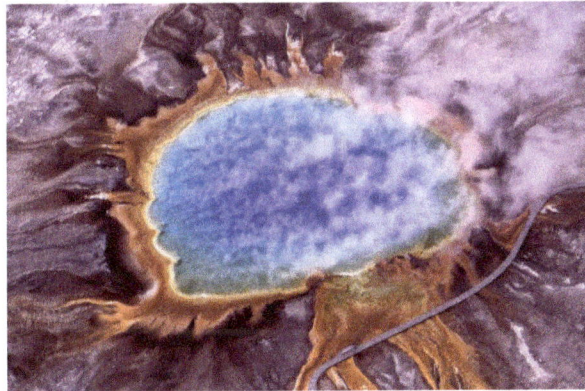

Archaea were first found in extreme environments, such as volcanic hot springs. Pictured here is Grand Prismatic Spring of Yellowstone National Park.

A superphylum - TACK - has been proposed that includes the Thaumarchaeota, Aigarchaeota, Crenarchaeota, and Korarchaeota. This superphylum may be related to the origin of eukaryotes.

Concept of Species

The classification of archaea into species is also controversial. Biology defines a species as a group of related organisms. The familiar exclusive breeding criterion (organisms that can breed with each other but not with others) is of no help here because archaea reproduce asexually.

Archaea show high levels of horizontal gene transfer between lineages. Some researchers suggest that individuals can be grouped into species-like populations given highly similar genomes and infrequent gene transfer to/from cells with less-related genomes, as in the genus *Ferroplasma*. On the other hand, studies in *Halorubrum* found significant genetic transfer to/from less-related populations, limiting the criterion's applicability. A second concern is to what extent such species designations have practical meaning.

The ARMAN are a new group of archaea recently discovered in acid mine drainage.

Current knowledge on genetic diversity is fragmentary and the total number of archaeal species cannot be estimated with any accuracy. Estimates of the number of phyla range from 18 to 23, of which only 8 have representatives that have been cultured and studied directly. Many of these hypothesized groups are known from a single rRNA sequence, indicating that the diversity among these organisms remains obscure. The Bacteria also contain many uncultured microbes with similar implications for characterization.

Origin and Evolution

The age of the Earth is about 4.54 billion years old. Scientific evidence suggests that life began on Earth at least 3.5 billion years ago. The earliest evidence for life on Earth is graphite found to be biogenic in 3.7 billion-year-old metasedimentary rocks discovered in Western Greenland and microbial mat fossils found in 3.48 billion-year-old sandstone discovered in Western Australia. More recently, in 2015, "remains of biotic life" were found in 4.1 billion-year-old rocks in Western Australia. An unrelated researcher, S. Blair Hedges, wrote in an email about the study saying, "If life arose relatively quickly on Earth ... then it could be common in the universe."

Although probable prokaryotic cell fossils date to almost 3.5 billion years ago, most prokaryotes do not have distinctive morphologies and fossil shapes cannot be used to identify them as archaea. Instead, chemical fossils of unique lipids are more informative because such compounds do not occur in other organisms. Some publications suggest that archaeal or eukaryotic lipid remains are present in shales dating from 2.7 billion years ago; such data have since been questioned. Such lipids have also been detected in even older rocks from west Greenland. The oldest such traces come from the Isua district, which include Earth's oldest known sediments, formed 3.8 billion years ago. The archaeal lineage may be the most ancient that exists on Earth.

Woese argued that the bacteria, archaea, and eukaryotes represent separate lines of descent that diverged early on from an ancestral colony of organisms. One possibility is that this occurred before the evolution of cells, when the lack of a typical cell membrane allowed unrestricted lateral gene transfer, and that the common ancestors of the three domains arose by fixation of specific subsets of genes. It is possible that the last common ancestor of the bacteria and archaea was a thermophile, which raises the possibility that lower temperatures are "extreme environments" in archaeal terms, and organisms that live in cooler environments appeared only later. Since the Archaea and Bacteria are no more related to each other than they are to eukaryotes, the term *prokaryote*'s only surviving meaning is "not a eukaryote", limiting its value.

Comparison to Other Domains

The following table compares some major characteristics of the three domains, to illustrate their similarities and differences. Many of these characteristics are also discussed below.

Property	Archaea	Bacteria	Eukarya
Cell Membrane	Ether-linked lipids, pseudopeptidoglycan	Ester-linked lipids, peptidoglycan	Ester-linked lipids, various structures
Gene Structure	Circular chromosomes, similar translation and transcription to Eukarya	Circular chromosomes, unique translation and transcription	Multiple, linear chromosomes, similar translation and transcription to Archaea
Internal Cell Structure	No membrane-bound organelles or nucleus	No membrane-bound organelles or nucleus	Membrane-bound organelles and nucleus
Metabolism	Various, with methanogenesis unique to Archaea	Various, including photosynthesis, aerobic and anaerobic respiration, fermentation, and autotrophy	Photosynthesis and cellular respiration
Reproduction	Asexual reproduction, horizontal gene transfer	Asexual reproduction, horizontal gene transfer	Sexual and asexual reproduction

Archaea were split off as a third domain because of the large differences in their ribosomal RNA structure. The particular RNA molecule sequenced, known as 16s rRNA, is present in all organisms and always has the same vital function, the production of proteins. Because this function is so central to life, organisms with mutations of its 16s rRNA are unlikely to survive, leading to great stability in the structure of this nucleotide over many generations. 16s rRNA is also large enough to retain organism-specific information, but small enough to be sequenced in a manageable amount of time. In 1977, Carl Woese, a microbiologist studying the genetic sequencing of organisms, developed a new sequencing method that involved splitting the RNA into fragments that could be sorted and compared to other fragments from other organisms. The more similar the patterns between species were, the more closely related the organisms.

Woese used his new rRNA comparison method to categorize and contrast different organisms. He sequenced a variety of different species and happened upon a group of methanogens that had vastly different patterns than any known prokaryotes or eukaryotes. These methanogens were much more similar to each other than they were to other organisms sequenced, leading Woese to propose the new domain of Archaea. One of the interesting results of his experiments was that the Archaea were more similar to eukaryotes than prokaryotes, even though they were more similar to prokaryotes in structure. This led to the conclusion that Archaea and Eukarya shared a more recent common ancestor than Eukarya and Bacteria in general. The development of the nucleus occurred after the split between Bacteria and this common ancestor. Although Archaea are pro-karyotic, they are more closely related to Eukarya and thus cannot be placed within either the Bacteria or Eukarya domains.

One property unique to Archaea is the abundant use of ether-linked lipids in their cell membranes. Ether linkages are more chemically stable than the ester linkages found in Bacteria and Eukarya, which may be a contributing factor to the ability of many Archaea to survive in extreme envi-ronments that place heavy stress on cell membranes, such as extreme heat and salinity. Another unique feature of Archaea is that no other known organisms are capable of methanogenesis (the metabolic production of methane). Methanogenic Archaea play a pivotal role in ecosystems with organisms that derive energy from oxidation of methane, many of which are Bacteria, as they are often a major source of methane in such environments and can play a role as primary producers. Methanogens also play a critical role in the carbon cycle, breaking down organic carbon into meth-ane, which is also a major greenhouse gas.

Relationship to Other Prokaryotes

The relationship between the three domains is of central importance for understanding the origin of life. Most of the metabolic pathways, which comprise the majority of an organism's genes, are common between Archaea and Bacteria, while most genes involved in genome expression are com-mon between Archaea and Eukarya. Within prokaryotes, archaeal cell structure is most similar to that of gram-positive bacteria, largely because both have a single lipid bilayer and usually contain a thick sacculus (exoskeleton) of varying chemical composition. In some phylogenetic trees based upon different gene/protein sequences of prokaryotic homologs, the archaeal homologs are more closely related to those of gram-positive bacteria. Archaea and gram-positive bacteria also share conserved indels in a number of important proteins, such as Hsp70 and glutamine synthetase I; however, the phylogeny of these genes was interpreted to reveal interdomain gene transfer, and might not reflect the organismal relationship(s).

It has been proposed that the archaea evolved from gram-positive bacteria in response to antibiot-ic selection pressure. This is suggested by the observation that archaea are resistant to a wide vari-ety of antibiotics that are primarily produced by gram-positive bacteria, and that these antibiotics primarily act on the genes that distinguish archaea from bacteria. The proposal is that the selective pressure towards resistance generated by the gram-positive antibiotics was eventually sufficient to cause extensive changes in many of the antibiotics' target genes, and that these strains represented the common ancestors of present-day Archaea. The evolution of Archaea in response to antibiotic selection, or any other competitive selective pressure, could also explain their adaptation to ex-treme environments (such as high temperature or acidity) as the result of a search for unoccupied niches to escape from antibiotic-producing organisms; Cavalier-Smith has made a similar sugges-tion. This proposal is also supported by other work investigating protein structural relationships and studies that suggest that gram-positive bacteria may constitute the earliest branching lineages within the prokaryotes.

Relation to Eukaryotes

The evolutionary relationship between archaea and eukaryotes remains unclear. Aside from the similarities in cell structure and function that are discussed below, many genetic trees group the two.

Complicating factors include claims that the relationship between eukaryotes and the archaeal phylum Crenarchaeota is closer than the relationship between the Euryarchaeota and the phylum Crenarchaeota and the presence of archaea-like genes in certain bacteria, such as *Thermotoga maritima*, from horizontal gene transfer. The standard hypothesis states that the ancestor of the eukaryotes diverged early from the Archaea, and that eukaryotes arose through fusion of an archaean and eubacterium, which became the nucleus and cytoplasm; this explains various genetic similarities but runs into difficulties explaining cell structure. An alternative hypothesis, the eocyte hypothesis, posits that Eukaryota emerged relatively late from the Archaea.

A recently discovered lineage of archaea, *Lokiarchaeum*, named for a hydrothermal vent called Loki's Castle in the Arctic Ocean, has been found to be most closely related to eukaryotes. It has been called a transitional organism between prokaryotes and eukaryotes.

Morphology

Individual archaea range from 0.1 micrometers (μm) to over 15 μm in diameter, and occur in various shapes, commonly as spheres, rods, spirals or plates. Other morphologies in the Crenarchaeota include irregularly shaped lobed cells in *Sulfolobus*, needle-like filaments that are less than half a micrometer in diameter in *Thermofilum*, and almost perfectly rectangular rods in *Thermoproteus* and *Pyrobaculum*. *Haloquadratum walsbyi* are flat, square archaea that live in hypersaline pools. These unusual shapes are probably maintained both by their cell walls and a prokaryotic cytoskeleton. Proteins related to the cytoskeleton components of other organisms exist in archaea, and filaments form within their cells, but in contrast to other organisms, these cellular structures are poorly understood. In *Thermoplasma* and *Ferroplasma* the lack of a cell wall means that the cells have irregular shapes, and can resemble amoebae.

Some species form aggregates or filaments of cells up to 200 μm long. These organisms can be prominent in biofilms. Notably, aggregates of *Thermococcus coalescens* cells fuse together in culture, forming single giant cells. Archaea in the genus *Pyrodictium* produce an elaborate multicell colony involving arrays of long, thin hollow tubes called *cannulae* that stick out from the cells' surfaces and connect them into a dense bush-like agglomeration. The function of these cannulae is not settled, but they may allow communication or nutrient exchange with neighbors. Multi-species colonies exist, such as the "string-of-pearls" community that was discovered in 2001 in a German swamp. Round whitish colonies of a novel Euryarchaeota species are spaced along thin filaments that can range up to 15 centimetres (5.9 in) long; these filaments are made of a particular bacteria species.

Structure, Composition Development, and Operation

Archaea and bacteria have generally similar cell structure, but cell composition and organization set the archaea apart. Like bacteria, archaea lack interior membranes and organelles. Like bacteria, archaea cell membranes are usually bounded by a cell wall and they swim using one or more flagella. Structurally, archaea are most similar to gram-positive bacteria. Most have a single plasma membrane and cell wall, and lack a periplasmic space; the exception to this general rule is

Ignicoccus, which possess a particularly large periplasm that contains membrane-bound vesicles and is enclosed by an outer membrane.

Membranes

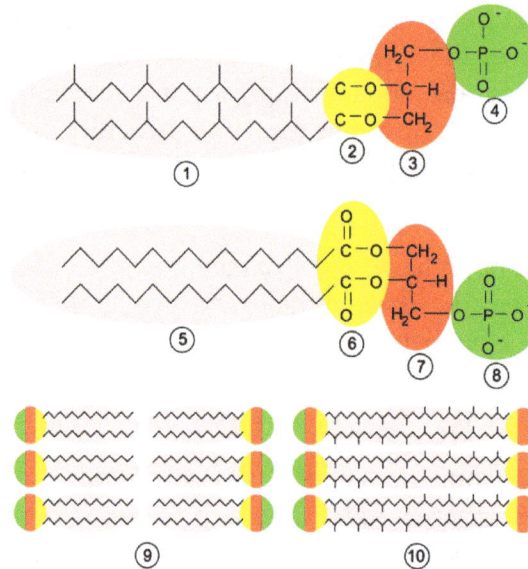

Membrane structures. Top, an archaeal phospholipid: 1, isoprene chains; 2, ether linkages; 3, L-glycerol moiety; 4, phosphate group. Middle, a bacterial or eukaryotic phospholipid: 5, fatty acid chains; 6, ester linkages; 7, D-glycerol moiety; 8, phosphate group. Bottom: 9, lipid bilayer of bacteria and eukaryotes; 10, lipid monolayer of some archaea.

Archaeal membranes are made of molecules that differ strongly from those in other life forms, showing that archaea are related only distantly to bacteria and eukaryotes. In all organisms, cell membranes are made of molecules known as phospholipids. These molecules possess both a polar part that dissolves in water (the phosphate "head"), and a "greasy" non-polar part that does not (the lipid tail). These dissimilar parts are connected by a glycerol moiety. In water, phospholipids cluster, with the heads facing the water and the tails facing away from it. The major structure in cell membranes is a double layer of these phospholipids, which is called a lipid bilayer.

These phospholipids are unusual in four ways:

- Bacteria and eukaryotes have membranes composed mainly of glycerol-ester lipids, whereas archaea have membranes composed of glycerol-ether lipids. The difference is the type of bond that joins the lipids to the glycerol moiety; the two types are shown in yellow in the figure at the right. In ester lipids this is an ester bond, whereas in ether lipids this is an ether bond. Ether bonds are chemically more resistant than ester bonds. This stability might help archaea to survive extreme temperatures and very acidic or alkaline environments. Bacteria and eukaryotes do contain some ether lipids, but in contrast to archaea these lipids are not a major part of their membranes.

- The stereochemistry of the glycerol moiety is the reverse of that found in other organisms. The glycerol moiety can occur in two forms that are mirror images of one another, called the right-handed and left-handed forms; in chemistry these are called *enantiomers*. Just as a right hand does not fit easily into a left-handed glove, a right-handed phospholipid generally cannot be used or made by enzymes adapted for the left-handed form. This suggests that archaea use entirely different enzymes for synthesizing phospholipids than do bacteria and eukaryotes. Such enzymes developed very early in life's history, suggesting an early split from the other two domains.

- Archaeal lipid tails are chemically different from other organisms. Archaeal lipids are based upon the isoprenoid sidechain and are long chains with multiple side-branches and sometimes even cyclopropane or cyclohexane rings. This is in contrast to the fatty acids found in other organisms' membranes, which have straight chains with no branches or rings. Although isoprenoids play an important role in the biochemistry of many organisms, only the archaea use them to make phospholipids. These branched chains may help prevent archaeal membranes from leaking at high temperatures.

- In some archaea, the lipid bilayer is replaced by a monolayer. In effect, the archaea fuse the tails of two independent phospholipid molecules into a single molecule with two polar heads (a bolaamphiphile); this fusion may make their membranes more rigid and better able to resist harsh environments. For example, the lipids in *Ferroplasma* are of this type, which is thought to aid this organism's survival in its highly acidic habitat.

Cell Wall and Flagella

Most archaea (but not *Thermoplasma* and *Ferroplasma*) possess a cell wall. In most archaea the wall is assembled from surface-layer proteins, which form an S-layer. An S-layer is a rigid array of protein molecules that cover the outside of the cell (like chain mail). This layer provides both chemical and physical protection, and can prevent macromolecules from contacting the cell membrane. Unlike bacteria, archaea lack peptidoglycan in their cell walls. Methanobacteriales do have cell walls containing pseudopeptidoglycan, which resembles eubacterial peptidoglycan in morphology, function, and physical structure, but pseudopeptidoglycan is distinct in chemical structure; it lacks D-amino acids and N-acetylmuramic acid.

Archaea flagella operate like bacterial flagella—their long stalks are driven by rotatory motors at the base. These motors are powered by the proton gradient across the membrane. However, archaeal flagella are notably different in composition and development. The two types of flagella evolved from different ancestors. The bacterial flagellum shares a common ancestor with the type III secretion system, while archaeal flagella appear to have evolved from bacterial type IV pili. In contrast to the bacterial flagellum, which is hollow and is assembled by subunits moving up the central pore to the tip of the flagella, archaeal flagella are synthesized by adding subunits at the base.

Metabolism

Archaea exhibit a great variety of chemical reactions in their metabolism and use many sources of energy. These reactions are classified into nutritional groups, depending on energy and carbon sources. Some archaea obtain energy from inorganic compounds such as sulfur or ammonia (they are lithotrophs). These include nitrifiers, methanogens and anaerobic methane oxidisers. In these reactions one compound passes electrons to another (in a redox reaction), releasing energy to fuel the cell's activities. One compound acts as an electron donor and one as an electron acceptor. The energy released generates adenosine triphosphate (ATP) through chemiosmosis, in the same basic process that happens in the mitochondrion of eukaryotic cells.

Archaea that grow in the hot water of the Morning Glory Hot Spring in Yellowstone National Park produce a bright colour

Other groups of archaea use sunlight as a source of energy (they are phototrophs). However, oxygen–generating photosynthesis does not occur in any of these organisms. Many basic metabolic pathways are shared between all forms of life; for example, archaea use a modified form of glycolysis (the Entner–Doudoroff pathway) and either a complete or partial citric acid cycle. These similarities to other organisms probably reflect both early origins in the history of life and their high level of efficiency.

Nutritional types in archaeal metabolism			
Nutritional type	**Source of energy**	**Source of carbon**	**Examples**
Phototrophs	Sunlight	Organic compounds	*Halobacterium*
Lithotrophs	Inorganic compounds	Organic compounds or carbon fixation	*Ferroglobus, Methanobacteria* or *Pyrolobus*
Organotrophs	Organic compounds	Organic compounds or carbon fixation	*Pyrococcus, Sulfolobus* or *Methanosarcinales*

Some Euryarchaeota are methanogens living in anaerobic environments, such as swamps. This form of metabolism evolved early, and it is even possible that the first free-living organism was a methanogen. A common reaction involves the use of carbon dioxide as an electron acceptor to oxidize hydrogen. Methanogenesis involves a range of coenzymes that are unique to these archaea, such as coenzyme M and methanofuran. Other organic compounds such as alcohols, acetic acid or formic acid are used as alternative electron acceptors by methanogens. These reactions are common in gut-dwelling archaea. Acetic acid is also broken down into methane and carbon dioxide directly, by *acetotrophic* archaea. These acetotrophs are archaea in the order Methanosarcinales, and are a major part of the communities of microorganisms that produce biogas.

Other archaea use CO_2 in the atmosphere as a source of carbon, in a process called carbon fixation (they are autotrophs). This process involves either a highly modified form of the Calvin cycle or a recently discovered metabolic pathway called the 3-hydroxypropionate/4-hydroxybutyrate cycle. The Crenarchaeota also use the reverse Krebs cycle while the Euryarchaeota also use the reductive acetyl-CoA pathway. Carbon–fixation is powered by inorganic energy sources. No known archaea carry out photosynthesis. Archaeal energy sources are extremely diverse, and range from the oxidation of ammonia by the Nitrosopumilales to the oxidation of hydrogen sulfide or elemental sulfur by species of *Sulfolobus*, using either oxygen or metal ions as electron acceptors.

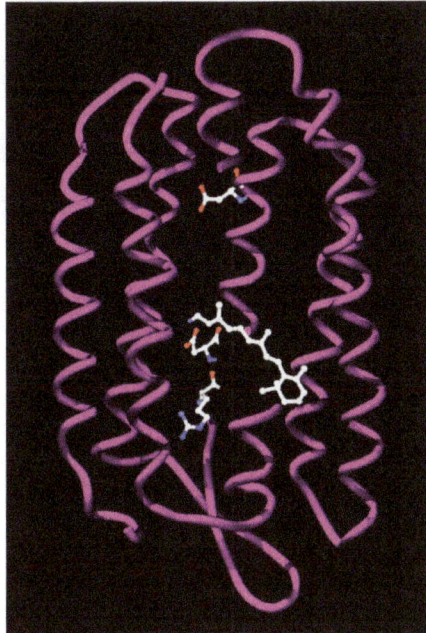

Bacteriorhodopsin from Halobacterium salinarum. The retinol cofactor and residues involved in proton transfer are shown as ball-and-stick models.

Phototrophic archaea use light to produce chemical energy in the form of ATP. In the Halobacteria, light-activated ion pumps like bacteriorhodopsin and halorhodopsin generate ion gradients by pumping ions out of the cell across the plasma membrane. The energy stored in these electrochemical gradients is then converted into ATP by ATP synthase. This process is a form of photophosphorylation. The ability of these light-driven pumps to move ions across membranes depends on light-driven changes in the structure of a retinol cofactor buried in the center of the protein.

Genetics

Archaea usually have a single circular chromosome, the size of which may be as great as 5,751,492 base pairs in *Methanosarcina acetivorans*, the largest known archaeal genome. The tiny 490,885 base-pair genome of *Nanoarchaeum equitans* is one-tenth of this size and the smallest archaeal genome known; it is estimated to contain only 537 protein-encoding genes. Smaller independent pieces of DNA, called *plasmids*, are also found in archaea. Plasmids may be transferred between cells by physical contact, in a process that may be similar to bacterial conjugation.

Archaea can be infected by double-stranded DNA viruses that are unrelated to any other form of virus and have a variety of unusual shapes, including bottles, hooked rods, or teardrops. These viruses have been studied in most detail in thermophilics, particularly the orders Sulfolobales and

Thermoproteales. Two groups of single-stranded DNA viruses that infect archaea have been recently isolated. One group is exemplified by the *Halorubrum pleomorphic virus 1* ("Pleolipoviridae") infecting halophilic archaea and the other one by the Aeropyrum coil-shaped virus ("Spiraviridae") infecting a hyperthermophilic (optimal growth at 90–95 °C) host. Notably, the latter virus has the largest currently reported ssDNA genome. Defenses against these viruses may involve RNA interference from repetitive DNA sequences that are related to the genes of the viruses.

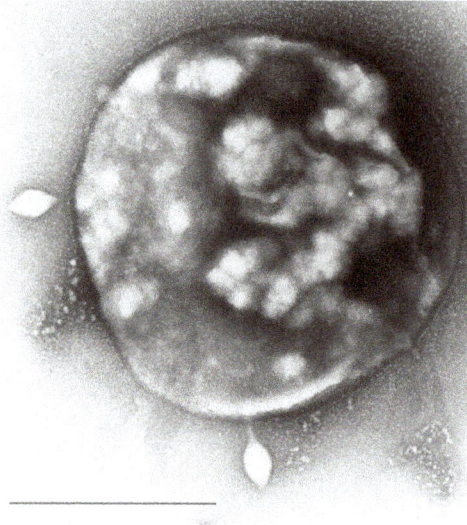

Sulfolobus infected with the DNA virus STSV1. Bar is 1 micrometer.

Archaea are genetically distinct from bacteria and eukaryotes, with up to 15% of the proteins encoded by any one archaeal genome being unique to the domain, although most of these unique genes have no known function. Of the remainder of the unique proteins that have an identified function, most belong to the Euryarchaea and are involved in methanogenesis. The proteins that archaea, bacteria and eukaryotes share form a common core of cell function, relating mostly to transcription, translation, and nucleotide metabolism. Other characteristic archaeal features are the organization of genes of related function—such as enzymes that catalyze steps in the same metabolic pathway into novel operons, and large differences in tRNA genes and their aminoacyl tRNA synthetases.

Transcription in archaea more closely resembles eukaryotic than bacterial transcription, with the archaeal RNA polymerase being very close to its equivalent in eukaryotes; while archaeal translation shows signs of both bacterial and eukaryal equivalents. Although archaea only have one type of RNA polymerase, its structure and function in transcription seems to be close to that of the eukaryotic RNA polymerase II, with similar protein assemblies (the general transcription factors) directing the binding of the RNA polymerase to a gene's promoter. However, other archaeal transcription factors are closer to those found in bacteria. Post-transcriptional modification is simpler than in eukaryotes, since most archaeal genes lack introns, although there are many introns in their transfer RNA and ribosomal RNA genes, and introns may occur in a few protein-encoding genes.

Gene Transfer and Genetic Exchange

Halobacterium volcanii, an extreme halophilic archaeon, forms cytoplasmic bridges between cells that appear to be used for transfer of DNA from one cell to another in either direction.

When the hyperthermophilic archaea *Sulfolobus solfataricus* and *Sulfolobus acidocaldarius* are exposed to the DNA damaging agents UV irradiation, bleomycin or mitomycin C, species-specific cellular aggregation is induced. Aggregation in *S. solfataricus* could not be induced by other physical stressors, such as pH or temperature shift, suggesting that aggregation is induced specifically by DNA damage. Ajon et al. showed that UV-induced cellular aggregation mediates chromosomal marker exchange with high frequency in *S. acidocaldarius*. Recombination rates exceeded those of uninduced cultures by up to three orders of magnitude. Frols et al. and Ajon et al. hypothesized that cellular aggregation enhances species specific DNA transfer between *Sulfolobus* cells in order to provide increased repair of damaged DNA by means of homologous recombination. This response may be a primitive form of sexual interaction similar to the more well-studied bacterial transformation systems that are also associated with species specific DNA transfer between cells leading to homologous recombinational repair of DNA damage.

Reproduction

Archaea reproduce asexually by binary or multiple fission, fragmentation, or budding; meiosis does not occur, so if a species of archaea exists in more than one form, all have the same genetic material. Cell division is controlled in a cell cycle; after the cell's chromosome is replicated and the two daughter chromosomes separate, the cell divides. In the genus *Sulfolobus*, the cycle has characteristics that are similar to both bacterial and eukaryotic systems. The chromosomes replicate from multiple starting-points (origins of replication) using DNA polymerases that resemble the equivalent eukaryotic enzymes.

In euryarchaea the cell division protein FtsZ, which forms a contracting ring around the cell, and the components of the septum that is constructed across the center of the cell, are similar to their bacterial equivalents. In cren- and thaumarchaea, however, the cell division machinery Cdv fulfills a similar role. This machinery is related to the eukaryotic ESCRT-III machinery which, while best known for its role in cell sorting, also has been seen to fulfill a role in separation between divided cell, suggesting an ancestral role in cell division.

Both bacteria and eukaryotes, but not archaea, make spores. Some species of Haloarchaea undergo phenotypic switching and grow as several different cell types, including thick-walled structures that are resistant to osmotic shock and allow the archaea to survive in water at low salt concentrations, but these are not reproductive structures and may instead help them reach new habitats.

Ecology

Habitats

Archaea exist in a broad range of habitats, and as a major part of global ecosystems, may represent about 20% of microbial cells in the oceans. The first-discovered archaeans were extremophiles. Indeed, some archaea survive high temperatures, often above 100 °C (212 °F), as found in geysers, black smokers, and oil wells. Other common habitats include very cold habitats and highly saline, acidic, or alkaline water. However, archaea include mesophiles that grow in mild conditions, in swamps and marshland, sewage, the oceans, the intestinal tract of animals, and soils.

Extremophile archaea are members of four main physiological groups. These are the halophiles, thermophiles, alkaliphiles, and acidophiles. These groups are not comprehensive or phylum-specific, nor are they mutually exclusive, since some archaea belong to several groups. Nonetheless, they are a useful starting point for classification.

Halophiles, including the genus *Halobacterium*, live in extremely saline environments such as salt lakes and outnumber their bacterial counterparts at salinities greater than 20–25%. Thermophiles grow best at temperatures above 45 °C (113 °F), in places such as hot springs; *hyperthermophilic* archaea grow optimally at temperatures greater than 80 °C (176 °F). The archaeal *Methanopyrus kandleri* Strain 116 can even reproduce at 122 °C (252 °F), the highest recorded temperature of any organism.

Other archaea exist in very acidic or alkaline conditions. For example, one of the most extreme archaean acidophiles is *Picrophilus torridus*, which grows at pH 0, which is equivalent to thriving in 1.2 molar sulfuric acid.

This resistance to extreme environments has made archaea the focus of speculation about the possible properties of extraterrestrial life. Some extremophile habitats are not dissimilar to those on Mars, leading to the suggestion that viable microbes could be transferred between planets in meteorites.

Recently, several studies have shown that archaea exist not only in mesophilic and thermophilic environments but are also present, sometimes in high numbers, at low temperatures as well. For example, archaea are common in cold oceanic environments such as polar seas. Even more significant are the large numbers of archaea found throughout the world's oceans in non-extreme habitats among the plankton community (as part of the picoplankton). Although these archaea can be present in extremely high numbers (up to 40% of the microbial biomass), almost none of these species have been isolated and studied in pure culture. Consequently, our understanding of the role of archaea in ocean ecology is rudimentary, so their full influence on global biogeochemical cycles remains largely unexplored. Some marine Crenarchaeota are capable of nitrification, suggesting these organisms may affect the oceanic nitrogen cycle, although these oceanic Crenarchaeota may also use other sources of energy. Vast numbers of archaea are also found in the sediments that cover the sea floor, with these organisms making up the majority of living cells at depths over 1 meter below the ocean bottom.

Role in Chemical Cycling

Archaea recycle elements such as carbon, nitrogen and sulfur through their various habitats. Although these activities are vital for normal ecosystem function, archaea can also contribute to human-made changes, and even cause pollution.

Archaea carry out many steps in the nitrogen cycle. This includes both reactions that remove nitrogen from ecosystems (such as nitrate-based respiration and denitrification) as well as processes that introduce nitrogen (such as nitrate assimilation and nitrogen fixation). Researchers recently discovered archaeal involvement in ammonia oxidation reactions. These reactions are particularly important in the oceans. The archaea also appear crucial for ammonia oxidation in soils. They produce nitrite, which other microbes then oxidize to nitrate. Plants and other organisms consume the latter.

In the sulfur cycle, archaea that grow by oxidizing sulfur compounds release this element from rocks, making it available to other organisms. However, the archaea that do this, such as *Sulfolobus*, produce sulfuric acid as a waste product, and the growth of these organisms in abandoned mines can contribute to acid mine drainage and other environmental damage.

In the carbon cycle, methanogen archaea remove hydrogen and play an important role in the decay of organic matter by the populations of microorganisms that act as decomposers in anaerobic ecosystems, such as sediments, marshes and sewage-treatment works.

Global methane levels in 2011 had increased by a factor of 2.5 since pre-industrial times: from 722 ppb to 1800 ppb, the highest value in at least 800,000 years. Methane has an anthropogenic global warming potential (AGWP) of 29, which means that it's 29 times stronger in heat-trapping than carbon dioxide is, over a 100-year time scale.

Interactions With Other Organisms

The well-characterized interactions between archaea and other organisms are either mutual or commensal. There are no clear examples of known archaeal pathogens or parasites. However, some species of methanogens have been suggested to be involved in infections in the mouth, and *Nanoarchaeum equitans* may be a parasite of another species of archaea, since it only survives and reproduces within the cells of the Crenarchaeon *Ignicoccus hospitalis*, and appears to offer no benefit to its host. In contrast, Archaeal Richmond Mine Acidophilic Nanoorganisms (ARMAN) occasionally connect with other archaeal cells in acid mine drainage biofilms. The nature of this relationship is unknown. However, it is distinct from that of *Nanarchaeaum–Ignicoccus* in that the ultrasmall ARMAN cells are usually seen independent of the Thermoplasmatales cells.

Methanogenic archaea form a symbiosis with termites.

Mutualism

One well-understood example of mutualism is the interaction between protozoa and methanogenic archaea in the digestive tracts of animals that digest cellulose, such as ruminants and termites. In these anaerobic environments, protozoa break down plant cellulose to obtain energy. This process releases hydrogen as a waste product, but high levels of hydrogen reduce energy production. When methanogens convert hydrogen to methane, protozoa benefit from more energy.

In anaerobic protozoa, such as *Plagiopyla frontata*, archaea reside inside the protozoa and consume hydrogen produced in their hydrogenosomes. Archaea also associate with larger organisms. For example, the marine archaean *Cenarchaeum symbiosum* lives within (is an endosymbiont of) the sponge *Axinella mexicana*.

Commensalism

Archaea can also be commensals, benefiting from an association without helping or harming the other organism. For example, the methanogen *Methanobrevibacter smithii* is by far the most common archaean in the human flora, making up about one in ten of all the prokaryotes in the human gut. In termites and in humans, these methanogens may in fact be mutualists, interacting with other microbes in the gut to aid digestion. Archaean communities also associate with a range of other organisms, such as on the surface of corals, and in the region of soil that surrounds plant roots (the rhizosphere).

Significance in Technology and Industry

Extremophile archaea, particularly those resistant either to heat or to extremes of acidity and alkalinity, are a source of enzymes that function under these harsh conditions. These enzymes have found many uses. For example, thermostable DNA polymerases, such as the Pfu DNA polymerase from *Pyrococcus furiosus*, revolutionized molecular biology by allowing the polymerase chain reaction to be used in research as a simple and rapid technique for cloning DNA. In industry, amylases, galactosidases and pullulanases in other species of *Pyrococcus* that function at over 100 °C (212 °F) allow food processing at high temperatures, such as the production of low lactose milk and whey. Enzymes from these thermophilic archaea also tend to be very stable in organic solvents, allowing their use in environmentally friendly processes in green chemistry that synthesize organic compounds. This stability makes them easier to use in structural biology. Consequently, the counterparts of bacterial or eukaryotic enzymes from extremophile archaea are often used in structural studies.

In contrast to the range of applications of archaean enzymes, the use of the organisms themselves in biotechnology is less developed. Methanogenic archaea are a vital part of sewage treatment, since they are part of the community of microorganisms that carry out anaerobic digestion and produce biogas. In mineral processing, acidophilic archaea display promise for the extraction of metals from ores, including gold, cobalt and copper.

Archaea host a new class of potentially useful antibiotics. A few of these archaeocins have been characterized, but hundreds more are believed to exist, especially within *Haloarchaea* and *Sulfolobus*. These compounds differ in structure from bacterial antibiotics, so they may have novel modes of action. In addition, they may allow the creation of new selectable markers for use in archaeal molecular biology.

Eukaryote

A eukaryote is any organism whose cells contain a nucleus and other organelles enclosed within membranes.

Eukaryotes belong to the taxon Eukarya or Eukaryota. The defining feature that sets eukaryotic cells apart from prokaryotic cells (Bacteria and Archaea) is that they have membrane-bound organelles, especially the nucleus, which contains the genetic material and is enclosed by the nuclear envelope. Eukaryotic cells also contain other membrane-bound organelles such as mitochondria and the Golgi apparatus. In addition, plants and algae contain chloroplasts. Eukaryotic organisms may be unicellular or multicellular. Only eukaryotes form multicellular organisms consisting of many kinds of tissue made up of different cell types.

Eukaryotes can reproduce both asexually through mitosis and sexually through meiosis and gamete fusion. In mitosis, one cell divides to produce two genetically identical cells. In meiosis, DNA replication is followed by two rounds of cell division to produce four daughter cells each with half the number of chromosomes as the original parent cell (haploid cells). These act as sex cells (gametes – each gamete has just one complement of chromosomes, each a unique mix of the corresponding pair of parental chromosomes) resulting from genetic recombination during meiosis.

The domain Eukaryota appears to be monophyletic, and so makes up one of the three domains of life. The two other domains, Bacteria and Archaea, are prokaryotes and have none of the above features. Eukaryotes represent a tiny minority of all living things. However, due to their much larger size, eukaryotes' collective worldwide biomass is estimated at about equal to that of prokaryotes. Eukaryotes first developed approximately 1.6–2.1 billion years ago (during the Proterozoic eon).

Cell Features

Eukaryotic cells are typically much larger than those of prokaryotes. They have a variety of internal membrane-bound structures, called organelles, and a cytoskeleton composed of microtubules, microfilaments, and intermediate filaments, which play an important role in defining the cell's organization and shape. Eukaryotic DNA is divided into several linear bundles called chromosomes, which are separated by a microtubular spindle during nuclear division.

Internal Membrane

Eukaryote cells include a variety of membrane-bound structures, collectively referred to as the endomembrane system. Simple compartments, called vesicles or vacuoles, can form by budding off other membranes. Many cells ingest food and other materials through a process of endocytosis, where the outer membrane invaginates and then pinches off to form a vesicle. It is probable that most other membrane-bound organelles are ultimately derived from such vesicles. Alternatively some products produced by the cell can leave in a vesicle through exocytosis.

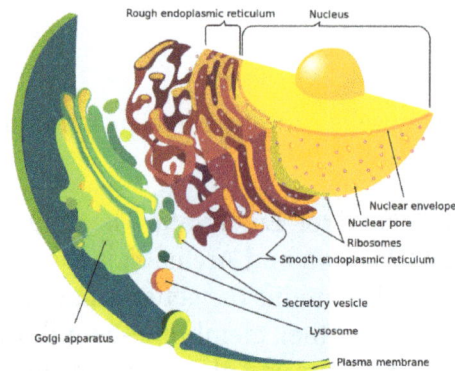

Detail of the endomembrane system and its components

A 3D rendering of an animal cell cut in half.

The nucleus is surrounded by a double membrane (commonly referred to as a nuclear membrane or nuclear envelope), with pores that allow material to move in and out. Various tube- and sheet-like extensions of the nuclear membrane form what is called the endoplasmic reticulum or ER, which is involved in protein transport and maturation. It includes the rough ER where ribosomes are attached to synthesize proteins, which enter the interior space or lumen. Subsequently, they generally enter vesicles, which bud off from the smooth ER. In most eukaryotes, these protein-carrying vesicles are released and further modified in stacks of flattened vesicles, called Golgi bodies or dictyosomes.

Vesicles may be specialized for various purposes. For instance, lysosomes contain digestive enzymes that break down the contents of food vacuoles, and peroxisomes are used to break down peroxide, which is toxic otherwise. Many protozoa have contractile vacuoles, which collect and expel excess water, and extrusomes, which expel material used to deflect predators or capture prey. In higher plants, most of a cell's volume is taken up by a central vacuole, which primarily maintains its osmotic pressure.

Mitochondria and Plastids

Mitochondria are organelles found in nearly all eukaryotes that provide energy to the cell by converting ingested sugars into ATP. They are surrounded by two membranes (each a phospholipid bi-layer), the inner of which is folded into invaginations called cristae, where aerobic respiration takes place. Mitochondria contain their own DNA. They are now generally held to have developed from endosymbiotic prokaryotes, probably proteobacteria. Protozoa and microbes that lack mitochondria, such as the amoebozoan Pelomyxa and metamonads such as Giardia and Trichomonas, have usually been found to contain mitochondrion-derived organelles, such as hydrogenosomes and mitosomes, and thus probably lost the mitochondria secondarily.

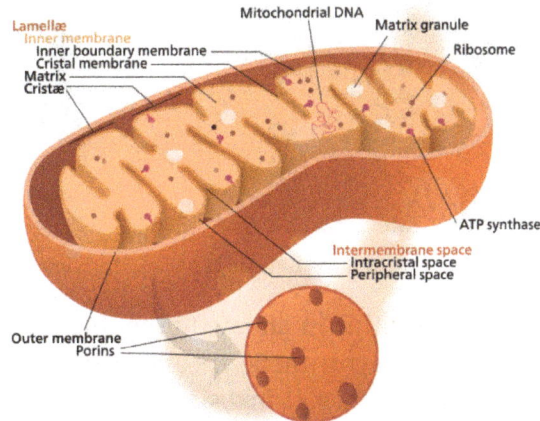

Simplified structure of a mitochondrion

In 2016, *Monocercomonoides*, a metamonad flagellate which resides in the intestines of the chinchilla, has been found to lack mitochondria entirely. *Monocercomonoides* obtains its energy by enzymatic action on nutrients absorbed from the environment. It has also acquired, by lateral gene transfer, a cytosolic sulphur mobilisation system which provides the clusters of iron and sulfur required for protein synthesis. The normal mitochondrial iron-sulphur cluster pathway is considered to have been lost secondarily.

Plants and various groups of algae also have plastids. Plastids have their own DNA and are developed from endosymbionts, in this case cyanobacteria. They usually take the form of chloroplasts, which like cyanobacteria contain chlorophyll and produce organic compounds (such as glucose) through photosynthesis. Others are involved in storing food. Although plastids probably had a single origin, not all plastid-containing groups are closely related. Instead, some eukaryotes have obtained them from others through secondary endosymbiosis or ingestion.

Endosymbiotic origins have also been proposed for the nucleus, and for eukaryotic flagella, supposed to have developed from spirochaetes. This is not generally accepted, both from a lack of cytological evidence and difficulty in reconciling this with cellular reproduction.

Cytoskeletal structures

Many eukaryotes have long slender motile cytoplasmic projections, called flagella, or similar structures called cilia. Flagella and cilia are sometimes referred to as undulipodia, and are variously involved in movement, feeding, and sensation. They are composed mainly of tubulin. These are entirely distinct from prokaryotic flagellae. They are supported by a bundle of microtubules arising from a basal body, also called a kinetosome or centriole, characteristically arranged as nine doublets surrounding two singlets. Flagella also may have hairs, or mastigonemes, and scales connecting membranes and internal rods. Their interior is continuous with the cell's cytoplasm.

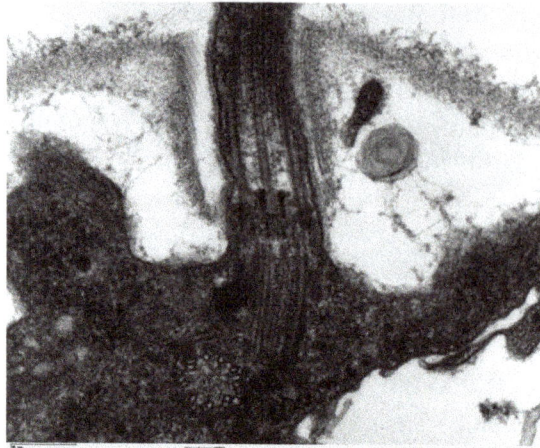

Longitudinal section through the flagellum of *Chlamydomonas reinhardtii*

Microfilamental structures composed of actin and actin binding proteins, e.g., α-actinin, fimbrin, filamin are present in submembraneous cortical layers and bundles, as well. Motor proteins of microtubules, e.g., dynein or kinesin and actin, e.g., myosins provide dynamic character of the network.

Centrioles are often present even in cells and groups that do not have flagella, but conifers and flowering plants have neither. They generally occur in groups of one or two, called kinetids, that give rise to various microtubular roots. These form a primary component of the cytoskeletal structure, and are often assembled over the course of several cell divisions, with one flagellum retained from the parent and the other derived from it. Centrioles may also be associated in the formation of a spindle during nuclear division.

The significance of cytoskeletal structures is underlined in the determination of shape of the cells, as well as their being essential components of migratory responses like chemotaxis and chemokinesis. Some protists have various other microtubule-supported organelles. These include the radiolaria and heliozoa, which produce axopodia used in flotation or to capture prey, and the haptophytes, which have a peculiar flagellum-like organelle called the haptonema.

Cell Wall

The cells of plants, fungi, and most chromalveolates have a cell wall, a layer outside the cell membrane, providing the cell with structural support, protection, and a filtering mechanism. The cell wall also prevents over-expansion when water enters the cell.

The major polysaccharides making up the primary cell wall of land plants are cellulose, hemicellulose, and pectin. The cellulose microfibrils are linked via hemicellulosic tethers to form the cellulose-hemicellulose network, which is embedded in the pectin matrix. The most common hemicellulose in the primary cell wall is xyloglucan.

Differences Among Eukaryotic Cells

There are many different types of eukaryotic cells, though animals and plants are the most familiar eukaryotes, and thus provide an excellent starting point for understanding eukaryotic structure. Fungi and many protists have some substantial differences, however.

Animal Cell

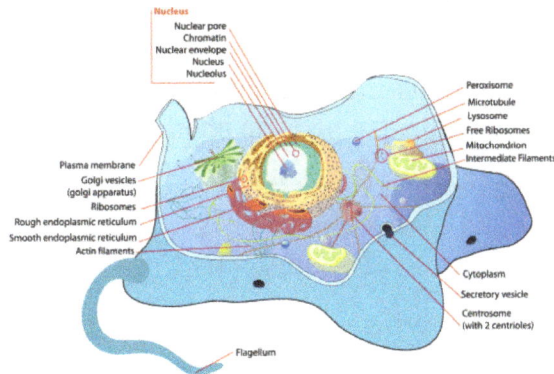

Structure of a typical animal cell

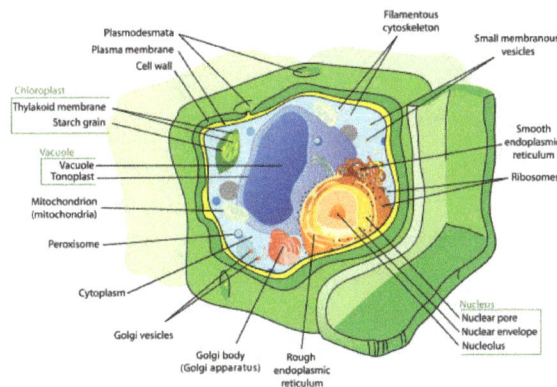

Structure of a typical plant cell

All animals consist of eukaryotic cells. Animal cells are distinct from those of other eukaryotes, most notably plants, as they lack cell walls and chloroplasts and have smaller vacuoles. Due to the lack of a cell wall, animal cells can adopt a variety of shapes. A phagocytic cell can even engulf other structures.

There are many other types of cell. For instance, there are approximately 210 distinct cell types in the adult human body.

Plant Cell

Plant cells are quite different from the cells of the other eukaryotic organisms. Their distinctive features are:

- A large central vacuole (enclosed by a membrane, the tonoplast), which maintains the cell's turgor and controls movement of molecules between the cytosol and sap

- A primary cell wall containing cellulose, hemicellulose and pectin, deposited by the protoplast on the outside of the cell membrane; this contrasts with the cell walls of fungi, which contain chitin, and the cell envelopes of prokaryotes, in which peptidoglycans are the main structural molecules

- The plasmodesmata, linking pores in the cell wall that allow each plant cell to communicate with other adjacent cells; this is different from the functionally analogous system of gap junctions between animal cells.

- Plastids, especially chloroplasts that contain chlorophyll, the pigment that gives plants their green color and allows them to perform photosynthesis

- Bryophytes and seedless vascular plants lack flagellae and centrioles except in the sperm cells. Sperm of cycads and *Ginkgo* are large, complex cells that swim with hundreds to thousands of flagellae.

- Conifers (Pinophyta) and flowering plants (Angiospermae) lack the flagellae and centrioles that are present in animal cells.

Fungal Cell

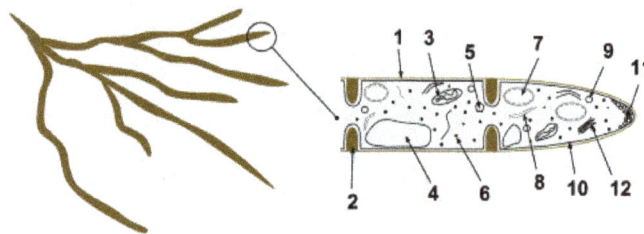

Fungal Hyphae Cells
1- Hyphal wall 2- Septum 3- Mitochondrion 4- Vacuole 5- Ergosterol crystal 6- Ribosome 7- Nucleus 8- Endoplasmic reticulum 9- Lipid body 10- Plasma membrane 11- Spitzenkörper 12- Golgi apparatus

Fungal cells are most similar to animal cells, with the following exceptions:

- A cell wall that contains chitin

- Less definition between cells; the hyphae of higher fungi have porous partitions called septa, which allow the passage of cytoplasm, organelles, and, sometimes, nuclei. Primitive fungi have few or no septa, so each organism is essentially a giant multinucleate supercell; these fungi are described as coenocytic.

- Only the most primitive fungi, chytrids, have flagella.

Other Eukaryotic Cells

Eukaryotes are a very diverse group, and their cell structures are equally diverse. Many have cell walls; many do not. Many have chloroplasts, derived from primary, secondary, or even tertiary endosymbiosis; and many do not. Some groups have unique structures, such as the cyanelles of the glaucophytes, the haptonema of the haptophytes, or the ejectisomes of the cryptomonads. Other structures, such as pseudopods, are found in various eukaryote groups in different forms, such as the lobose amoebozoans or the reticulose foraminiferans.

Reproduction

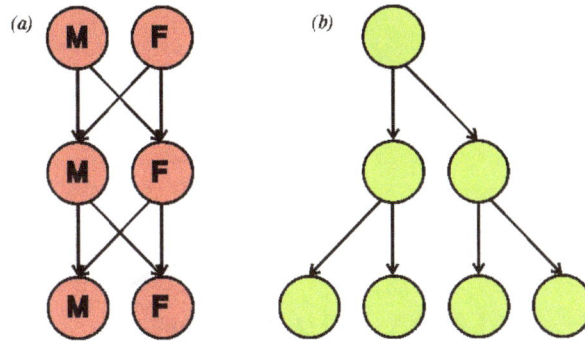

This diagram illustrates the twofold cost of sex. If each individual were to contribute to the same number of offspring (two), (a) the sexual population remains the same size each generation, where the (b) asexual population doubles in size each generation.

Cell division generally takes place asexually by mitosis, a process that allows each daughter nucleus to receive one copy of each chromosome. In most eukaryotes, there is also a process of sexual reproduction, typically involving an alternation between haploid generations, wherein only one copy of each chromosome is present, and diploid generations, wherein two copies of each chromosome are present, occurring through meiosis. There is considerable variation in this pattern.

Eukaryotes have a smaller surface area to volume ratio than prokaryotes, and thus have lower metabolic rates and longer generation times. In some multicellular organisms, cells specialized for metabolism will have enlarged surface areas, such as intestinal vili.

The evolution of sexual reproduction may be a primordial and fundamental characteristic of eukaryotes. Based on a phylogenetic analysis, Dacks and Roger proposed that facultative sex was present in the common ancestor of all eukaryotes. A core set of genes that function in meiosis is present in both *Trichomonas vaginalis* and *Giardia intestinalis*, two organisms previously thought to be asexual. Since these two species are descendants of lineages that diverged early from the eukaryotic evolutionary tree, it was inferred that core meiotic genes, and hence sex, were likely present in a common ancestor of all eukaryotes. Other studies on eukaryotic species once thought to be asexual have revealed evidence for a sexual cycle. For instance, parasitic protozoa of the genus Leishmania have recently been shown to have a sexual cycle. Also, evidence now indicates that amoeba, that were previously regarded as asexual, are anciently sexual and that the majority of present-day asexual groups likely arose recently and independently.

Classification

In antiquity, the two clades of animals and plants were recognized. They were given the taxonomic rank of Kingdom by Linnaeus. Though he included the fungi with plants with some reservations, it was later realized that they are quite distinct and warrant a separate kingdom, the composition of which was not entirely clear until the 1980s. The various single-cell eukaryotes were originally placed with plants or animals when they became known. In 1830, the German biologist Georg A. Goldfuss coined the word *protozoa* to refer to organisms such as ciliates, and this group was expanded until it encompassed all single-celled eukaryotes, and given their own kingdom, the Protista, by Ernst Haeckel in 1866. The eukaryotes thus came to be composed of four kingdoms:

- Kingdom Protista

- Kingdom Plantae

- Kingdom Fungi

- Kingdom Animalia

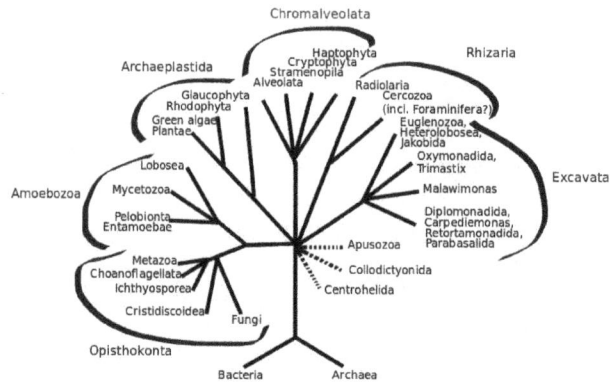

One hypothesis of eukaryotic relationships. The Opisthokonta group includes both animals (Metazoa) and fungi. Plants (Plantae) are placed in Archaeplastida.

The protists were understood to be "primitive forms", and thus an evolutionary grade, united by their primitive unicellular nature. The disentanglement of the deep splits in the tree of life only really got going with DNA sequencing, leading to a system of domains rather than kingdoms as top level rank being put forward by Carl Woese, uniting all the eukaryote kingdoms under the eukaryote domain. At the same time, work on the protist tree intensified, and is still actively going on today. Several alternative classifications have been forwarded, though there is no consensus in the field.

A classification produced in 2005 for the International Society of Protistologists, which reflected the consensus of the time, divided the eukaryotes into six supposedly monophyletic 'supergroups'. However, in the same year (2005), doubts were expressed as to whether some of these supergroups were monophyletic, particularly the Chromalveolata, and a review in 2006 noted the lack of evidence for several of the supposed six supergroups. A revised classification in 2012 recognizes five supergroups.

Archaeplastida (or Primo-plantae)	Land plants, green algae, red algae, and glaucophytes
SAR supergroup	Stramenopiles (brown algae, diatoms, etc.), Alveolata, and Rhizaria (Foraminifera, Radiolaria, and various other amoeboid protozoa).
Excavata	Various flagellate protozoa
Amoebozoa	Most lobose amoeboids and slime molds
Opisthokonta	Animals, fungi, choanoflagellates, etc.

There are also smaller groups of eukaryotes whose position is uncertain or seems to fall outside the major groups. In particular, Haptophyta, Cryptophyta, Centrohelida, Telonemia, Picozoa, Apusomonadida, Ancyromonadida, Breviatea, and the genus *Collodictyon*. Overall, it seems that, although progress has been made, there are still very significant uncertainties in the evolutionary history and classification of eukaryotes. As Roger & Simpson said in 2009 "with the current pace of change in our understanding of the eukaryote tree of life, we should proceed with caution."

In an article published in *Nature Microbiology* in April 2016 the authors, "reinforced once again that the life we see around us – plants, animals, humans and other so-called eukaryotes – represent a tiny percentage of the world's biodiversity." They classified eukaryote "based on the inheritance of their information systems as opposed to lipid or other cellular structures." Jillian F. Banfield of the University of California, Berkeley and fellow scientists used a super computer to generate a diagram of a new tree of life based on DNA from 3000 species including 2,072 known species and 1,011 newly reported microbial organisms, whose DNA they had gathered from diverse environments. As the capacity to sequence DNA became easier, Banfield and team were able to do metagenomic sequencing—"sequencing whole communities of organisms at once and picking out the individual groups based on their genes alone."

Phylogeny

The rRNA trees constructed during the 1980s and 1990s left most eukaryotes in an unresolved "crown" group (not technically a true crown), which was usually divided by the form of the mitochondrial cristae. The few groups that lack mitochondria branched separately, and so the absence was believed to be primitive; but this is now considered an artifact of long-branch attraction, and they are known to have lost them secondarily.

As of 2011, there is widespread agreement that the Rhizaria belong with the Stramenopiles and the Alveolata, in a clade dubbed the SAR supergroup, so that Rhizaria is not one of the main eukaryote groups; also that the Amoebozoa and Opisthokonta are each monophyletic and form a clade, often called the unikonts. Beyond this, there does not appear to be a consensus.

It has been estimated that there may be 75 distinct lineages of eukaryotes. Most of these lineages are protists.

The known eukaryote genome sizes vary from 8.2 megabases (Mb) in *Babesia bovis* to 112,000– 220,050 Mb in the dinoflagellate *Prorocentrum micans*, suggesting that the genome of the ancestral eukaryote has undergone considerable variation during its evolution. The last common ancestor of all eukaryotes is believed to have been a phagotrophic protist with a nucleus, at least one centriole and cilium, facultatively aerobic mitochondria, sex (meiosis and syngamy), a dormant cyst with a cell wall of chitin and/or cellulose and peroxisomes. Later endosymbiosis led to the spread of plastids in some lineages.

Five Supergroups

A global tree of eukaryotes from a consensus of phylogenetic evidence (in particular, phylogenomics), rare genomic signatures, and morphological characteristics is presented in Adl *et al.* 2012 and Burki 2014.

In some analyses, the Hacrobia group (Haptophyta + Cryptophyta) is placed next to Archaeplastida, but in other ones it is nested inside the Archaeplastida. However, several recent studies have concluded that Haptophyta and Cryptophyta do not form a monophyletic group.. The former could be a sister group to the SAR group, the latter cluster with the Archaeplastida (plants in the broad sense). As of February 2012, it remains unclear whether the Hacrobia forms a monophyletic group.

The division of the eukaryotes into two primary clades, bikonts (Archaeplastida + SAR + Excavata) and unikonts (Amoebozoa + Opisthokonta), derived from an ancestral biflagellar organism and an ancestral uniflagellar organism, respectively, had been suggested earlier. A 2012 study produced a somewhat similar division, although noting that the terms "unikonts" and "bikonts" were not used in the original sense.

Cavalier-Smith's Tree

Cavalier-Smith 2010, 2013, and 2014 places the eukaryotic tree's root between Excavata (with ventral feeding groove supported by a microtubular root) and the grooveless Euglenozoa:

Alternative Views

Other analyses place the SAR supergroup within an expanded Chromalveolata, although they differ on the placement of the resulting five groups. Rogozin *et al.* in 2009 produced the tree shown below, where the primary division is between the Archaeplastida and all other eukaryotes.

A paper published in 2009, which re-examined the data used in some of the analyses presented above as well as performing new ones, strongly suggested that the Archaeplastida are polyphyletic. The phylogeny finally proposed in the paper is shown below.

There are also the cladistic analyses of Diana Lipscomb based on classical data which have red algae as basal and Archeoplastida as paraphyletic. In the survey by Parfrey et al. it is recovered in only 42.6% of the molecular analyses that include it, that is, 26 out of 61. It is not recovered in Goloboff et al.'s combined analysis, and is mostly weakly supported in other molecular analyses. Laura Parfrey et al. point out that Archeoplastida support comes primarily from phylogenomic analyses and these may be picking up misleading endosymbiotic gene transfer signal of genes independently transferred from the plastid to the host nucleus in the 3 archeoplastid clades. And Stiller and Harrell emphasize that the group can be explained by "short-branch exclusion" and "subtle and easily overlooked biases can dominate the overall results of molecular phylogenetic analyses of ancient eukaryotic relationships. Sources of potential phylogenetic artifact should be investigated routinely, not just when obvious 'long-branch attraction' is encountered."

Red algae as basal is also supported by molecular evidence. *Cyanidioschyzon*, a red alga, is considered basal (sister group to the rest of eukaryotes) by Nagashima et al. and Seckbach. It has the most primitive chloroplast, only 1 mitochondrion, no vacuoles, no trienoic acids, and the smallest eukaryotic genome at 8 Mbp.

Rhizaria are only moderately supported (65.5% of the studies, i.e., 19 of 29), statistical support for them is inconsistent in multigene genealogies with larger taxon sampling and the group is ambiguously supported in Goloboff et al.

Origin of Eukaryotes

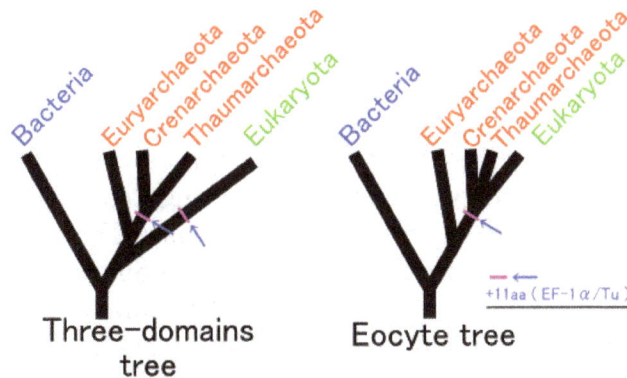

The three-domains tree and the Eocyte hypothesis.

Fossils

The origin of the eukaryotic cell is considered a milestone in the evolution of life, since eukaryotes include all complex cells and almost all multicellular organisms. The timing of this series of events is hard to determine; Knoll (2006) suggests they developed approximately 1.6–2.1 billion years ago. Some acritarchs are known from at least 1.65 billion years ago, and the possible alga *Grypania* has been found as far back as 2.1 billion years ago. The *Geosiphon*-like fossil fungus *Diskagma* has been found in paleosols 2.2 billion years old

Organized living structures have been found in the black shales of the Palaeoproterozoic Francevillian B Formation in Gabon, dated at 2.1 billion years old. Eukaryotic life could have evolved at that time. Fossils that are clearly related to modern groups start appearing an estimated 1.2 billion years ago, in the form of a red alga, though recent work suggests the existence of fossilized filamentous algae in the Vindhya basin dating back perhaps to 1.6 to 1.7 billion years ago.

Biomarkers suggest that at least stem eukaryotes arose even earlier. The presence of steranes in Australian shales indicates that eukaryotes were present in these rocks dated at 2.7 billion years old.

Relationship to Archaea

Eukaryotes are more closely related to Archaea than Bacteria, at least in terms of nuclear DNA and genetic machinery, and one controversial idea is to place them with Archaea in the clade Neomura. However, in other respects, such as membrane composition, eukaryotes are similar to Bacteria. Three main explanations for this have been proposed:

- Eukaryotes resulted from the complete fusion of two or more cells, wherein the cytoplasm formed from a eubacterium, and the nucleus from an archaeon, from a virus, or from a pre-cell.

- Eukaryotes developed from Archaea, and acquired their eubacterial characteristics from the proto-mitochondrion.

- Eukaryotes and Archaea developed separately from a modified eubacterium.

The chronocyte hypothesis for the origin of the eukaryotic cell postulates that a primitive eukaryotic cell was formed by the endosymbiosis of both archaea and bacteria by a third type of cell, termed a chronocyte.

Endomembrane System and Mitochondria

The origins of the endomembrane system and mitochondria are also unclear. The phagotrophic hypothesis proposes that eukaryotic-type membranes lacking a cell wall originated first, with the development of endocytosis, whereas mitochondria were acquired by ingestion as endosymbionts. The syntrophic hypothesis proposes that the proto-eukaryote relied on the proto-mitochondrion for food, and so ultimately grew to surround it. Here the membranes originated after the engulfment of the mitochondrion, in part thanks to mitochondrial genes (the hydrogen hypothesis is one particular version).

In a study using genomes to construct supertrees, Pisani *et al.* (2007) suggest that, along with evidence that there was never a mitochondrion-less eukaryote, eukaryotes evolved from a syntrophy between an archaea closely related to Thermoplasmatales and an α-proteobacterium, likely a symbiosis driven by sulfur or hydrogen. The mitochondrion and its genome is a remnant of the α-proteobacterial endosymbiont.

Hypotheses for the Origin of Eukaryotes

Different hypotheses have been proposed as to how eukaryotic cells came into existence. These hypotheses can be classified into two distinct classes – autogenous models and chimeric models.

Autogenous Models

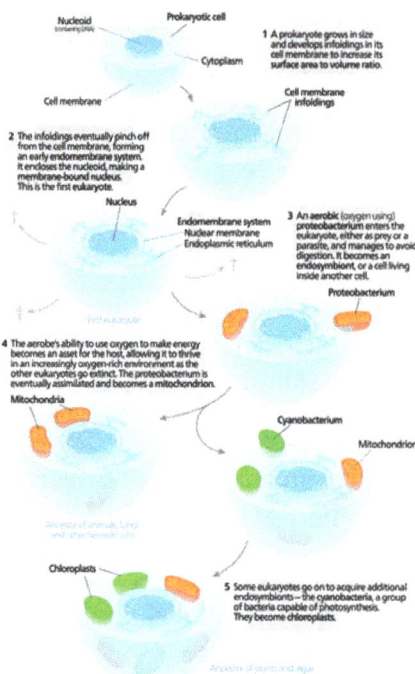

An autogenous model for the origin of eukaryotes.

Autogenous models propose that a proto-eukaryotic cell containing a nucleus existed first, and later acquired mitochondria. According to this model, a large prokaryote developed invaginations in its plasma membrane in order to obtain enough surface area to service its cytoplasmic volume. As the invaginations differentiated in function, some became separate compartments—giving rise to the endomembrane system, including the endoplasmic reticulum, golgi apparatus, nuclear membrane, and single membrane structures such as lysosomes. Mitochondria are proposed to come from the endosymbiosis of an aerobic proteobacterium, and it's assumed that all the eukaryotic lineages that did not acquire mitochondria became extinct. Chloroplasts came about from another endosymbiotic event involving cyanobacteria. Since all eukaryotes have mitochondria, but not all have chloroplasts, mitochondria are thought to have come first. This is the serial endosymbiosis theory.

Some models propose that the origins of double layered organelles, such as mitochondria and chloroplasts, in the proto-eukaryotic cell is due to the compartmentalization of DNA vesicles that were formed from the secondary invaginations or more detailed infoldings of cellular membrane.

Chimeric Models

Chimeric models claim that two prokaryotic cells existed initially – an archaeon and a bacterium. These cells underwent a merging process, either by a physical fusion or by endosymbiosis, thereby leading to the formation of a eukaryotic cell. Within these chimeric models, some studies further claim that mitochondria originated from a bacterial ancestor while others emphasize the role of endosymbiotic processes behind the origin of mitochondria.

Based on the process of mutualistic symbiosis, the hypotheses can be categorized as – the serial endosymbiotic theory (SET), the hydrogen hypothesis (mostly a process of symbiosis where hydrogen transfer takes place among different species), and the syntrophy hypothesis.

According to serial endosymbiotic theory (championed by Dr. Lynn Margulis), a union between a motile anaerobic bacterium (like *Spirochaeta*) and a thermoacidophilic crenarchaeon (like *Thermoplasma* which is sulfidogenic in nature) gave rise to the present day eukaryotes. This union established a motile organism capable of living in the already existing acidic and sulfurous waters. Oxygen is known to cause toxicity to organisms that lack the required metabolic machinery. Thus, the archaeon provided the bacterium with a highly beneficial reduced environment (sulfur and sulfate were reduced to sulfide). In microaerophilic conditions, oxygen was reduced to water thereby creating a mutual benefit platform. The bacterium on the other hand, contributed the necessary fermentation products and electron acceptors along with its motility feature to the archaeon thereby gaining a swimming motility for the organism. From a consortium of bacterial and archaeal DNA originated the nuclear genome of eukaryotic cells. Spirochetes gave rise to the motile features of eukaryotic cells. Endosymbiotic unifications of the ancestors of alpha-proteobacteria and cyanobacteria, led to the origin of mitochondria and plastids respectively. For example, *Thiodendron* has been known to have originated via an ectosymbiotic process based on a similar syntrophy of sulfur existing between the two types of bacteria – *Desulphobacter* and *Spirochaeta*. However, such an association based on motile symbiosis have never been observed practically. Also there is no evidence of archaeans and spirochetes adapting to intense acid-based environments.

In the hydrogen hypothesis, the symbiotic linkage of an anaerobic and autotrophic methanogenic archaeon (host) with an alpha-proteobacterium (the symbiont) gave rise to the eukaryotes. The host utilized hydrogen (H_2) and carbon dioxide (CO_2) to produce methane while the symbiont, capable of aerobic respiration, expelled H_2 and CO_2 as byproducts of anaerobic fermentation process. The host's methanogenic environment worked as a sink for H_2, which resulted in heightened bacterial fermentation. Endosymbiotic gene transfer (EGT) acted as a catalyst for the host to acquire the symbionts' carbohydrate metabolism and turn heterotrophic in nature. Subsequently, the host's methane forming capability was lost. Thus, the origins of the heterotrophic organelle (symbiont) are identical to the origins of the eukaryotic lineage. In this hypothesis, the presence of H_2 represents the selective force that forged eukaryotes out of prokaryotes.

The syntrophy hypothesis was developed in contrast to the hydrogen hypothesis and proposes the existence of two symbiotic events. According to this theory, eukaryogenesis (i.e. origin of eukaryotic cells) occurred based on metabolic symbiosis (syntrophy) between a methanogenic archaeon and a delta-proteobacterium. This syntrophic symbiosis was initially facilitated by H_2 transfer between different species under anaerobic environments. In earlier stages, an alpha-proteobacterium became a member of this integration, and later developed into the mitochondrion. Gene transfer from a delta-proteobacterium to an archaeon led to the methanogenic archaeon developing into a nucleus. The archaeon constituted the genetic apparatus while the delta-proteobacterium contributed towards the cytoplasmic features. This theory incorporates two selective forces that were needed to be considered during the time of nucleus evolution – (a) presence of metabolic partitioning in order to avoid the harmful effects of the co-existence of anabolic and catabolic cellular pathways, and (b) prevention of abnormal biosynthesis of proteins that occur due to a vast spread of introns in the archaeal genes after acquiring the mitochondrion and the loss of methanogenesis.

Thus, the origin of eukaryotes by endosymbiotic processes has been broadly recognized and accepted so far. Mitochondria and plastids have been known to originate from a bacterial ancestor during parallel adaptation to anaerobiosis. However, there still remains a greater need in assessing the question of how much eukaryotic complexity is being originated via an implementation of these symbiogenetic theories.

Protist

Protists (/ˈproʊtɪst/) are the members of an informal grouping of diverse eukaryotic organisms that are not animals, plants or fungi. They do not form a natural group, or clade, but are often grouped together for convenience, like algae or invertebrates. In some systems of biological classification, such as the popular 5-kingdom scheme proposed by Robert Whittaker in 1969, the protists make up a kingdom called Protista, composed of "organisms which are unicellular or unicellular-colonial and which form no tissues."[A]

Besides their relatively simple levels of organization, protists do not necessarily have much in common. When used, the term "protists" is now considered to mean similar-appearing but diverse phyla that are not related through an exclusive common ancestor, and that have different life cycles, trophic levels, modes of locomotion, and cellular structures. In the classification system of Lynn Margulis, the term *protist* is reserved for microscopic organisms, while the more inclusive term *Protoctista* is applied to a biological kingdom which includes certain large multicellular eukaryotes, such as kelp, red algae and slime molds. Other workers use the term *protist* more broadly, to encompass both microbial eukaryotes and macroscopic organisms that do not fit into the other traditional kingdoms.

In cladistic systems, there are no equivalents to the taxa Protista or Protoctista, both terms referring to a paraphyletic group which spans the entire eukaryotic tree of life. In cladistic classification, the contents of Protista are distributed among various *supergroups* (SAR, Archaeplastida, Excavata, Opisthokonta, etc.) and "Protista", "Protoctista" and "Protozoa" are considered obsolete. However, there still remains some ambiguity about the position in the cladistic tree of some taxa – such as most excavata (metamonads, jakobids, Malawimonas and Collodictyon) and the term "protist" continues to be used informally as a catch-all term for Eukaryotic microorganisms – for example "protist pathogen" is used to denote any disease causing microbe which is not bacteria, virus, viroid or metazoa.[B]

Subdivisions

The term *protista* was first used by Ernst Haeckel in 1866. Protists were traditionally subdivided into several groups based on similarities to the "higher" kingdoms such as:

Protozoa the unicellular "animal-like (*heterotrophic*/parasitic)" protozoa which was further sub-divided based on motility such as (flagellated) *Flagellata*, (ciliated) *Ciliophora*, (phagocytic) *amoeba* and spore-forming *Sporozoans*

Protophyta the "plant-like" (*autotrophic*) protophyta (mostly unicellular algae)

Molds the "fungus-like" (*saprophytic*) slime molds and water molds.

Some protists, sometimes called ambiregnal protists, have been considered to be both protozoa and algae or fungi (e.g., slime molds and flagellated algae), and names for these have been published under either or both of the *ICN* and the *ICZN*. Conflicts, such as these – for example the dual-classification of Euglenids and Dinobryons, which are mixotrophic – is an example of why the kingdom Protista was adopted.

These traditional subdivisions, largely based on superficial commonalities, have been replaced by classifications based on phylogenetics (evolutionary relatedness among organisms). Molecular analyses in modern taxonomy have been used to redistribute former members of this group into diverse and sometimes distantly related phyla. For instance, the water molds are now considered to be closely related to photosynthetic organisms such as Brown algae and Diatoms, the slime molds are grouped mainly under Amoebozoa, and the Amoebozoa itself includes only a subset of "Amoeba" group, and significant number of erstwhile "Amoeboid" genera are distributed among Rhizarians and other Phyla.

However, the older terms are still used as informal names to describe the morphology and ecology of various protists. For example, the term *protozoa* is used to refer to heterotrophic species of protists that do not form filaments.

Classification

Historical Classifications

Among the pioneers in the study of the protists, which were almost ignored by Linnaeus except for some genera (e.g., *Vorticella, Chaos, Volvox, Corallina, Conferva, Ulva, Chara, Fucus*) were Leeuwenhoek, O. F. Müller, C. G. Ehrenberg and Félix Dujardin. The first groups used to classify microscopic organism were the Animalcules and the Infusoria. In 1817, the German naturalist Georg August Goldfuss introduced the word *Protozoa* to refer to organisms such as ciliates and corals. After the cell theory of Schwann and Schleiden (1838–39), this group was modified in 1848 by Carl von Siebold to include only animal-like unicellular organisms, such as foraminifera and amoebae. The formal taxonomic category *Protoctista* was first proposed in the early 1860s by John Hogg, who argued that the protists should include what he saw as primitive unicellular forms of both plants and animals. He defined the Protoctista as a "fourth kingdom of nature", in addition to the then-traditional kingdoms of plants, animals and minerals. The kingdom of minerals was later removed from taxonomy in 1866 by Ernst Haeckel, leaving plants, animals, and the protists (*Protista*), defined as a "kingdom of primitive forms".

In 1938, Herbert Copeland resurrected Hogg's label, arguing that Haeckel's term *Protista* included anucleated microbes such as bacteria, which the term "Protoctista" (literally meaning "first established beings") did not. In contrast, Copeland's term included nucleated eukaryotes such as diatoms, green algae and fungi. This classification was the basis for Whittaker's later definition of Fungi, Animalia, Plantae and Protista as the four kingdoms of life. The kingdom Protista was later modified to separate prokaryotes into the separate kingdom of Monera, leaving the protists as a group of eukaryotic microorganisms. These five kingdoms remained the accepted classification until the development of molecular phylogenetics in the late 20th century, when it became apparent that neither protists nor monera were single groups of related organisms (they were not monophyletic groups).

Modern Classifications

Many systematists today do not treat Protista as a formal taxon, but the term "protist" is still commonly used for convenience in two ways. The most popular contemporary definition is a phylogenetic one, that identifies a paraphyletic group: a protist is any eukaryote that is not an animal, (land) plant, or (true) fungus; this definition excludes many unicellular groups, like the Microsporidia (fungi), many Chytridiomycetes (fungi), and yeasts (fungi), and also a non-unicellular group included in Protista in the past, the Myxozoa (animal). Some systematists judge paraphyletic taxa acceptable, and use Protista in this sense as a formal taxon (as found in some secondary textbooks, for pedagogical purpose).

The other definition describes protists primarily by functional or biological criteria: protists are essentially those eukaryotes that are never multicellular, that either exist as independent cells, or if they occur in colonies, do not show differentiation into tissues (but vegetative cell differentiation may occur restricted to sexual reproduction, alternate vegetative morphology, and quiescent or resistant stages, such as cysts); this definition excludes many brown, many red and some green algae, which may have tissues.

The taxonomy of protists is still changing. Newer classifications attempt to present monophyletic groups based on morphological (especially ultrastructural), biochemical (chemotaxonomy) and DNA sequence (molecular research) information. However, there are sometimes discordances between molecular and morphological investigations; these can be categorized as two types: (i) one morphology, multiple lineages (e.g. morphological convergence, cryptic species) and (ii) one lineage, multiple morphologies (e.g. phenotypic plasticity, multiple life-cycle stages).

Because the protists as a whole are paraphyletic, new systems often split up or abandon the kingdom, instead treating the protist groups as separate lines of eukaryotes. The recent scheme by Adl *et al.* (2005) does not recognize formal ranks (phylum, class, etc.) and instead treats groups as clades of phylogenetically related organisms. This is intended to make the classification more stable in the long term and easier to update. Some of the main groups of protists, which may be treated as phyla, are listed in the taxobox, upper right. Many are thought to be monophyletic, though there is still uncertainty. For instance, the excavates are probably not monophyletic and the chromalveolates are probably only monophyletic if the haptophytes and cryptomonads are excluded.

Metabolism

Nutrition can vary according to the type of protist. Most eukaryotic algae are autotrophic, but the pigments were lost in some groups. Other protists are heterotrophic, and may present phagotrophy, osmotrophy, saprotrophy or parasitism. Some are mixotrophic. Some protists that do not have / lost chloroplasts/mitochondria have entered into endosymbiontic relationship with other bacteria/algae to replace the missing functionality. For ex. Paramecium bursaria & Paulinella have captured a green algae (Zoochlorella) & a cyanobacterium respectively that act as a replacement for chloroplast. Meanwhile, a protist, Mixotricha paradoxa that has lost its mitochondria uses endosymbiontic bacteria as mitochondria and ectosymbiontic hair-like bacteria (Treponema spirochetes) for locomotion.

Many protists are flagellate, for example, and filter feeding can take place where the flagella find prey. Other protists can engulf bacteria and other food particles, by extending their cell membrane around them to form a food vacuole and digest them internally, in a process termed phagocytosis.

Nutritional types in protist metabolism			
Nutritional type	**Source of energy**	**Source of carbon**	**Examples**
Photoautotrophs	Sunlight	Organic compounds or carbon fixation	Most algae
Chemoheterotrophs	Organic compounds	Organic compounds	Apicomplexa, Trypanosomes or Amoebae

Reproduction

Some protists reproduce sexually using gametes, while others reproduce asexually by binary fission.

Some species, for example *Plasmodium falciparum*, have extremely complex life cycles that involve multiple forms of the organism, some of which reproduce sexually and others asexually. However, it is unclear how frequently sexual reproduction causes genetic exchange between different strains of *Plasmodium* in nature and most populations of parasitic protists may be clonal lines that rarely exchange genes with other members of their species.

Eukaryotes emerged in evolution more than 1.5 billion years ago. The earliest eukaryotes were likely protists. Although sexual reproduction is widespread among extant eukaryotes, it seemed unlikely until recently, that sex could be a primordial and fundamental characteristic of eukaryotes. A principal reason for this view was that sex appeared to be lacking in certain pathogenic protists whose ancestors branched off early from the eukaryotic family tree. However, several of these protists are now known to be capable of, or to recently have had the capability for, meiosis and hence sexual reproduction. For example, the common intestinal parasite *Giardia lamblia* was once considered to be a descendant of a protist lineage that predated the emergence of meiosis and sex. However, *G. lamblia* was recently found to have a core set of genes that function in meiosis and that are widely present among sexual eukaryotes. These results suggested that *G. lamblia* is capable of meiosis and thus sexual reproduction. Furthermore, direct evidence for meiotic recombination, indicative of sex, was also found in *G. lamblia*.

The pathogenic parasitic protists of the genus Leishmania have been shown to be capable of a sexual cycle in the invertebrate vector, likened to the meiosis undertaken in the trypanosomes.

Trichomonas vaginalis, a parasitic protist, is not known to undergo meiosis, but when Malik et al. tested for 29 genes that function in meiosis, they found 27 to be present, including 8 of 9 genes specific to meiosis in model eukaryotes. These findings suggest that *T. vaginalis* may be capable of meiosis. Since 21 of the 29 meiotic genes were also present in G. lamblia, it appears that most of these meiotic genes were likely present in a common ancestor of *T. vaginalis* and *G. lamblia*. These two species are descendants of protist lineages that are highly divergent among eukaryotes, leading Malik et al. to suggest that these meiotic genes were likely present in a common ancestor of all eukaryotes.

Based on a phylogenetic analysis, Dacks and Roger proposed that facultative sex was present in the common ancestor of all eukaryotes.

This view was further supported by a study of amoebae by Lahr et al. Amoeba have generally been regarded as asexual protists. However these authors describe evidence that most amoeboid lineages are anciently sexual, and that the majority of asexual groups likely arose recently and independently. It should be noted that early researchers (e.g., Calkins) have interpreted phenomena related to chromidia (chromatin granules free in the cytoplasm) in amoeboid organisms as sexual reproduction.

Protists generally reproduce asexually under favorable environmental conditions, but tend to reproduce sexually under stressful conditions, such as starvation or heat shock. Oxidative stress, which is associated with the production of reactive oxygen species leading to DNA damage, also appears to be an important factor in the induction of sex in protists.

Ecology

Protists live in almost any environment that contains liquid water. Many protists, such as algae, are photosynthetic and are vital primary producers in ecosystems, particularly in the ocean as part of the plankton. Other protists include pathogenic species such as the kinetoplastid *Trypanosoma brucei*, which causes sleeping sickness and species of the apicomplexan *Plasmodium* which cause malaria.

Parasitism: Role as Pathogens

Some protists are significant parasites of animals (e.g., five species of the parasitic genus *Plasmodium* cause malaria in humans and many others cause similar diseases in other vertebrates), plants (the oomycete *Phytophthora infestans* causes late blight in potatoes) or even of other protists. Protist pathogens share many metabolic pathways with their Eukaryotic hosts. This makes therapeutic target development extremely difficult – a drug that harms a protist parasite is also likely to harm its animal/plant host. A more thorough understanding of protist biology may allow these diseases to be treated more efficiently. For example, the apicoplast (a nonphotosynthetic chloroplast but essential to carry out important functions other than photosynthesis) present in apicomplexans provides an attractive target for treating diseases caused by dangerous pathogens such as plasmodium.

Recent papers have proposed the use of viruses to treat infections caused by protozoa.

Researchers from the Agricultural Research Service are taking advantage of protists as pathogens to control red imported fire ant (*Solenopsis invicta*) populations in Argentina. Spore-producing protists such as *Kneallhazia solenopsae* (recognized as a sister clade or the closest relative to the fungus kingdom now) can reduce red fire ant populations by 53–100%. Researchers have also been able to infect phorid fly parasitoids of the ant with the protist without harming the flies. This turns the flies into a vector that can spreads the pathogenic protist between red fire ant colonies.

Fossil Record

Many protists have neither hard parts nor resistant spores, and their fossils are extremely rare or unknown. Examples of such groups include the apicomplexans, most ciliates, some green algae (the Klebsormidiales), choanoflagellates, oomycetes, brown algae, yellow-green algae, excavates (e.g., euglenids). Some of these have been found preserved in amber (fossilized tree resin) or under unusual conditions (e.g., *Paleoleishmania*, a kinetoplastid).

Others are relatively common in the fossil record, as the diatoms, golden algae, haptophytes (coccoliths), silicoflagellates, tintinnids (ciliates), dinoflagellates, green algae, red algae, heliozoans, radiolarians, foraminiferans, ebriids and testate amoebae (euglyphids, arcellaceans). Some are even used as paleoecological indicators to reconstruct ancient environments.

More probable eukaryote fossils begin to appear at about 1.8 billion years ago, the acritarchs, spherical fossils of likely algal protists. Another possible representant of early fossil eukaryotes are the Gabonionta.

Fungus

A fungus (/ˈfʌŋgəs/; plural: fungi or funguses) is any member of the group of eukaryotic organisms that includes unicellular microorganisms such as yeasts and molds, as well as multicellular fungi that produce familiar fruiting forms known as mushrooms. These organisms are classified as a kingdom, Fungi, which is separate from the other eukaryotic life kingdoms of plants and animals.

A characteristic that places fungi in a different kingdom from plants, bacteria and some protists, is chitin in their cell walls. Similar to animals, fungi are heterotrophs; they acquire their food by absorbing dissolved molecules, typically by secreting digestive enzymes into their environment. Fungi do not photosynthesise. Growth is their means of mobility, except for spores (a few of which are flagellated), which may travel through the air or water. Fungi are the principal decomposers in ecological systems. These and other differences place fungi in a single group of related organisms, named the *Eumycota* (*true fungi* or *Eumycetes*), which share a common ancestor (is a *monophyletic group*), an interpretation that is also strongly supported by molecular phylogenetics. This fungal group is distinct from the structurally similar myxomycetes (slime molds) and oomycetes (water molds). The discipline of biology devoted to the study of fungi is known as mycology. In the past, mycology was regarded as a branch of botany, although it is now known fungi are genetically more closely related to animals than to plants.

Abundant worldwide, most fungi are inconspicuous because of the small size of their structures, and their cryptic lifestyles in soil or on dead matter. Fungi include symbionts of plants, animals, or other fungi and also parasites. They may become noticeable when fruiting, either as mushrooms or as molds. Fungi perform an essential role in the decomposition of organic matter and have fundamental roles in nutrient cycling and exchange in the environment. They have long been used as a direct source of food, in the form of mushrooms and truffles; as a leavening agent for bread; and in the fermentation of various food products, such as wine, beer, and soy sauce. Since the 1940s, fungi have been used for the production of antibiotics, and, more recently, various enzymes produced by fungi are used industrially and in detergents. Fungi are also used as biological pesticides to control weeds, plant diseases and insect pests. Many species produce bioactive compounds called mycotoxins, such as alkaloids and polyketides, that are toxic to animals including humans. The fruiting structures of a few species contain psychotropic compounds and are consumed recreationally or in traditional spiritual ceremonies. Fungi can break down manufactured materials and buildings, and become significant pathogens of humans and other animals. Losses of crops due to fungal diseases (e.g., rice blast disease) or food spoilage can have a large impact on human food supplies and local economies.

The fungus kingdom encompasses an enormous diversity of taxa with varied ecologies, life cycle strategies, and morphologies ranging from unicellular aquatic chytrids to large mushrooms. However, little is known of the true biodiversity of Kingdom Fungi, which has been estimated at 1.5 million to 5 million species, with about 5% of these having been formally classified. Ever since the pioneering 18th and 19th century taxonomical works of Carl Linnaeus, Christian Hendrik Persoon, and Elias Magnus Fries, fungi have been classified according to their morphology (e.g., characteristics such as spore color or microscopic features) or physiology. Advances in molecular genetics have opened the way for DNA analysis to be incorporated into taxonomy, which has sometimes challenged the historical groupings based on morphology and other traits. Phylogenetic studies published in the last decade have helped reshape the classification within Kingdom Fungi, which is divided into one subkingdom, seven phyla, and ten subphyla.

Characteristics

Before the introduction of molecular methods for phylogenetic analysis, taxonomists considered fungi to be members of the plant kingdom because of similarities in lifestyle: both fungi and plants are mainly immobile, and have similarities in general morphology and growth habitat. Like plants, fungi often grow in soil and, in the case of mushrooms, form conspicuous fruit bodies, which sometimes resemble plants, such as mosses. The fungi are now considered a separate kingdom, distinct from both plants and animals, from which they appear to have diverged around one billion years ago. Some morphological, biochemical, and genetic features are shared with other organisms, while others are unique to the fungi, clearly separating them from the other kingdoms:

Shared features:

- With other eukaryotes: Fungal cells contain membrane-bound nuclei with chromosomes that contain DNA with noncoding regions called introns and coding regions called exons. Fungi have membrane-bound cytoplasmic organelles such as mitochondria, sterol-containing membranes, and ribosomes of the 80S type. They have a characteristic range of soluble carbohydrates and storage compounds, including sugar alcohols (e.g., mannitol), disaccharides, (e.g., trehalose), and polysaccharides (e.g., glycogen, which is also found in animals).

- With animals: Fungi lack chloroplasts and are heterotrophic organisms and so require preformed organic compounds as energy sources.

- With plants: Fungi have a cell wall and vacuoles. They reproduce by both sexual and asexual means, and like basal plant groups (such as ferns and mosses) produce spores. Similar to mosses and algae, fungi typically have haploid nuclei.

- With euglenoids and bacteria: Higher fungi, euglenoids, and some bacteria produce the amino acid L-lysine in specific biosynthesis steps, called the α-aminoadipate pathway.

- The cells of most fungi grow as tubular, elongated, and thread-like (filamentous) structures called hyphae, which may contain multiple nuclei and extend by growing at their tips. Each tip contains a set of aggregated vesicles—cellular structures consisting of proteins, lipids, and other organic molecules—called the Spitzenkörper. Both fungi and oomycetes grow as filamentous hyphal cells. In contrast, similar-looking organisms, such as filamentous green algae, grow by repeated cell division within a chain of cells. There are also single-celled fungi (yeasts) that do not form hyphae, and fungi with both hyphal and yeast forms.

- In common with some plant and animal species, more than 70 fungal species display bioluminescence.

Unique features:

- Some species grow as unicellular yeasts that reproduce by budding or binary fission. Dimorphic fungi can switch between a yeast phase and a hyphal phase in response to environmental conditions.

- The fungal cell wall is composed of glucans and chitin; while glucans are also found in plants and chitin in the exoskeleton of arthropods, fungi are the only organisms that combine these two structural molecules in their cell wall. Unlike those of plants and oomycetes, fungal cell walls do not contain cellulose.

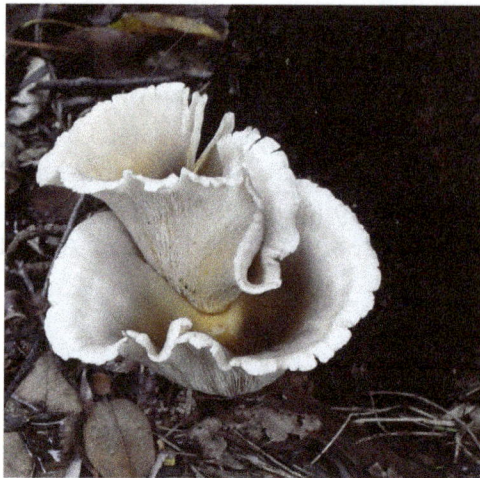

Omphalotus nidiformis, a bioluminescent mushroom

Most fungi lack an efficient system for the long-distance transport of water and nutrients, such as the xylem and phloem in many plants. To overcome this limitation, some fungi, such as *Armillaria*, form rhizomorphs, which resemble and perform functions similar to the roots of plants. As eukaryotes, fungi possess a biosynthetic pathway for producing terpenes that uses mevalonic acid and pyrophosphate as chemical building blocks. Plants and some other organisms have an additional terpene biosynthesis pathway in their chloroplasts, a structure fungi and animals do not have. Fungi produce several secondary metabolites that are similar or identical in structure to those made by plants. Many of the plant and fungal enzymes that make these compounds differ from each other in sequence and other characteristics, which indicates separate origins and evolution of these enzymes in the fungi and plants.

Diversity

Bracket fungi on a tree stump

Fungi have a worldwide distribution, and grow in a wide range of habitats, including extreme environments such as deserts or areas with high salt concentrations or ionizing radiation, as well as in deep sea sediments. Some can survive the intense UV and cosmic radiation encountered during space travel. Most grow in terrestrial environments, though several species live partly or solely in aquatic habitats, such as the chytrid fungus *Batrachochytrium dendrobatidis*, a parasite that has been responsible for a worldwide decline in amphibian populations. This organism spends part of its life cycle as a motile zoospore, enabling it to propel itself through water and enter its amphibian host. Other examples of aquatic fungi include those living in hydrothermal areas of the ocean.

Around 100,000 species of fungi have been formally described by taxonomists, but the global biodiversity of the fungus kingdom is not fully understood. On the basis of observations of the ratio of the number of fungal species to the number of plant species in selected environments, the fungal kingdom has been estimated to contain about 1.5 million species. A recent (2011) estimate suggests there may be over 5 million species. In mycology, species have historically been distinguished by a variety of methods and concepts. Classification based on morphological characteristics, such as the size and shape of spores or fruiting structures, has traditionally dominated fungal taxonomy. Species may also be distinguished by their biochemical and physiological characteristics, such as their ability to metabolize certain biochemicals, or their reaction to chemical tests. The biological species concept discriminates species based on their ability to mate. The application of molecular tools, such as DNA sequencing and phylogenetic analysis, to study diversity has greatly enhanced the resolution and added robustness to estimates of genetic diversity within various taxonomic groups.

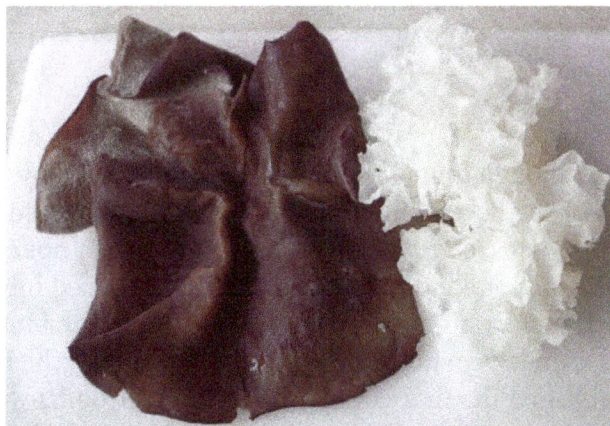

Two types of edible fungi

Mycology

Mycology is the branch of biology concerned with the systematic study of fungi, including their genetic and biochemical properties, their taxonomy, and their use to humans as a source of medicine, food, and psychotropic substances consumed for religious purposes, as well as their dangers, such as poisoning or infection. The field of phytopathology, the study of plant diseases, is closely related because many plant pathogens are fungi.

The use of fungi by humans dates back to prehistory; Ötzi the Iceman, a well-preserved mummy of a 5,300-year-old Neolithic man found frozen in the Austrian Alps, carried two species of polypore mushrooms that may have been used as tinder (*Fomes fomentarius*), or for medicinal purposes (*Piptoporus betulinus*). Ancient peoples have used fungi as food sources–often unknowingly–for millennia, in the preparation of leavened bread and fermented juices. Some of the oldest written records contain references to the destruction of crops that were probably caused by pathogenic fungi.

Pietro Antonio MICHELI
(1679 - 1737)

In 1729, Pier A. Micheli first published descriptions of fungi.

History

Mycology is a relatively new science that became systematic after the development of the micro-scope in the 16th century. Although fungal spores were first observed by Giambattista della Porta in 1588, the seminal work in the development of mycology is considered to be the publication of Pier Antonio Micheli's 1729 work *Nova plantarum genera*. Micheli not only observed spores but also showed that, under the proper conditions, they could be induced into growing into the same species of fungi from which they originated. Extending the use of the binomial system of nomen-clature introduced by Carl Linnaeus in his *Species plantarum* (1753), the Dutch Christian Hendrik Persoon (1761–1836) established the first classification of mushrooms with such skill so as to be considered a founder of modern mycology. Later, Elias Magnus Fries (1794–1878) further elabo-rated the classification of fungi, using spore color and various microscopic characteristics, methods still used by taxonomists today. Other notable early contributors to mycology in the 17th–19th and early 20th centuries include Miles Joseph Berkeley, August Carl Joseph Corda, Anton de Bary, the brothers Louis René and Charles Tulasne, Arthur H. R. Buller, Curtis G. Lloyd, and Pier Andrea Saccardo. The 20th century has seen a modernization of mycology that has come from advances in biochemistry, genetics, molecular biology, and biotechnology. The use of DNA sequencing tech-nologies and phylogenetic analysis has provided new insights into fungal relationships and biodi-versity, and has challenged traditional morphology-based groupings in fungal taxonomy.

Morphology

Microscopic Structures

Most fungi grow as hyphae, which are cylindrical, thread-like structures 2–10 µm in diameter and up to several centimeters in length. Hyphae grow at their tips (apices); new hyphae are typically formed by emergence of new tips along existing hyphae by a process called *branching*, or occa-sionally growing hyphal tips fork, giving rise to two parallel-growing hyphae. The combination of apical growth and branching/forking leads to the development of a mycelium, an interconnected network of hyphae. Hyphae can be either septate or coenocytic. Septate hyphae are divided into compartments separated by cross walls (internal cell walls, called septa, that are formed at right angles to the cell wall giving the hypha its shape), with each compartment containing one or more nuclei; coenocytic hyphae are not compartmentalized. Septa have pores that allow cytoplasm, organelles, and sometimes nuclei to pass through; an example is the dolipore septum in fungi of the phylum Basidiomycota. Coenocytic hyphae are in essence multinucleate supercells.

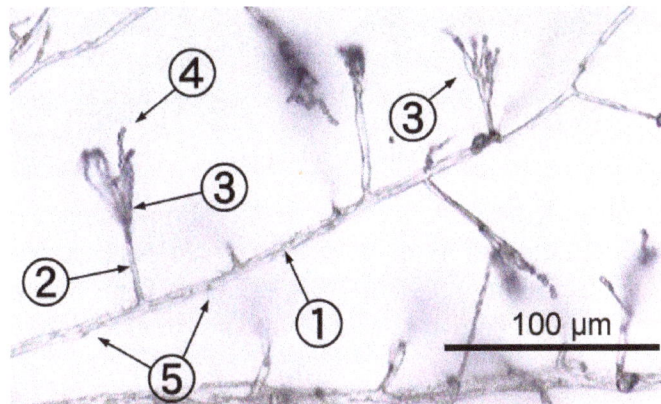

An environmental isolate of *Penicillium*

1. hypha

2. conidiophore

3. phialide

4. conidia

5. septa

Many species have developed specialized hyphal structures for nutrient uptake from living hosts; examples include haustoria in plant-parasitic species of most fungal phyla, and arbuscules of several mycorrhizal fungi, which penetrate into the host cells to consume nutrients.

Although fungi are opisthokonts—a grouping of evolutionarily related organisms broadly characterized by a single posterior flagellum—all phyla except for the chytrids have lost their posterior flagella. Fungi are unusual among the eukaryotes in having a cell wall that, in addition to glucans (e.g., β-1,3-glucan) and other typical components, also contains the biopolymer chitin.

Macroscopic Structures

Fungal mycelia can become visible to the naked eye, for example, on various surfaces and substrates, such as damp walls and spoiled food, where they are commonly called molds. Mycelia grown on solid agar media in laboratory petri dishes are usually referred to as colonies. These colonies can exhibit growth shapes and colors (due to spores or pigmentation) that can be used as diagnostic features in the identification of species or groups. Some individual fungal colonies can reach extraordinary dimensions and ages as in the case of a clonal colony of *Armillaria solidipes*, which extends over an area of more than 900 ha (3.5 square miles), with an estimated age of nearly 9,000 years.

Armillaria solidipes

The apothecium—a specialized structure important in sexual reproduction in the ascomycetes—is a cup-shaped fruit body that holds the hymenium, a layer of tissue containing the spore-bearing cells. The fruit bodies of the basidiomycetes (basidiocarps) and some ascomycetes can sometimes grow very large, and many are well known as mushrooms.

Growth and Physiology

Mold growth covering a decaying peach. The frames were taken approximately 12 hours apart over a period of six days.

The growth of fungi as hyphae on or in solid substrates or as single cells in aquatic environments is adapted for the efficient extraction of nutrients, because these growth forms have high surface area to volume ratios. Hyphae are specifically adapted for growth on solid surfaces, and to invade substrates and tissues. They can exert large penetrative mechanical forces; for example, the plant pathogen *Magnaporthe grisea* forms a structure called an appressorium that evolved to puncture plant tissues. The pressure generated by the appressorium, directed against the plant epidermis, can exceed 8 megapascals (1,200 psi). The filamentous fungus *Paecilomyces lilacinus* uses a similar structure to penetrate the eggs of nematodes.

The mechanical pressure exerted by the appressorium is generated from physiological processes that increase intracellular turgor by producing osmolytes such as glycerol. Adaptations such as these are complemented by hydrolytic enzymes secreted into the environment to digest large organic molecules—such as polysaccharides, proteins, and lipids—into smaller molecules that may then be absorbed as nutrients. The vast majority of filamentous fungi grow in a polar fashion—i.e., by extension into one direction—by elongation at the tip (apex) of the hypha. Other forms of fungal growth include intercalary extension (longitudinal expansion of hyphal compartments that are below the apex) as in the case of some endophytic fungi, or growth by volume expansion during the development of mushroom stipes and other large organs. Growth of fungi as multicellular structures consisting of somatic and reproductive cells—a feature independently evolved in animals and plants—has several functions, including the development of fruit bodies for dissemination of sexual spores and biofilms for substrate colonization and intercellular communication.

The fungi are traditionally considered heterotrophs, organisms that rely solely on carbon fixed by other organisms for metabolism. Fungi have evolved a high degree of metabolic versatility that allows them to use a diverse range of organic substrates for growth, including simple compounds such as nitrate, ammonia, acetate, or ethanol. In some species the pigment melanin may play a role in extracting energy from ionizing radiation, such as gamma radiation. This form of "radiotrophic" growth has been described for only a few species, the effects on growth rates are small, and the underlying biophysical and biochemical processes are not well known. This process might bear similarity to CO_2 fixation via visible light, but instead uses ionizing radiation as a source of energy.

Reproduction

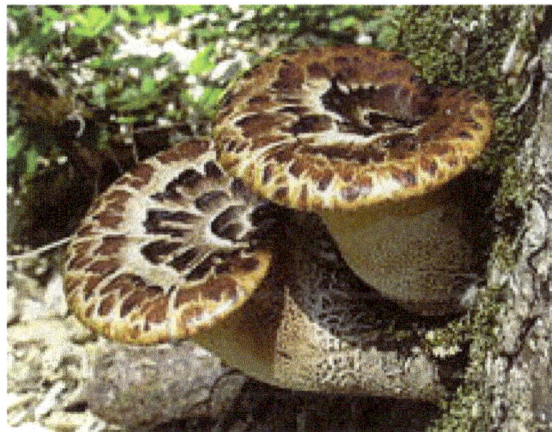

Polyporus squamosus

Fungal reproduction is complex, reflecting the differences in lifestyles and genetic makeup within this diverse kingdom of organisms. It is estimated that a third of all fungi reproduce using more than one method of propagation; for example, reproduction may occur in two well-differentiated stages within the life cycle of a species, the teleomorph and the anamorph. Environmental conditions trigger genetically determined developmental states that lead to the creation of specialized structures for sexual or asexual reproduction. These structures aid reproduction by efficiently dispersing spores or spore-containing propagules.

Asexual Reproduction

Asexual reproduction occurs via vegetative spores (conidia) or through mycelial fragmentation.

Mycelial fragmentation occurs when a fungal mycelium separates into pieces, and each component grows into a separate mycelium. Mycelial fragmentation and vegetative spores maintain clonal populations adapted to a specific niche, and allow more rapid dispersal than sexual reproduction. The "Fungi imperfecti" (fungi lacking the perfect or sexual stage) or Deuteromycota comprise all the species that lack an observable sexual cycle.

Sexual Reproduction

Sexual reproduction with meiosis exists in all fungal phyla except Glomeromycota. It differs in many aspects from sexual reproduction in animals or plants. Differences also exist between fungal groups and can be used to discriminate species by morphological differences in sexual structures and reproductive strategies. Mating experiments between fungal isolates may identify species on the basis of biological species concepts. The major fungal groupings have initially been delineated based on the morphology of their sexual structures and spores; for example, the spore-containing structures, asci and basidia, can be used in the identification of ascomycetes and basidiomycetes, respectively. Some species may allow mating only between individuals of opposite mating type, whereas others can mate and sexually reproduce with any other individual or itself. Species of the former mating system are called heterothallic, and of the latter homothallic.

Most fungi have both a haploid and a diploid stage in their life cycles. In sexually reproducing fungi, compatible individuals may combine by fusing their hyphae together into an interconnected network; this process, anastomosis, is required for the initiation of the sexual cycle. Ascomycetes and basidiomycetes go through a dikaryotic stage, in which the nuclei inherited from the two parents do not combine immediately after cell fusion, but remain separate in the hyphal cells.

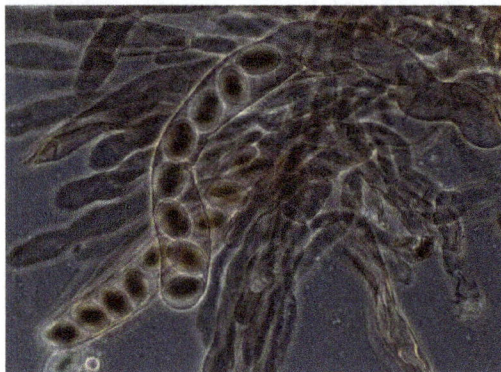

The 8-spore asci of *Morchella elata*, viewed with phase contrast microscopy

In ascomycetes, dikaryotic hyphae of the hymenium (the spore-bearing tissue layer) form a characteristic *hook* at the hyphal septum. During cell division, formation of the hook ensures proper distribution of the newly divided nuclei into the apical and basal hyphal compartments. An ascus (plural *asci*) is then formed, in which karyogamy (nuclear fusion) occurs. Asci are embedded in an ascocarp, or fruiting body. Karyogamy in the asci is followed immediately by meiosis and the production of ascospores. After dispersal, the ascospores may germinate and form a new haploid mycelium.

Sexual reproduction in basidiomycetes is similar to that of the ascomycetes. Compatible haploid hyphae fuse to produce a dikaryotic mycelium. However, the dikaryotic phase is more extensive in the basidiomycetes, often also present in the vegetatively growing mycelium. A specialized anatomical structure, called a clamp connection, is formed at each hyphal septum. As with the structurally similar hook in the ascomycetes, the clamp connection in the basidiomycetes is required for controlled transfer of nuclei during cell division, to maintain the dikaryotic stage with two genetically different nuclei in each hyphal compartment. A basidiocarp is formed in which club-like structures known as basidia generate haploid basidiospores after karyogamy and meiosis. The most commonly known basidiocarps are mushrooms, but they may also take other forms.

In glomeromycetes (formerly zygomycetes), haploid hyphae of two individuals fuse, forming a gametangium, a specialized cell structure that becomes a fertile gamete-producing cell. The gametangium develops into a zygospore, a thick-walled spore formed by the union of gametes. When the zygospore germinates, it undergoes meiosis, generating new haploid hyphae, which may then form asexual sporangiospores. These sporangiospores allow the fungus to rapidly disperse and germinate into new genetically identical haploid fungal mycelia.

Spore dispersal

Both asexual and sexual spores or sporangiospores are often actively dispersed by forcible ejection from their reproductive structures. This ejection ensures exit of the spores from the reproductive structures as well as traveling through the air over long distances.

The bird's nest fungus *Cyathus stercoreus*

Specialized mechanical and physiological mechanisms, as well as spore surface structures (such as hydrophobins), enable efficient spore ejection. For example, the structure of the spore-bearing cells in some ascomycete species is such that the buildup of substances affecting cell volume and fluid balance enables the explosive discharge of spores into the air. The forcible discharge of single spores termed *ballistospores* involves formation of a small drop of water (Buller's drop), which upon contact with the spore leads to its projectile release with an initial acceleration of more than 10,000 g; the net result is that the spore is ejected 0.01–0.02 cm, sufficient distance for it to fall through the gills or pores into the air below. Other fungi, like the puffballs, rely on alternative mechanisms for spore release, such as external mechanical forces. The bird's nest fungi use the force of falling water drops to liberate the spores from cup-shaped fruiting bodies. Another strategy is seen in the stinkhorns, a group of fungi with lively colors and putrid odor that attract insects to disperse their spores.

Other Sexual Processes

Besides regular sexual reproduction with meiosis, certain fungi, such as those in the genera *Penicillium* and *Aspergillus*, may exchange genetic material via parasexual processes, initiated by anastomosis between hyphae and plasmogamy of fungal cells. The frequency and relative importance of parasexual events is unclear and may be lower than other sexual processes. It is known to play a role in intraspecific hybridization and is likely required for hybridization between species, which has been associated with major events in fungal evolution.

Evolution

In contrast to plants and animals, the early fossil record of the fungi is meager. Factors that likely contribute to the under-representation of fungal species among fossils include the nature of fungal fruiting bodies, which are soft, fleshy, and easily degradable tissues and the microscopic dimensions of most fungal structures, which therefore are not readily evident. Fungal fossils are difficult to distinguish from those of other microbes, and are most easily identified when they resemble extant fungi. Often recovered from a permineralized plant or animal host, these samples are typically studied by making thin-section preparations that can be examined with light microscopy or transmission electron microscopy. Researchers study compression fossils by dissolving the surrounding matrix with acid and then using light or scanning electron microscopy to examine surface details.

The earliest fossils possessing features typical of fungi date to the Proterozoic eon, some 1,430 million years ago (Ma); these multicellular benthic organisms had filamentous structures with septa, and were capable of anastomosis. More recent studies (2009) estimate the arrival of fungal organisms at about 760–1060 Ma on the basis of comparisons of the rate of evolution in closely related groups. For much of the Paleozoic Era (542–251 Ma), the fungi appear to have been aquatic and consisted of organisms similar to the extant chytrids in having flagellum-bearing spores. The evolutionary adaptation from an aquatic to a terrestrial lifestyle necessitated a diversification of ecological strategies for obtaining nutrients, including parasitism, saprobism, and the development of mutualistic relationships such as mycorrhiza and lichenization. Recent (2009) studies suggest that the ancestral ecological state of the Ascomycota was saprobism, and that independent lichenization events have occurred multiple times.

It is presumed that the fungi colonized the land during the Cambrian (542–488.3 Ma), long before land plants. Fossilized hyphae and spores recovered from the Ordovician of Wisconsin (460 Ma) resemble modern-day Glomerales, and existed at a time when the land flora likely consisted of only non-vascular bryophyte-like plants. Prototaxites, which was probably a fungus or lichen, would have been the tallest organism of the late Silurian. Fungal fossils do not become common and uncontroversial until the early Devonian (416–359.2 Ma), when they occur abundantly in the Rhynie chert, mostly as Zygomycota and Chytridiomycota. At about this same time, approximately 400 Ma, the Ascomycota and Basidiomycota diverged, and all modern classes of fungi were present by the Late Carboniferous (Pennsylvanian, 318.1–299 Ma).

Lichen-like fossils have been found in the Doushantuo Formation in southern China dating back to 635–551 Ma. Lichens formed a component of the early terrestrial ecosystems, and the estimated age of the oldest terrestrial lichen fossil is 400 Ma; this date corresponds to the age of the oldest known sporocarp fossil, a *Paleopyrenomycites* species found in the Rhynie Chert. The oldest fossil with microscopic features resembling modern-day basidiomycetes is *Palaeoancistrus*, found permineralized with a fern from the Pennsylvanian. Rare in the fossil record are the Homobasidiomycetes (a taxon roughly equivalent to the mushroom-producing species of the Agaricomycetes). Two amber-preserved specimens provide evidence that the earliest known mushroom-forming fungi (the extinct species *Archaeomarasmius leggetti*) appeared during the late Cretaceous, 90 Ma.

Some time after the Permian–Triassic extinction event (251.4 Ma), a fungal spike (originally thought to be an extraordinary abundance of fungal spores in sediments) formed, suggesting that fungi were the dominant life form at this time, representing nearly 100% of the available fossil record for this period. However, the relative proportion of fungal spores relative to spores formed by algal species is difficult to assess, the spike did not appear worldwide, and in many places it did not fall on the Permian–Triassic boundary.

Taxonomic Groups

The major phyla (sometimes called divisions) of fungi have been classified mainly on the basis of characteristics of their sexual reproductive structures. Currently, seven phyla are proposed: Microsporidia, Chytridiomycota, Blastocladiomycota, Neocallimastigomycota, Glomeromycota, Ascomycota, and Basidiomycota.

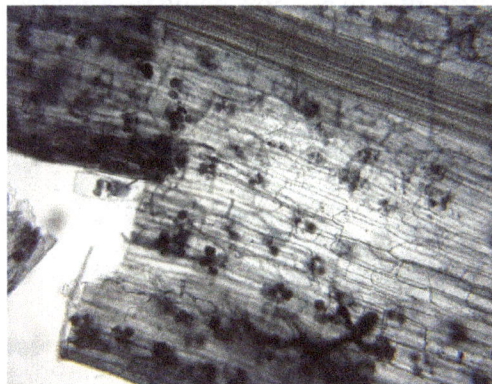

Arbuscular mycorrhiza seen under microscope. Flax root cortical cells containing paired arbuscules.

Phylogenetic analysis has demonstrated that the Microsporidia, unicellular parasites of animals and protists, are fairly recent and highly derived endobiotic fungi (living within the tissue of another species). One 2006 study concludes that the Microsporidia are a sister group to the true fungi; that is, they are each other's closest evolutionary relative. Hibbett and colleagues suggest that this analysis does not clash with their classification of the Fungi, and although the Microsporidia are elevated to phylum status, it is acknowledged that further analysis is required to clarify evolutionary relationships within this group.

The Chytridiomycota are commonly known as chytrids. These fungi are distributed worldwide. Chytrids produce zoospores that are capable of active movement through aqueous phases with a single flagellum, leading early taxonomists to classify them as protists. Molecular phylogenies, inferred from rRNA sequences in ribosomes, suggest that the Chytrids are a basal group divergent from the other fungal phyla, consisting of four major clades with suggestive evidence for paraphyly or possibly polyphyly.

The Blastocladiomycota were previously considered a taxonomic clade within the Chytridiomycota. Recent molecular data and ultrastructural characteristics, however, place the Blastocladiomycota as a sister clade to the Zygomycota, Glomeromycota, and Dikarya (Ascomycota and Basidiomycota). The blastocladiomycetes are saprotrophs, feeding on decomposing organic matter, and they are parasites of all eukaryotic groups. Unlike their close relatives, the chytrids, most of which exhibit zygotic meiosis, the blastocladiomycetes undergo sporic meiosis.

The Neocallimastigomycota were earlier placed in the phylum Chytridomycota. Members of this small phylum are anaerobic organisms, living in the digestive system of larger herbivorous mammals and in other terrestrial and aquatic environments enriched in cellulose (e.g., domestic waste landfill sites). They lack mitochondria but contain hydrogenosomes of mitochondrial origin. As the related chrytrids, neocallimastigomycetes form zoospores that are posteriorly uniflagellate or polyflagellate.

Members of the Glomeromycota form arbuscular mycorrhizae, a form of symbiosis wherein fungal hyphae invade plant root cells and both species benefit from the resulting increased supply of nutrients. All known Glomeromycota species reproduce asexually. The symbiotic association between the Glomeromycota and plants is ancient, with evidence dating to 400 million years ago. Formerly part of the Zygomycota (commonly known as 'sugar' and 'pin' molds), the Glomeromycota were elevated to phylum status in 2001 and now replace the older phylum Zygomycota. Fungi that were placed in the Zygomycota are now being reassigned to the Glomeromycota, or the subphyla incertae sedis Mucoromycotina, Kickxellomycotina, the Zoopagomycotina and the Entomophthoromycotina. Some well-known examples of fungi formerly in the Zygomycota include black bread mold (*Rhizopus stolonifer*), and *Pilobolus* species, capable of ejecting spores several meters through the air. Medically relevant genera include *Mucor*, *Rhizomucor*, and *Rhizopus*.

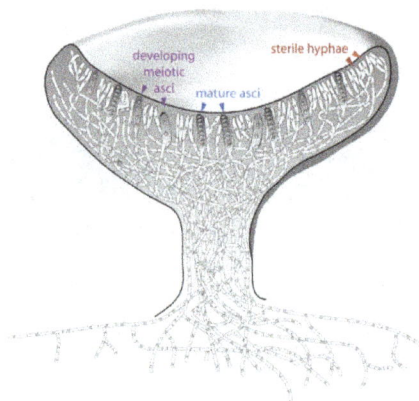

Diagram of an apothecium (the typical cup-like reproductive structure of Ascomycetes) showing sterile tissues as well as developing and mature asci.

The Ascomycota, commonly known as sac fungi or ascomycetes, constitute the largest taxonomic group within the Eumycota. These fungi form meiotic spores called ascospores, which are enclosed in a special sac-like structure called an ascus. This phylum includes morels, a few mushrooms and truffles, unicellular yeasts (e.g., of the genera *Saccharomyces*, *Kluyveromyces*, *Pichia*, and *Candida*), and many filamentous fungi living as saprotrophs, parasites, and mutualistic symbionts. Prominent and important genera of filamentous ascomycetes include *Aspergillus*, *Penicillium*, *Fusarium*, and *Claviceps*. Many ascomycete species have only been observed undergoing asexual reproduction (called anamorphic species), but analysis of molecular data has often been able to identify their closest teleomorphs in the Ascomycota. Because the products of meiosis are retained within the sac-like ascus, ascomycetes have been used for elucidating principles of genetics and heredity (e.g., *Neurospora crassa*).

Members of the Basidiomycota, commonly known as the club fungi or basidiomycetes, produce meiospores called basidiospores on club-like stalks called basidia. Most common mushrooms belong to this group, as well as rust and smut fungi, which are major pathogens of grains. Other important basidiomycetes include the maize pathogen *Ustilago maydis*, human commensal species of the genus *Malassezia*, and the opportunistic human pathogen, *Cryptococcus neoformans*.

Fungus-like Organisms

Because of similarities in morphology and lifestyle, the slime molds (mycetozoans, plasmodiophorids, acrasids, *Fonticula* and labyrinthulids, now in Amoebozoa, Rhizaria, Excavata, Opisthokonta and Stramenopiles, respectively), water molds (oomycetes) and hyphochytrids (both Stramenopiles) were formerly classified in the kingdom Fungi, in groups like Mastigomycotina, Gymnomycota and Phycomycetes. The slime molds were studied also as protozoans, leading to a ambiregnal, duplicated taxonomy.

Unlike true fungi, the cell walls of oomycetes contain cellulose and lack chitin. Hyphochytrids have both chitin and cellulose. Slime molds lack a cell wall during the assimilative phase (except labyrinthulids, which have a wall of scales), and ingest nutrients by ingestion (phagocytosis, except labyrinthulids) rather than absorption (osmotrophy, as fungi, labyrinthulids, oomycetes and hyphochytrids). Neither water molds nor slime molds are closely related to the true fungi, and, therefore, taxonomists no longer group them in the kingdom Fungi. Nonetheless, studies of the oomycetes and myxomycetes are still often included in mycology textbooks and primary research literature.

The Eccrinales and Amoebidiales are opisthokont protists, previously thought to be zygomycete fungi. Other groups now in Opisthokonta (e.g., *Corallochytrium*, Ichthyosporea) were also at given time classified as fungi. The genus *Blastocystis*, now in Stramenopiles, was originally classified as a yeast. *Ellobiopsis*, now in Alveolata, was considered a chytrid. The bacteria were also included in fungi in some classifications, as the group Schizomycetes.

The Rozellida clade, including the "ex-chytrid" *Rozella*, is a genetically disparate group known mostly from environmental DNA sequences that is a sister group to fungi. Members of the group that have been isolated lack the chitinous cell wall that is characteristic of fungi.

The nucleariids, protists currently grouped in the Choanozoa (Opisthokonta), may be the next sister group to the eumycete clade, and as such could be included in an expanded fungal kingdom.

Many Actinomycetales (Actinobacteria), a group with many filamentous bacteria, were also long believed to be fungi.

Ecology

Although often inconspicuous, fungi occur in every environment on Earth and play very important roles in most ecosystems. Along with bacteria, fungi are the major decomposers in most terrestrial (and some aquatic) ecosystems, and therefore play a critical role in biogeochemical cycles and in many food webs. As decomposers, they play an essential role in nutrient cycling, especially as saprotrophs and symbionts, degrading organic matter to inorganic molecules, which can then re-enter anabolic metabolic pathways in plants or other organisms.

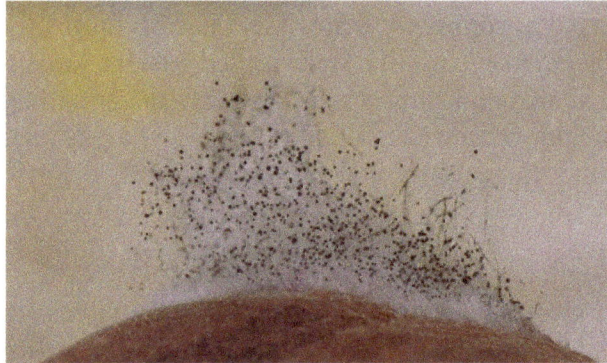

A pin mold decomposing a peach

Symbiosis

Many fungi have important symbiotic relationships with organisms from most if not all Kingdoms. These interactions can be mutualistic or antagonistic in nature, or in the case of commensal fungi are of no apparent benefit or detriment to the host.

With Plants

Mycorrhizal symbiosis between plants and fungi is one of the most well-known plant–fungus associations and is of significant importance for plant growth and persistence in many ecosystems; over 90% of all plant species engage in mycorrhizal relationships with fungi and are dependent upon this relationship for survival.

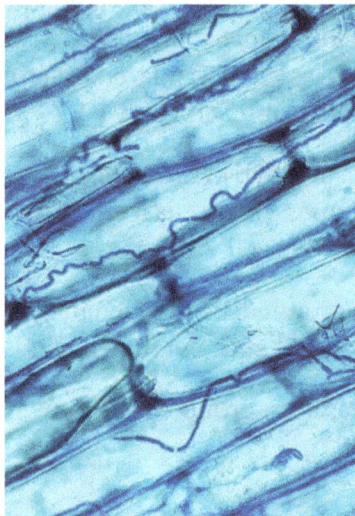

The dark filaments are hyphae of the endophytic fungus Neotyphodium coenophialum in the intercellular spaces of tall fescue leaf sheath tissue

The mycorrhizal symbiosis is ancient, dating to at least 400 million years ago. It often increases the plant's uptake of inorganic compounds, such as nitrate and phosphate from soils having low concentrations of these key plant nutrients. The fungal partners may also mediate plant-to-plant transfer of carbohydrates and other nutrients. Such mycorrhizal communities are called "common mycorrhizal networks". A special case of mycorrhiza is myco-heterotrophy, whereby the plant parasitizes the fungus, obtaining all of its nutrients from its fungal symbiont. Some fungal species inhabit the tissues inside roots, stems, and leaves, in which case they are called endophytes. Similar to mycorrhiza, endophytic colonization by fungi may benefit both symbionts; for example, endophytes of grasses impart to their host increased resistance to herbivores and other environmental stresses and receive food and shelter from the plant in return.

With Algae and Cyanobacteria

Lichens are a symbiotic relationship between fungi and algae or cyanobacteria. The algae partner in the relationship is referred to in lichen terminology as a "photobiont". The fungi part of the relationship are composed mostly of various species of ascomycetes and a few basidiomycetes. Lichens occur in every ecosystem on all continents, play a key role in soil formation and the initiation of biological succession, and are the dominating life forms in extreme environments, including polar, alpine, and semiarid desert regions. They are able to grow on inhospitable surfaces, including bare soil, rocks, tree bark, wood, shells, barnacles and leaves. As in mycorrhizas, the photobiont provides sugars and other carbohydrates via photosynthesis to the fungus, while the fungus provides minerals and water to the photobiont. The functions of both symbiotic organisms are so closely intertwined that they function almost as a single organism; in most cases the resulting organism differs greatly from the individual components. Lichenization is a common mode of nutrition for fungi; around 20% of fungi—between 17,500 and 20,000 described species—are lichenized. Characteristics common to most lichens include obtaining organic carbon by photosynthesis, slow growth, small size, long life, long-lasting (seasonal) vegetative reproductive structures, mineral nutrition obtained largely from airborne sources, and greater tolerance of desiccation than most other photosynthetic organisms in the same habitat.

The lichen Lobaria pulmonaria, a symbiosis of fungal, algal, and cyanobacterial species

With Insects

Many insects also engage in mutualistic relationships with fungi. Several groups of ants cultivate fungi in the order Agaricales as their primary food source, while ambrosia beetles cultivate various species of fungi in the bark of trees that they infest. Likewise, females of several wood wasp species (genus *Sirex*) inject their eggs together with spores of the wood-rotting fungus *Amylostereum areolatum* into the sapwood of pine trees; the growth of the fungus provides ideal nutritional conditions for the development of the wasp larvae. At least one species of stingless bee has a relationship with a fungus in the genus *Monascus*, where the larvae consume and depend on fungus transferred from old to new nests. Termites on the African savannah are also known to cultivate fungi, and yeasts of the genera *Candida* and *Lachancea* inhabit the gut of a wide range of insects, including neuropterans, beetles, and cockroaches; it is not known whether these fungi benefit their hosts. Fungi ingrowing dead wood are essential for xylophagous insects (e.g. wood-boring beetles). They deliver nutrients needed by xylophages to nutritionally scarce dead wood. Thanks to this nutritional enrichment the larvae of woodboring insect is able to grow and develop to adulthood. The larvae of many families of fungicolous flies, particularly those within the superfamily Sciaroidea such as the Mycetophilidae and some Keroplatidae feed on fungal fruiting bodies and sterile mycorrhizae.

As Pathogens and Parasites

Many fungi are parasites on plants, animals (including humans), and other fungi. Serious pathogens of many cultivated plants causing extensive damage and losses to agriculture and forestry include the rice blast fungus *Magnaporthe oryzae*, tree pathogens such as *Ophiostoma ulmi* and *Ophiostoma novo-ulmi* causing Dutch elm disease, and *Cryphonectria parasitica* responsible for chestnut blight, and plant pathogens in the genera *Fusarium, Ustilago, Alternaria*, and *Cochliobolus*. Some carnivorous fungi, like *Paecilomyces lilacinus*, are predators of nematodes, which they capture using an array of specialized structures such as constricting rings or adhesive nets.

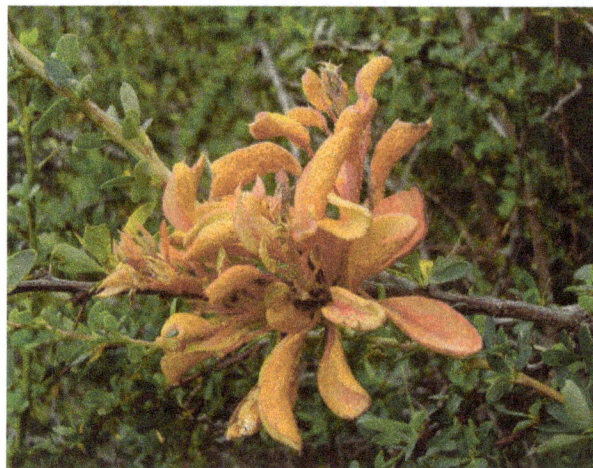

The plant pathogen Aecidium magellanicum causes calafate rust, seen here on a Berberis shrub in Chile.

Some fungi can cause serious diseases in humans, several of which may be fatal if untreated. These include aspergillosis, candidiasis, coccidioidomycosis, cryptococcosis, histoplasmosis, mycetomas, and paracoccidioidomycosis. Furthermore, persons with immuno-deficiencies are particularly susceptible to disease by genera such as *Aspergillus, Candida, Cryptoccocus, Histoplasma*, and *Pneumocystis*. Other fungi can attack eyes, nails, hair, and especially skin, the so-called dermatophytic and keratinophilic fungi, and cause local infections such as ringworm and athlete's foot. Fungal spores are also a cause of allergies, and fungi from different taxonomic groups can evoke allergic reactions.

Mycotoxins

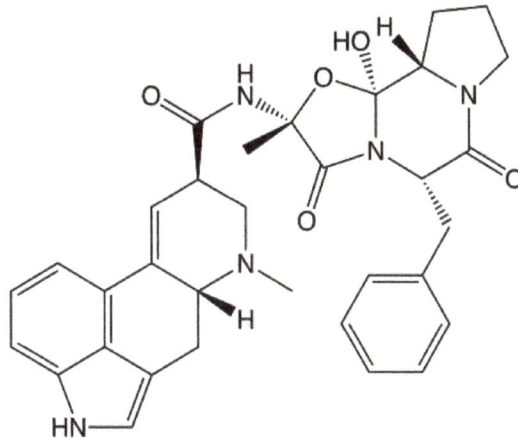

Ergotamine, a major mycotoxin produced by Claviceps species, which if ingested can cause gangrene, convulsions, and hallucinations

Many fungi produce biologically active compounds, several of which are toxic to animals or plants and are therefore called mycotoxins. Of particular relevance to humans are mycotoxins produced by molds causing food spoilage, and poisonous mushrooms. Particularly infamous are the lethal amatoxins in some *Amanita* mushrooms, and ergot alkaloids, which have a long history of causing serious epidemics of ergotism (St Anthony's Fire) in people consuming rye or related cereals contaminated with sclerotia of the ergot fungus, *Claviceps purpurea*. Other notable mycotoxins include the aflatoxins, which are insidious liver toxins and highly carcinogenic metabolites produced by certain *Aspergillus* species often growing in or on grains and nuts consumed by humans, ochratoxins, patulin, and trichothecenes (e.g., T-2 mycotoxin) and fumonisins, which have significant impact on human food supplies or animal livestock.

Mycotoxins are secondary metabolites (or natural products), and research has established the existence of biochemical pathways solely for the purpose of producing mycotoxins and other natural products in fungi. Mycotoxins may provide fitness benefits in terms of physiological adaptation, competition with other microbes and fungi, and protection from consumption (fungivory).

Pathogenic Mechanisms

Ustilago maydis is a pathogenic plant fungus that causes smut disease in maize and teosinte. Plants have evolved efficient defense systems against pathogenic microbes such as *U. maydis*. A rapid defense reaction after pathogen attack is the oxidative burst where the plant produces reactive oxygen species at the site of the attempted invasion. *U. maydis* can respond to the oxidative burst with an oxidative stress response, regulated by the gene *YAP1*. The response protects *U. maydis* from the host defense, and is necessary for the pathogen's virulence. Furthermore, *U. maydis* has a well-established recombinational DNA repair system which acts during mitosis and meiosis. The system may assist the pathogen in surviving DNA damage arising from the host plant's oxidative defensive response to infection.

Cryptococcus neoformans is an encapsulated yeast that can live in both plants and animals. *C. neoformans* usually infects the lungs, where it is phagocytosed by alveolar macrophages. Some

C. neoformans can survive inside macrophages, which appears to be the basis for latency, disseminated disease, and resistance to antifungal agents. One mechanism by which *C. neoformans* survives the hostile macrophage environment is by up-regulating the expression of genes involved in the oxidative stress response. Another mechanism involves meiosis. The majority of *C. neoformans* are mating "type a". Filaments of mating "type a" ordinarily have haploid nuclei, but they can become diploid (perhaps by endoduplication or by stimulated nuclear fusion) to form blastospores. The diploid nuclei of blastospores can undergo meiosis, including recombination, to form haploid basidiospores that can be dispersed. This process is referred to as monokaryotic fruiting. this process requires a gene called *DMC1*, which is a conserved homologue of genes *recA* in bacteria and *RAD51* in eukaryotes, that mediates homologous chromosome pairing during meiosis and repair of DNA double-strand breaks. Thus, *C. neoformans* can undergo a meiosis, monokaryotic fruiting, that promotes recombinational repair in the oxidative, DNA damaging environment of the host macrophage, and the repair capability may contribute to its virulence.

Human Use

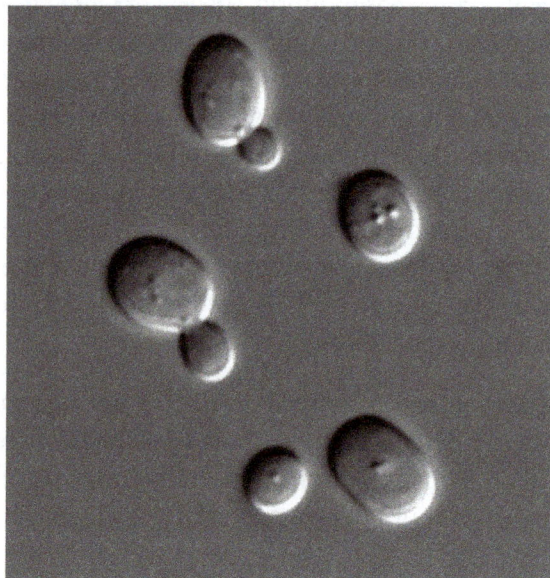

Saccharomyces cerevisiae cells shown with DIC microscopy

The human use of fungi for food preparation or preservation and other purposes is extensive and has a long history. Mushroom farming and mushroom gathering are large industries in many countries. The study of the historical uses and sociological impact of fungi is known as ethnomycology. Because of the capacity of this group to produce an enormous range of natural products with antimicrobial or other biological activities, many species have long been used or are being developed for industrial production of antibiotics, vitamins, and anti-cancer and cholesterol-lowering drugs. More recently, methods have been developed for genetic engineering of fungi, enabling metabolic engineering of fungal species. For example, genetic modification of yeast species—which are easy to grow at fast rates in large fermentation vessels—has opened up ways of pharmaceutical production that are potentially more efficient than production by the original source organisms.

Therapeutic Uses

Modern Chemotherapeutics

Many species produce metabolites that are major sources of pharmacologically active drugs. Particularly important are the antibiotics, including the penicillins, a structurally related group of β-lactam antibiotics that are synthesized from small peptides. Although naturally occurring penicillins such as penicillin G (produced by *Penicillium chrysogenum*) have a relatively narrow spectrum of biological activity, a wide range of other penicillins can be produced by chemical modification of the natural penicillins. Modern penicillins are semisynthetic compounds, obtained initially from fermentation cultures, but then structurally altered for specific desirable properties. Other antibiotics produced by fungi include: ciclosporin, commonly used as an immunosuppressant during transplant surgery; and fusidic acid, used to help control infection from methicillin-resistant *Staphylococcus aureus* bacteria. Widespread use of antibiotics for the treatment of bacterial diseases, such as tuberculosis, syphilis, leprosy, and others began in the early 20th century and continues to date. In nature, antibiotics of fungal or bacterial origin appear to play a dual role: at high concentrations they act as chemical defense against competition with other microorganisms in species-rich environments, such as the rhizosphere, and at low concentrations as quorum-sensing molecules for intra- or interspecies signaling. Other drugs produced by fungi include griseofulvin isolated from *Penicillium griseofulvum*, used to treat fungal infections, and statins (HMG-CoA reductase inhibitors), used to inhibit cholesterol synthesis. Examples of statins found in fungi include mevastatin from *Penicillium citrinum* and lovastatin from *Aspergillus terreus* and the oyster mushroom.

Traditional and Folk Medicine

The medicinal fungi *Ganoderma lucidum* (left) and *Ophiocordyceps sinensis* (right)

Certain mushrooms enjoy usage as therapeutics in folk medicines, such as Traditional Chinese medicine. Notable medicinal mushrooms with a well-documented history of use include *Agaricus subrufescens*, *Ganoderma lucidum*, and *Ophiocordyceps sinensis*. Research has identified compounds produced by these and other fungi that have inhibitory biological effects against viruses and cancer cells. Specific metabolites, such as polysaccharide-K, ergotamine, and β-lactam antibiotics, are routinely used in clinical medicine. The shiitake mushroom is a source of lentinan, a clinical drug approved for use in cancer treatments in several countries, including Japan. In Europe and Japan, polysaccharide-K (brand name Krestin), a chemical derived from *Trametes versicolor*, is an approved adjuvant for cancer therapy.

Cultured Foods

Baker's yeast or *Saccharomyces cerevisiae*, a unicellular fungus, is used to make bread and other wheat-based products, such as pizza dough and dumplings. Yeast species of the genus *Saccharomyces* are also used to produce alcoholic beverages through fermentation. Shoyu koji mold (*Aspergillus oryzae*) is an essential ingredient in brewing Shoyu (soy sauce) and sake, and the preparation of miso, while *Rhizopus* species are used for making tempeh. Several of these fungi are domesticated species that were bred or selected according to their capacity to ferment food without producing harmful mycotoxins, which are produced by very closely related *Aspergilli*. Quorn, a meat substitute, is made from *Fusarium venenatum*.

Edible and Poisonous Species

Edible mushrooms are well-known examples of fungi. Many are commercially raised, but others must be harvested from the wild. *Agaricus bisporus*, sold as button mushrooms when small or Portobello mushrooms when larger, is a commonly eaten species, used in salads, soups, and many other dishes. Many Asian fungi are commercially grown and have increased in popularity in the West. They are often available fresh in grocery stores and markets, including straw mushrooms (*Volvariella volvacea*), oyster mushrooms (*Pleurotus ostreatus*), shiitakes (*Lentinula edodes*), and enokitake (*Flammulina* spp.).

Amanita phalloides accounts for the majority of fatal mushroom poisonings worldwide.

There are many more mushroom species that are harvested from the wild for personal consumption or commercial sale. Milk mushrooms, morels, chanterelles, truffles, black trumpets, and *porcini* mushrooms (*Boletus edulis*) (also known as king boletes) demand a high price on the market. They are often used in gourmet dishes.

Certain types of cheeses require inoculation of milk curds with fungal species that impart a unique flavor and texture to the cheese. Examples include the blue color in cheeses such as Stilton or Roquefort, which are made by inoculation with *Penicillium roqueforti*. Molds used in cheese production are non-toxic and are thus safe for human consumption; however, mycotoxins (e.g., aflatoxins, roquefortine C, patulin, or others) may accumulate because of growth of other fungi during cheese ripening or storage.

Stilton cheese veined with *Penicillium roqueforti*

Many mushroom species are poisonous to humans, with toxicities ranging from slight digestive problems or allergic reactions as well as hallucinations to severe organ failures and death. Genera with mushrooms containing deadly toxins include *Conocybe, Galerina, Lepiota,* and, the most infamous, *Amanita*. The latter genus includes the destroying angel *(A. virosa)* and the death cap *(A. phalloides)*, the most common cause of deadly mushroom poisoning. The false morel (*Gyromitra esculenta*) is occasionally considered a delicacy when cooked, yet can be highly toxic when eaten raw. *Tricholoma equestre* was considered edible until it was implicated in serious poisonings causing rhabdomyolysis. Fly agaric mushrooms (*Amanita muscaria*) also cause occasional non-fatal poisonings, mostly as a result of ingestion for its hallucinogenic properties. Historically, fly agaric was used by different peoples in Europe and Asia and its present usage for religious or shamanic purposes is reported from some ethnic groups such as the Koryak people of north-eastern Siberia.

As it is difficult to accurately identify a safe mushroom without proper training and knowledge, it is often advised to assume that a wild mushroom is poisonous and not to consume it.

Pest Control

In agriculture, fungi may be useful if they actively compete for nutrients and space with pathogenic microorganisms such as bacteria or other fungi via the competitive exclusion principle, or if they are parasites of these pathogens. For example, certain species may be used to eliminate or suppress the growth of harmful plant pathogens, such as insects, mites, weeds, nematodes, and other fungi that cause diseases of important crop plants. This has generated strong interest in practical applications that use these fungi in the biological control of these agricultural pests. Entomopathogenic fungi can be used as biopesticides, as they actively kill insects. Examples that have been used as biological insecticides are *Beauveria bassiana*, *Metarhizium* spp, *Hirsutella* spp, *Paecilomyces* (*Isaria*) spp, and *Lecanicillium lecanii*. Endophytic fungi of grasses of the genus *Neotyphodium*, such as *N. coenophialum*, produce alkaloids that are toxic to a range of invertebrate and vertebrate herbivores. These alkaloids protect grass plants from herbivory, but several endophyte alkaloids can poison grazing animals, such as cattle and sheep. Infecting cultivars of pasture or forage grasses with *Neotyphodium* endophytes is one approach being used in grass breeding programs; the fungal strains are selected for producing only alkaloids that increase resistance to herbivores such as insects, while being non-toxic to livestock.

Grasshoppers killed by *Beauveria bassiana*

Bioremediation

Certain fungi, in particular "white rot" fungi, can degrade insecticides, herbicides, pentachlorophenol, creosote, coal tars, and heavy fuels and turn them into carbon dioxide, water, and basic elements. Fungi have been shown to biomineralize uranium oxides, suggesting they may have application in the bioremediation of radioactively polluted sites.

Model Organisms

Several pivotal discoveries in biology were made by researchers using fungi as model organisms, that is, fungi that grow and sexually reproduce rapidly in the laboratory. For example, the one gene-one enzyme hypothesis was formulated by scientists using the bread mold *Neurospora crassa* to test their biochemical theories. Other important model fungi are *Aspergillus nidulans* and the yeasts *Saccaromyces cerevisiae* and *Schizosaccharomyces pombe*, each of which with a long history of use to investigate issues in eukaryotic cell biology and genetics, such as cell cycle regulation, chromatin structure, and gene regulation. Other fungal models have more recently emerged that address specific biological questions relevant to medicine, plant pathology, and industrial uses; examples include *Candida albicans*, a dimorphic, opportunistic human pathogen, *Magnaporthe grisea*, a plant pathogen, and *Pichia pastoris*, a yeast widely used for eukaryotic protein production.

Others

Fungi are used extensively to produce industrial chemicals like citric, gluconic, lactic, and malic acids, and industrial enzymes, such as lipases used in biological detergents, cellulases used in making cellulosic ethanol and stonewashed jeans, and amylases, invertases, proteases and xylanases. Several species, most notably *Psilocybin mushrooms* (colloquially known as *magic mushrooms*), are ingested for their psychedelic properties, both recreationally and religiously.

References

- Madigan M; Martinko J, eds. (2006). Brock Biology of Microorganisms (13th ed.). Pearson Education. p. 1096. ISBN 0-321-73551-X.

- Mahavira is dated 599 BCE - 527 BCE. See. Dundas, Paul; John Hinnels, eds. (2002). The Jains. London: Routledge. ISBN 0-415-26606-8. p. 24

- Gillen, Alan L. (2007). The Genesis of Germs: The Origin of Diseases and the Coming Plagues. New Leaf Publishing Group. p. 10. ISBN 0-89051-493-3.

- Tickell, Joshua; et al. (2000). From the Fryer to the Fuel Tank: The Complete Guide to Using Vegetable Oil as an Alternative Fuel. Biodiesel America. p. 53. ISBN 0-9707227-0-2.

- Inslee, Jay; et al. (2008). Apollo's Fire: Igniting America's Clean Energy Economy. Island Press. p. 157. ISBN 1-59726-175-0.

- Langford, Roland E. (2004). Introduction to Weapons of Mass Destruction: Radiological, Chemical, and Biological. Wiley-IEEE. p. 140. ISBN 0-471-46560-7.

- Bauman, Robert W.; Tizard, Ian R.; Machunis-Masouka, Elizabeth (2006). Microbiology. San Francisco: Pearson Benjamin Cummings. ISBN 0-8053-7693-3.

- Madigan, Michael T. (2012). Brock biology of microorganisms (13th ed.). San Francisco: Benjamin Cummings. ISBN 9780321649638.

- "Direct Observations". The Biofilm Primer. Springer Series on Biofilms. 1. 2007. pp. 3–4. doi:10.1007/978-3-540-68022-2_2. ISBN 978-3-540-68021-5.

- Balaban, N.; Ren, D.; Givskov, M.; Rasmussen, T. B. (2008). "Introduction". Control of Biofilm Infections by Signal Manipulation. Springer Series on Biofilms. 2. p. 1. doi:10.1007/7142_2007_006. ISBN 978-3-540-73852-7.

- The Molecular Biology of the Cell, fourth edition. Bruce Alberts, et al. Garland Science (2002) pg. 808 ISBN 0-8153-3218-1

- Dusenbery, David B. (2009). Living at Micro Scale, pp. 20–25. Harvard University Press, Cambridge, Mass. ISBN 978-0-674-03116-6.

3

Essential Topics of Microbiology

Some of the essential topics of microbiology are sterilization, endosymbiont, microbiota, microbial cyst, colony–forming unit and viral eukaryogenesis. Sterilization refers to the process that eliminates biological agents and viruses. Sterilization is majorly used to sanitize and pasteurize, and is usually achieved by using chemicals, irradiation and filtration. The section serves as a source to understand the major concepts related to microbiology.

Sterilization (Microbiology)

Sterilization (or sterilisation) referring to any process that eliminates (removes) or kills (deactivates) all forms of life and other biological agents (such as prions, as well as viruses which some do not consider to be alive but are biological pathogens nonetheless), including transmissible agents (such as fungi, bacteria, viruses, prions, spore forms, unicellular eukaryotic organisms such as Plasmodium, etc.) present in a specified region, such as a surface, a volume of fluid, medication, or in a compound such as biological culture media. Sterilization can be achieved with one or more of the following: heat, chemicals, irradiation, high pressure, and filtration. Sterilization is distinct from disinfection, sanitization, and pasteurization in that sterilization kills, deactivates, or eliminates all forms of life and other biological agents.Sterilization can be achieved by physical, chemical and physiochemical means. Chemicals used as sterilizing agents are called chemosterilants.

Applications

Foods

One of the first steps toward sterilization was made by Nicolas Appert who discovered that thorough application of heat over a suitable period slowed the decay of foods and various liquids, preserving them for safe consumption for a longer time than was typical. Canning of foods is an extension of the same principle, and has helped to reduce food borne illness ("food poisoning"). Other methods of sterilizing foods include food irradiation and high pressure (pascalization).

Medicine and Surgery

In general, surgical instruments and medications that enter an already aseptic part of the body (such as the bloodstream, or penetrating the skin) must be sterile. Examples of such instruments include scalpels, hypodermic needles and artificial pacemakers. This is also essential in the manufacture of parenteral pharmaceuticals.

Preparation of injectable medications and intravenous solutions for fluid replacement therapy requires not only sterility but also well-designed containers to prevent entry of adventitious agents after initial product sterilization.

Joseph Lister was a pioneer of antiseptic surgery.

Apparatus to sterilize surgical instruments, Verwaltungsgebäude der Schweiz. Kranken- und Hilfsanstalt, 1914-1918

Most medical and surgical devices used in healthcare facilities are made of materials that are able to go under steam sterilization. However, since 1950, there has been an increase in medical devices and instruments made of materials (e.g., plastics) that require low-temperature sterilization. Ethylene oxide gas has been used since the 1950s for heat- and moisture-sensitive medical devices. Within the past 15 years, a number of new, low-temperature sterilization systems (e.g., hydrogen peroxide gas plasma, peracetic acid immersion, ozone) have been developed and are being used to sterilize medical devices.] Steam sterilization is the most widely used and the most dependable. Steam sterilization is nontoxic, inexpensive, rapidly microbicidal, sporicidal, and rapidly heats and penetrates fabrics.

Spacecraft

There are strict international rules to protect the contamination of Solar System bodies from biological material from Earth. Standards vary depending on both the type of mission and its destination; the more likely a planet is considered to be habitable, the stricter the requirements are.

Many components of instruments used on spacecraft cannot withstand very high temperatures, so techniques not requiring excessive temperatures are used as tolerated, including heating to at least 120 °C, chemical sterilization, oxidization, ultraviolet, and irradiation.

Quantification

The aim of sterilization is the reduction of initially present microorganisms or other potential pathogens.The degree of sterilization is commonly expressed by multiples of the decimal reduction time, or D-value, denoting the time needed to reduce the initial number N_0 to one tenth (10^{-1}) of its original value. Then the number of microorganisms N after sterilization time t is given by:

$$\frac{N}{N_0} = 10^{\left(-\frac{t}{D}\right)}.$$

The D-value is a function of sterilization conditions and varies with the type of microorganism, temperature, water activity, pH etc.. For steam sterilization typically the temperature (in °Celsius) is given as index.

Theoretically, the likelihood of survival of an individual microorganism is never zero. To compensate for this, the overkill method is often used. Using the overkill method, sterilization is performed by sterilizing for longer than is required to kill the bioburden present on or in the item being sterilized. This provides a sterility assurance level (SAL) equal to the probability of a non-sterile unit.

For high-risk applications such as medical devices and injections, a sterility assurance level of at least 10^{-6} is required by the United States Food and Drug Administration (FDA)

Heat

Steam

Front-loading autoclave

A widely used method for heat sterilization is the autoclave, sometimes called a converter or steam sterilizer. Autoclaves use steam heated to 121-134 °C under pressure. To achieve sterility, the article is heated in a chamber by injected steam until the article reaches a time and temperature setpoint. The article is then held at that setpoint for a period of time which varies depending on the bioburden present on the article being sterilized and its resistance (D-value) to steam sterilization. A general cycle would be anywhere between 3 and 15 minutes, (depending on the generated heat) at 121 °C at 100 kPa, which is sufficient to provide a sterility assurance level of 10^{-4} for a product with a bioburden of 10^6 and a D-value of 2.0 minutes. Following sterilization, liquids in a pressurized autoclave must be cooled slowly to avoid boiling over when the pressure is released. This may be achieved by gradually depressurizing the sterilization chamber and allowing liquids to evaporate under a negative pressure, while cooling the contents.

Proper autoclave treatment will inactivate all resistant bacterial spores in addition to fungi, bacteria, and viruses, but is not expected to eliminate all prions, which vary in their resistance. For prion elimination, various recommendations state 121-132 °C for 60 minutes or 134 °C for at least 18 minutes. The 263K scrapie prion is inactivated relatively quickly by such sterilization procedures; however, other strains of scrapie, and strains of CJD and BSE are more resistant. Using mice as test animals, one experiment showed that heating BSE positive brain tissue at 134-138 °C for 18 minutes resulted in only a 2.5 log decrease in prion infectivity.

Most autoclaves have meters and charts that record or display information, particularly temperature and pressure as a function of time. The information is checked to ensure that the conditions required for sterilization have been met. Indicator tape is often placed on packages of products prior to autoclaving, and some packaging incorporates indicators. The indicator changes color when exposed to steam, providing a visual confirmation.

Biological indicators can also be used to independently confirm autoclave performance. Simple bioindicator devices are commercially available based on microbial spores. Most contain spores of the heat resistant microbe *Geobacillus stearothermophilus* (formerly *Bacillus stearothermophilus*), which is extremely resistant to steam sterilization. Biological indicators may take the form of glass vials of spores and liquid media, or as spores on strips of paper inside glassine envelopes. These indicators are placed in locations where it is difficult for steam to reach to verify that steam is penetrating there.

For autoclaving, cleaning is critical. Extraneous biological matter or grime may shield organisms from steam penetration. Proper cleaning can be achieved through physical scrubbing, sonication, ultrasound or pulsed air. Pressure cooking and canning are analogous to autoclaving, and when performed correctly renders food sterile.

Moist heat causes destruction of micro-organisms by denaturation of macromolecules, primarily proteins. This method is a faster process than dry heat sterilization.

Dry Heat

Dry heat sterilizer

Dry heat was the first method of sterilization, and is a longer process than moist heat sterilization. The destruction of microorganisms through the use of dry heat is a gradual phenomenon. With longer exposure to lethal temperatures, the number of killed microorganisms increases. Forced ventilation of hot air can be used to increase the rate at which heat is transferred to an organism and reduce the temperature and amount of time needed to achieve sterility. At higher temperatures, shorter exposure times are required to kill organisms. This can reduce heat-induced damage to food products.

The standard setting for a hot air oven is at least two hours at 160 °C. A rapid method heats air to 190 °C for 6 minutes for unwrapped objects and 12 minutes for wrapped objects. Dry heat has the advantage that it can be used on powders and other heat-stable items that are adversely affected by steam (e.g. it does not cause rusting of steel objects).

Flaming

Flaming is done to loops and straight-wires in microbiology labs. Leaving the loop in the flame of a Bunsen burner or alcohol lamp until it glows red ensures that any infectious agent gets inactivated. This is commonly used for small metal or glass objects, but not for large objects. However, during the initial heating infectious material may be "sprayed" from the wire surface before it is killed, contaminating nearby surfaces and objects. Therefore, special heaters have been developed that surround the inoculating loop with a heated cage, ensuring that such sprayed material does not further contaminate the area. Another problem is that gas flames may leave carbon or other residues on the object if the object is not heated enough. A variation on flaming is to dip the object in 70% or higher ethanol, then briefly touch the object to a Bunsen burner flame. The ethanol will ignite and burn off rapidly, leaving less residue than a gas flame.

Incineration

Incineration is a waste treatment process that involves the combustion of organic substances contained in waste materials. This method also burns any organism to ash. It is used to sterilize medical and other biohazardous waste before it is discarded with non-hazardous waste. Bacteria incinerators are mini furnaces used to incinerate and kill off any micro organisms that may be on an inoculating loop or wire.

Tyndallization

Named after John Tyndall, Tyndallization is an obsolete and lengthy process designed to reduce the level of activity of sporulating bacteria that are left by a simple boiling water method. The process involves boiling for a period (typically 20 minutes) at atmospheric pressure, cooling, incubating for a day, then repeating the process a total of three to four times. The incubation periods are to allow heat-resistant spores surviving the previous boiling period to germinate to form the heat-sensitive vegetative (growing) stage, which can be killed by the next boiling step. This is effective because many spores are stimulated to grow by the heat shock. The procedure only works for media that can support bacterial growth, and will not sterilize non-nutritive substrates like water. Tyndallization is also ineffective against prions.

Glass Bead Sterilizers

Glass bead sterilizers work by heating glass beads to 250 °C. Instruments are then quickly doused in these glass beads, which heat the object while physically scraping contaminants off their surface. Glass bead sterilizers were once a common sterilization method employed in dental offices as well as biologic laboratories, but are not approved by the U.S. Food and Drug Administration (FDA) and Centers for Disease Control and Prevention (CDC) to be used as a sterilizers since 1997. They are still popular in European as well as Israeli dental practices although there are no current evidence-based guidelines for using this sterilizer.

Chemical Sterilization

Chemiclav

Chemicals are also used for sterilization. Heating provides a reliable way to rid objects of all transmissible agents, but it is not always appropriate if it will damage heat-sensitive materials such as biological materials, fiber optics, electronics, and many plastics. In these situations chemicals, either as gases or in liquid form, can be used as sterilants. While the use of gas and liquid chemical sterilants avoids the problem of heat damage, users must ensure that article to be sterilized is chemically compatible with the sterilant being used. In addition, the use of chemical sterilants poses new challenges for workplace safety, as the properties that make chemicals effective sterilants usually make them harmful to humans.

Ethylene Oxide

Ethylene oxide (EO,EtO) gas treatment is one of the common methods used to sterilize, pasteurize, or disinfect items because of its wide range of material compatibility. It is also used to process items that are sensitive to processing with other methods, such as radiation (gamma, electron beam, X-ray), heat (moist or dry), or other chemicals. Ethylene oxide treatment is the most common sterilization method, used for approximately 70% of total sterilizations, and for over 50% of all disposable medical devices.

Ethylene oxide treatment is generally carried out between 30 °C and 60 °C with relative humidity above 30% and a gas concentration between 200 and 800 mg/l. Typically, the process lasts for several hours. Ethylene oxide is highly effective, as it penetrates all porous materials, and it can penetrate through some plastic materials and films. Ethylene oxide kills all known microorganisms such as bacteria (including spores), viruses, and fungi (including yeasts and molds), and is compatible with almost all materials even when repeatedly applied. It is flammable, toxic and carcinogenic, however, with a reported potential for some adverse health effects when not used in compliance with published requirements. Ethylene oxide sterilizers and processes require biological validation after sterilizer installation, significant repairs or process changes.

The traditional process consists of a preconditioning phase (in a separate room or cell), a processing phase (more commonly in a vacuum vessel and sometimes in a pressure rated vessel), and an aeration phase (in a separate room or cell) to remove ethylene oxide residues and lower by-products such as ethylene chlorohydrin (EC or ECH) and, of lesser importance, ethylene glycol (EG). An alternative process, known as all-in-one processing, also exists for some products whereby all three phases are performed in the vacuum or pressure rated vessel. This latter option can facilitate faster overall processing time and residue dissipation.

The most common ethylene oxide processing method is the gas chamber method. To benefit from economies of scale, ethylene oxide has traditionally been delivered by filling a large chamber with a combination of gaseous ethylene oxide either as pure ethylene oxide, or with other gases used as diluents (chlorofluorocarbons (CFCs), hydrochlorofluorocarbons (HCFCs), or carbon dioxide).

Ethylene oxide is still widely used by medical device manufacturers . Since ethylene oxide is explosive at concentrations above 3%, ethylene oxide was traditionally supplied with an inert carrier gas such as a CFC or HCFC. The use of CFCs or HCFCs as the carrier gas was banned because of concerns of ozone depletion. These halogenated hydrocarbons are being replaced by systems using 100% ethylene oxide because of regulations and the high cost of the blends. In hospitals, most ethylene oxide sterilizers use single use cartridges because of the convenience and ease of use compared to the former plumbed gas cylinders of ethylene oxide blends.

It is important to adhere to patient and healthcare personnel government specified limits of ethylene oxide residues in and/or on processed products, operator exposure after processing, during storage and handling of ethylene oxide gas cylinders, and environmental emissions produced when using ethylene oxide.

The U.S. Occupational Safety and Health Administration (OSHA) has set the permissible exposure limit (PEL) at 1 ppm calculated as an eight-hour time weighted average (TWA) [29 CFR 1910.1047] and 5 ppm as a 15-minute excursion limit (EL). The National Institute for Occupational Safety and Health (NIOSH) immediately dangerous to life and health limit (IDLH) for ethylene oxide is 800 ppm. The odor threshold is around 500 ppm, so ethylene oxide is imperceptible until concentrations well above the OSHA PEL. Therefore, OSHA recommends that continuous gas monitoring systems be used to protect workers using ethylene oxide for processing. Employees' health records must be maintained during employment and after termination of employment for 30 years.

Nitrogen Dioxide

Nitrogen dioxide (NO_2) gas is a rapid and effective sterilant for use against a wide range of microorganisms, including common bacteria, viruses, and spores. The unique physical properties of

NO_2 gas allow for sterilant dispersion in an enclosed environment at room temperature and ambient pressure. The mechanism for lethality is the degradation of DNA in the spore core through nitration of the phosphate backbone, which kills the exposed organism as it absorbs NO_2. This degradation occurs at even very low concentrations of the gas. NO_2 has a boiling point of 21 °C at sea level, which results in a relatively high saturated vapor pressure at ambient temperature. Because of this, liquid NO_2 may be used as a convenient source for the sterilant gas. Liquid NO_2 is often referred to by the name of its dimer, dinitrogen tetroxide (N_2O_4). Additionally, the low levels of concentration required, coupled with the high vapor pressure, assures that no condensation occurs on the devices being sterilized. This means that no aeration of the devices is required immediately following the sterilization cycle. NO_2 is also less corrosive than other sterilant gases, and is compatible with most medical materials and adhesives.

The most-resistant organism (MRO) to sterilization with NO_2 gas is the spore of Geobacillus stearothermophilus, which is the same MRO for both steam and hydrogen peroxide sterilization processes. The spore form of G. stearothermophilus has been well characterized over the years as a biological indicator in sterilization applications. Microbial inactivation of G. stearothermophilus with NO_2 gas proceeds rapidly in a log-linear fashion, as is typical of other sterilization processes. Noxilizer, Inc. has commercialized this technology to offer contract sterilization services for medical devices at its Baltimore, MD facility. This has been demonstrated in Noxilizer's lab in multiple studies and is supported by published reports from other labs. These same properties also allow for quicker removal of the sterilant and residuals through aeration of the enclosed environment. The combination of rapid lethality and easy removal of the gas allows for shorter overall cycle times during the sterilization (or decontamination) process and a lower level of sterilant residuals than are found with other sterilization methods.

Ozone

Ozone is used in industrial settings to sterilize water and air, as well as a disinfectant for surfaces. It has the benefit of being able to oxidize most organic matter. On the other hand, it is a toxic and unstable gas that must be produced on-site, so it is not practical to use in many settings.

Ozone offers many advantages as a sterilant gas; ozone is a very efficient sterilant because of its strong oxidizing properties (E = 2.076 vs SHE) capable of destroying a wide range of pathogens, including prions without the need for handling hazardous chemicals since the ozone is generated within the sterilizer from medical grade oxygen. The high reactivity of ozone means that waste ozone can be destroyed by passing over a simple catalyst that reverts it to oxygen and ensures that the cycle time is relatively short. The disadvantage of using ozone is that the gas is very reactive and very hazardous. The NIOSH immediately dangerous to life and health limit for ozone is 5 ppm, 160 times smaller than the 800 ppm IDLH for ethylene oxide. Documentation for Immediately Dangerous to Life or Health Concentrations (IDLH): NIOSH Chemical Listing and Documentation of Revised IDLH Values (as of 3/1/95) and OSHA has set the PEL for ozone at 0.1 ppm calculated as an 8 hour time weighted average (29 CFR 1910.1000, Table Z-1). The Canadian Center for Occupation Health and Safety provides an excellent summary of the health effects of exposure to ozone. The sterilant gas manufacturers include many safety features in their products but prudent practice is to provide continuous monitoring to below the OSHA PEL to provide a rapid warning in the event of a leak. Monitors for determining workplace exposure to ozone are commercially available.

Glutaraldehyde and Formaldehyde

Glutaraldehyde and formaldehyde solutions (also used as fixatives) are accepted liquid sterilizing agents, provided that the immersion time is sufficiently long. To kill all spores in a clear liquid can take up to 22 hours with glutaraldehyde and even longer with formaldehyde. The presence of solid particles may lengthen the required period or render the treatment ineffective. Sterilization of blocks of tissue can take much longer, due to the time required for the fixative to penetrate. Glutaraldehyde and formaldehyde are volatile, and toxic by both skin contact and inhalation. Glutaraldehyde has a short shelf life (<2 weeks), and is expensive. Formaldehyde is less expensive and has a much longer shelf life if some methanol is added to inhibit polymerization to paraformaldehyde, but is much more volatile. Formaldehyde is also used as a gaseous sterilizing agent; in this case, it is prepared on-site by depolymerization of solid paraformaldehyde. Many vaccines, such as the original Salk polio vaccine, are sterilized with formaldehyde.

Hydrogen Peroxide

Hydrogen peroxide, in both liquid and as vaporized hydrogen peroxide (VHP), is another chemical sterilizing agent. Hydrogen peroxide is strong oxidant, which allows it to destroy a wide range of pathogens. Hydrogen peroxide is used to sterilize heat or temperature sensitive articles such as rigid endoscopes. In medical sterilization hydrogen peroxide is used at higher concentrations, ranging from around 35% up to 90%. The biggest advantage of hydrogen peroxide as a sterilant is the short cycle time. Whereas the cycle time for ethylene oxide may be 10 to 15 hours, some modern hydrogen peroxide sterilizers have a cycle time as short as 28 minutes.

Drawbacks of hydrogen peroxide include material compatibility, a lower capability for penetration and operator health risks. Products containing cellulose, such as paper, cannot be sterilized using VHP and products containing nylon may become brittle. The penetrating ability of hydrogen peroxide is not as good as ethylene oxide and so there are limitations on the length and diameter of lumens that can be effectively sterilized and guidance is available from the sterilizer manufacturers. Hydrogen peroxide is primary irritant and the contact of the liquid solution with skin will cause bleaching or ulceration depending on the concentration and contact time. It is relatively non-toxic when diluted to low concentrations, but is a dangerous oxidizer at high concentrations (> 10% w/w). The vapor is also hazardous, primarily affecting the eyes and respiratory system. Even short term exposures can be hazardous and NIOSH has set the Immediately Dangerous to Life and Health Level (IDLH) at 75 ppm, less than one tenth the IDLH for ethylene oxide (800 ppm). Prolonged exposure to lower concentrations can cause permanent lung damage and consequently OSHA has set the permissible exposure limit to 1.0 ppm, calculated as an 8-hour time weighted average. Sterilizer manufacturers go to great lengths to make their products safe through careful design and incorporation of many safety features, though there are still workplace exposures of hydrogen peroxide from gas sterilizers are documented in the FDA MAUDE database. When using any type of gas sterilizer, prudent work practices will include good ventilation, a continuous gas monitor for hydrogen peroxide and good work practices and training.

Vaporized hydrogen peroxide (VHP) is used to sterilize large enclosed and sealed areas such as entire rooms and aircraft interiors.

Peracetic Acid

Peracetic acid (0.2%) is a recognized sterilant by the FDA for use in sterilizing medical devices such as endoscopes.

Potential for Chemical Sterilization of Prions

Prions are highly resistant to chemical sterilization. Treatment with aldehydes such as formaldehyde have actually been shown to increase prion resistance. Hydrogen peroxide (3%) for one hour was shown to be ineffective, providing less than 3 logs (10^{-3}) reduction in contamination. Iodine, formaldehyde, glutaraldehyde and peracetic acid also fail this test (one hour treatment). Only chlorine, phenolic compounds, guanidinium thiocyanate, and sodium hydroxide (NaOH) reduce prion levels by more than 4 logs; chlorine (too corrosive to use on certain objects) and NaOH are the most consistent. Many studies have shown the effectiveness of sodium hydroxide.

Radiation Sterilization

Sterilization can be achieved using electromagnetic radiation such as electron beams, X-rays, gamma rays, or irradiation by subatomic particles. Electromagnetic or particulate radiation can be energetic enough to ionize atoms or molecules (ionizing radiation), or less energetic (non-ionizing radiation).

Non-ionizing Radiation Sterilization

Ultraviolet light irradiation (UV, from a germicidal lamp) is useful for sterilization of surfaces and some transparent objects. Many objects that are transparent to visible light absorb UV. UV irradiation is routinely used to sterilize the interiors of biological safety cabinets between uses, but is ineffective in shaded areas, including areas under dirt (which may become polymerized after prolonged irradiation, so that it is very difficult to remove). It also damages some plastics, such as polystyrene foam if exposed for prolonged periods of time.

Ionizing Radiation Sterilization

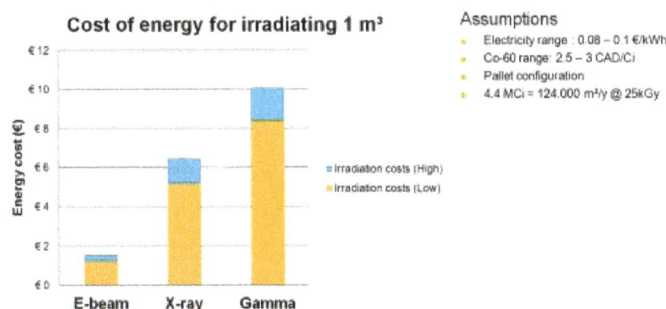

Efficiency illustration of the different radiation technologies (electron beam, X-ray, gamma rays)

The safety of irradiation facilities is regulated by the United Nations International Atomic Energy Agency and monitored by the different national Nuclear Regulatory Commissions. The incidents that have occurred in the past are documented by the agency and thoroughly analyzed to determine root cause and improvement potential. Such improvements are then mandated to retrofit existing facilities and future design.

Gamma radiation is very penetrating, and is commonly used for sterilization of disposable medical equipment, such as syringes, needles, cannulas and IV sets, and food. It is emitted by a radioisotope, usually Cobalt-60(^{60}Co) or caesium-137 (^{137}Cs).

Use of a radioisotope requires shielding for the safety of the operators while in use and in storage. With most designs the radioisotope is lowered into a water-filled source storage pool, which absorbs radiation and allows maintenance personnel to enter the radiation shield. One variant keeps the radioisotope under water at all times and lowers the product to be irradiated into the water towards the source in hermetic bells; no further shielding is required for such designs. Other uncommonly used designs use dry storage, providing movable shields that reduce radiation levels in areas of the irradiation chamber. An incident in Decatur Georgia, US, where water-soluble caesium-137 leaked into the source storage pool, requiring NRC intervention has led to use of this radioisotope being almost entirely discontinued in favour of the more costly, non-water-soluble cobalt-60. Cobalt-60 gamma photons have about twice the energy, and hence greater penetrating range, of Caesium-137 radiation.

Electron beam processing is also commonly used for sterilization. Electron beams use an on-off technology and provide a much higher dosing rate than gamma or x-rays. Due to the higher dose rate, less exposure time is needed and thereby any potential degradation to polymers is reduced. A limitation is that electron beams are less penetrating than either gamma or x-rays. Facilities rely on substantial concrete shields to protect workers and the environment from radiation exposure.

X-rays: high-energy X-rays (produced by bremsstrahlung) allow irradiation of large packages and pallet loads of medical devices. They are sufficiently penetrating to treat multiple pallet loads of low-density packages with very good dose uniformity ratios. X-ray sterilization does not require chemical or radioactive material: high-energy X-rays are generated at high intensity by an X-ray generator that does not require shielding when not in use. X-rays are generated by bombarding a dense material (target) such as tantalum or tungsten with high-energy electrons in a process known as bremsstrahlung conversion. These systems are energy-inefficient, requiring much more electrical energy than other systems for the same result.

Irradiation with X-rays or gamma rays, electromagnetic radiation rather than particles, does not make materials radioactive. Irradiation with particles may make materials radioactive, depending upon the type of particles and their energy, and the type of target material: neutrons and very high-energy particles can make materials radioactive, but have good penetration, whereas lower energy particles (other than neutrons) cannot make materials radioactive, but have poorer penetration.

Sterilization by irradiation with gamma rays may however in some cases affect material properties.

Irradiation is used by the United States Postal Service to sterilize mail in the Washington, D.C. area. Some foods (e.g. spices, ground meats) are sterilized by irradiation.

Subatomic particles may be more or less penetrating, and may be generated by a radioisotope or a device, depending upon the type of particle.

Sterile Filtration

Fluids that would be damaged by heat, irradiation or chemical sterilization, such as drug products, can be sterilized by microfiltration using membrane filters. This method is commonly used for heat labile pharmaceuticals and protein solutions in medicinal drug processing. A microfilter with pore size 0.2 µm will usually effectively remove microorganisms. In the processing of biologics, viruses must be removed or inactivated, requiring the use of nanofilters with a smaller pore size (20 -50 nm) are used. Smaller pore sizes lower the flow rate, so in order to achieve higher total throughput or to avoid premature blockage, pre-filters might be used to protect small pore membrane filters.

Membrane filters used in production processes are commonly made from materials such as mixed cellulose ester or polyethersulfone (PES). The filtration equipment and the filters themselves may be purchased as pre-sterilized disposable units in sealed packaging, or must be sterilized by the user, generally by autoclaving at a temperature that does not damage the fragile filter membranes. To ensure proper functioning of the filter, the membrane filters are integrity tested post-use and sometimes before use. The non-destructive integrity test assures the filter is undamaged, and is a regulatory requirement. Typically, terminal pharmaceutical sterile filtration is performed inside of a cleanroom to prevent contamination.

Preservation of Sterility

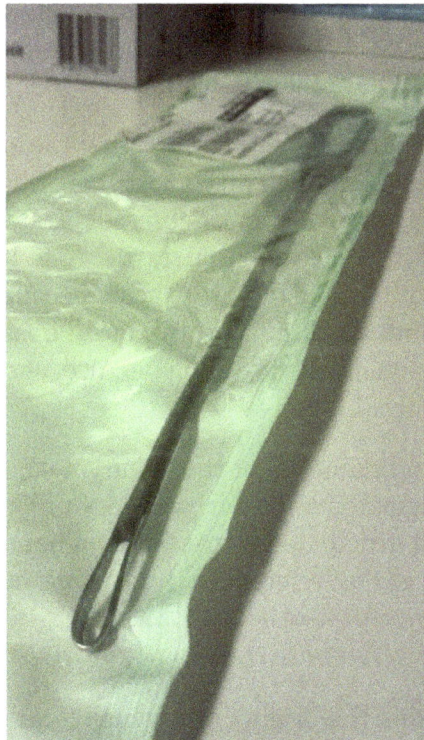

A curette in sterile packaging.

Instruments that have undergone sterilization can be maintained in such condition by containment in sealed packaging until use.

Aseptic technique is the act of maintaining sterility during procedures.

Endosymbiont

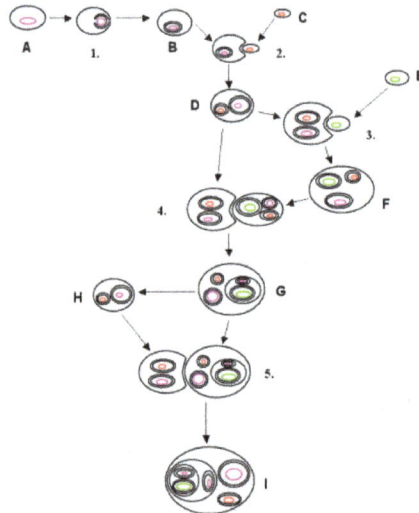

A representation of the endosymbiotic theory

An endosymbiont or endobiont is any organism that lives within the body or cells of another organism, i.e. forming an endosymbiosis. Examples are nitrogen-fixing bacteria (called rhizobia), which live in root nodules on legume roots, single-cell algae inside reef-building corals, and bacterial endosymbionts that provide essential nutrients to about 10–15% of insects.

Many instances of endosymbiosis are obligate; that is, either the endosymbiont or the host cannot survive without the other, such as the gutless marine worms of the genus *Riftia*, which get nutrition from their endosymbiotic bacteria. The most common examples of obligate endosymbioses are mitochondria and chloroplasts. Some human parasites, e.g. *Wuchereria bancrofti* and *Mansonella perstans*, thrive in their intermediate insect hosts because of an obligate endosymbiosis with *Wolbachia spp*. They can both be eliminated from said hosts by treatments that target this bacterium. However, not all endosymbioses are obligate. Also, some endosymbioses can be harmful to either of the organisms involved.

It is generally agreed that certain organelles of the eukaryotic cell, especially mitochondria and plastids such as chloroplasts, originated as bacterial endosymbionts. This theory is called the endosymbiotic theory, and was first articulated by the Russian botanist Konstantin Mereschkowski in 1910, even though the first paper that referenced this theory was published in 1905.

Endosymbiosis Theory and Mitochondria and Chloroplasts

The endosymbiosis theory explains the origins of organelles such as mitochondria and chloroplasts in eukaryotic cells. The theory proposes that chloroplasts and mitochondria evolved from certain types of bacteria that eukaryotic cells engulfed through endophagocytosis. These cells and the bacteria trapped inside them entered a symbiotic relationship, a close association between different types of organisms over an extended time. However, to be specific, the relationship was endosymbiotic, meaning that one of the organisms (the bacteria) lived within the other (the eukaryotic cells).

According to endosymbiosis theory, an anaerobic cell probably ingested an aerobic bacterium but failed to digest it. The aerobic bacterium flourished within the cell because the cell's cytoplasm was abundant in half-digested food molecules. The bacterium digested these molecules with oxygen and gained great amounts of energy. Because the bacterium had so much energy, it probably leaked some of it as adenosine triphosphate into the cell's cytoplasm. This benefited the anaerobic cell because it was now able to breathe aerobically, which means more potential for energy gain. Eventually, the aerobic bacterium could no longer live independently from the cell, and it, therefore, became a mitochondrion. The origin of the chloroplast is very similar to that of the mitochondrion. A cell must have captured a photosynthetic cyanobacterium and failed to digest it. The cyanobacterium thrived in the cell and eventually evolved into the first chloroplast. Other eukaryotic organelles may have also evolved through endosymbiosis; it has been proposed that cilia, flagella, centrioles, and microtubules may have originated from a symbiosis between a Spirochaete bacterium and an early eukaryotic cell, but this is not widely accepted among biologists.

There are several examples of evidence that support endosymbiosis theory. Mitochondria and chloroplasts contain their own small supply of DNA, which may be remnants of the genome the organelles had when they were independent aerobic bacteria. The single most convincing evidence of the descent of organelles from bacteria is the position of mitochondria and plastid DNA sequences in phylogenetic trees of bacteria. Mitochondria have sequences that clearly indicate origin from a group of bacteria called the alphaproteobacteria. Plastids have DNA sequences that indicate origin from the cyanobacteria (blue-green algae). In addition, there are organisms alive today, called living intermediates, that are in a similar endosymbiotic condition to the prokaryotic cells and the aerobic bacteria. Living intermediates show that the evolution proposed by the endosymbiont theory is possible. For example, the giant amoeba *Pelomyxa* lacks mitochondria but has aerobic bacteria that carry out a similar role. A variety of corals, clams, snails, and one species of *Paramecium* permanently host algae in their cells. Many of the insect endosymbionts have been shown to have ancient associations with their hosts, involving strictly vertical inheritance. In addition, these insect symbionts have similar patterns of genome evolution to those found in true organelles: genome reduction, rapid rates of gene evolution, and bias in nucleotide base composition favoring adenine and thymine, at the expense of guanine and cytosine.

Further evidence of endosymbiosis are the prokaryotic ribosomes found within chloroplasts and mitochondria as well as the double-membrane enclosing them. It used to be widely assumed that the inner membrane is the original membrane of the once independent prokaryote, while the outer one is the food vacuole (phagosomal membrane) it was enclosed in initially. However, this view neglects the fact that i) both modern cyanobacteria and alpha-proteobacteria are Gram-negative bacteria, which are surrounded by double membranes; ii) the outer membranes of the endosymbiotic organelles (chloroplasts and mitochondria) are very similar to those of these bacteria in their lipid and protein compositions. Accumulating biochemical data strongly suggests that the double-membrane-enclosing chloroplasts and mitochondria derived from those of the ancestral bacteria, and the phagosomal membrane disappeared during organelle evolution. Triple or quadruple membranes are found among certain algae, probably resulting from repeated endosymbiosis (although little else was retained of the engulfed cell).

These modern organisms with endosymbiotic relationships with aerobic bacteria have verified the endosymbiotic theory, which explains the origin of mitochondria and chloroplasts from bacteria.

Researchers in molecular and evolutionary biology no longer question this theory, although some of the details, such as the mechanisms for loss of genes from organelles to host nuclear genomes, are still being worked out.

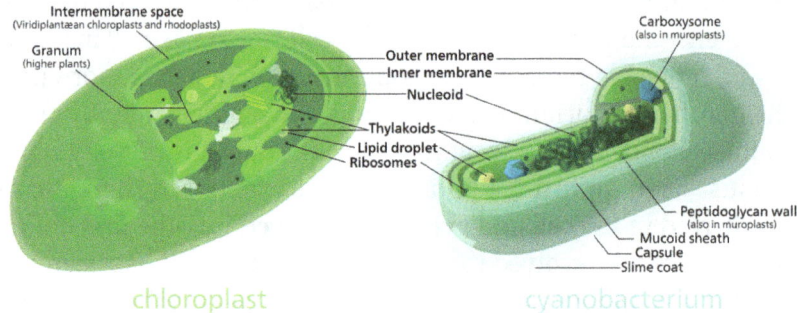

Comparison of chloroplasts and cyanobacteria showing their similarities.

Bacterial Endosymbionts in Marine Invertebrates

Extracellular endosymbionts are also represented in all four extant classes of Echinodermata (Crinoidea, Ophiuroidea, Echinoidea, and Holothuroidea). Little is known of the nature of the association (mode of infection, transmission, metabolic requirements, etc.) but phylogenetic analysis indicates that these symbionts belong to the alpha group of the class Proteobacteria, relating them to *Rhizobium* and *Thiobacillus*. Other studies indicate that these subcuticular bacteria may be both abundant within their hosts and widely distributed among the Echinoderms in general.

Some marine oligochaeta (e.g., Olavius or Inanidrillus) have obligate extracellular endosymbionts that fill the entire body of their host. These marine worms are nutritionally dependent on their symbiotic chemoautotrophic bacteria lacking any digestive or excretory system (no gut, mouth, or nephridia).

Symbiodinium Dinoflagellate Endosymbionts in Marine Metazoa and Protists

Dinoflagellate endosymbionts of the genus *Symbiodinium*, commonly known as zooxanthellae, are found in corals, mollusks (esp. giant clams, the *Tridacna*), sponges, and foraminifera. These endosymbionts drive the formation of coral reefs by capturing sunlight and providing their hosts with energy for carbonate deposition.

Previously thought to be a single species, molecular phylogenetic evidence over the past couple decades has shown there to be great diversity in *Symbiodinium*. In some cases, there is specificity between host and *Symbiodinium* clade. More often, however, there is an ecological distribution of *Symbiodinium*, the symbionts switching between hosts with apparent ease. When reefs become environmentally stressed, this distribution of symbionts is related to the observed pattern of coral bleaching and recovery. Thus, the distribution of *Symbiodinium* on coral reefs and its role in coral bleaching presents one of the most complex and interesting current problems in reef ecology.

In Protists

Mixotricha paradoxa is a protozoan that lacks mitochondria. However, spherical bacteria live inside the cell and serve the function of the mitochondria. *Mixotricha* also has three other species of symbionts that live on the surface of the cell.

Paramecium bursaria, a species of ciliate, has a mutualistic symbiotic relationship with green alga called Zoochlorella. The algae live inside the cell, in the cytoplasm.

Paulinella chromatophora is a freshwater amoeboid which has recently (evolutionarily speaking) taken on a cyanobacterium as an endosymbiont.

Bacterial Endosymbionts in Insects

Scientists classify insect endosymbionts in two broad categories, 'Primary' and 'Secondary'. Primary endosymbionts (sometimes referred to as P-endosymbionts) have been associated with their insect hosts for many millions of years (from 10 to several hundred million years in some cases). They form obligate associations, and display cospeciation with their insect hosts. Secondary endosymbionts exhibit a more recently developed association, are sometimes horizontally transferred between hosts, live in the hemolymph of the insects, and are not obligate.

Among primary endosymbionts of insects, the best-studied are the pea aphid (*Acyrthosiphon pisum*) and its endosymbiont *Buchnera sp.* APS, the tsetse fly *Glossina morsitans morsitans* and its endosymbiont *Wigglesworthia glossinidia brevipalpis* and the endosymbiotic protists in lower termites. As with endosymbiosis in other insects, the symbiosis is obligate in that neither the bacteria nor the insect is viable without the other. Scientists have been unable to cultivate the bacteria in lab conditions outside of the insect. With special nutritionally-enhanced diets, the insects can survive, but are unhealthy, and at best survive only a few generations.

In some insect groups, these endosymbionts live in specialized insect cells called bacteriocytes (also called *mycetocytes*), and are maternally-transmitted, i.e. the mother transmits her endosymbionts to her offspring. In some cases, the bacteria are transmitted in the egg, as in *Buchnera*; in others like *Wigglesworthia*, they are transmitted via milk to the developing insect embryo. In termites, the endosymbionts reside within the hindguts and are transmitted through trophallaxis among colony members.

The primary endosymbionts are thought to help the host either by providing nutrients that the host cannot obtain itself or by metabolizing insect waste products into safer forms. For example, the putative primary role of *Buchnera* is to synthesize essential amino acids that the aphid cannot acquire from its natural diet of plant sap. Likewise, the primary role of *Wigglesworthia*, it is presumed, is to synthesize vitamins that the tsetse fly does not get from the blood that it eats. In lower termites, the endosymbiotic protists play a major role in the digestion of lignocellulosic materials that constitute a bulk of the termites' diet.

Bacteria benefit from the reduced exposure to predators and competition from other bacterial species, the ample supply of nutrients and relative environmental stability inside the host.

Genome sequencing reveals that obligate bacterial endosymbionts of insects have among the smallest of known bacterial genomes and have lost many genes that are commonly found in closely related bacteria. Several theories have been put forth to explain the loss of genes. It is presumed that some of these genes are not needed in the environment of the host insect cell. A complementary theory suggests that the relatively small numbers of bacteria inside each insect decrease the efficiency of natural selection in 'purging' deleterious mutations and small mutations from the population, resulting in a loss of genes over many millions of years. Research in which a parallel phylogeny of bacteria and insects was inferred supports the belief that the primary endosymbionts are transferred only vertically (i.e., from the mother), and not horizontally (i.e., by escaping the host and entering a new host).

Attacking obligate bacterial endosymbionts may present a way to control their insect hosts, many of which pests or carriers of human disease. For example, aphids are crop pests and the tsetse fly carries the organism *Trypanosoma brucei* that causes African sleeping sickness. Other motivations for their study is to understand symbiosis, and to understand how bacteria with severely depleted genomes are able to survive, thus improving our knowledge of genetics and molecular biology.

Less is known about secondary endosymbionts. The pea aphid (*Acyrthosiphon pisum*) is known to contain at least three secondary endosymbionts, *Hamiltonella defensa*, *Regiella insecticola*, and *Serratia symbiotica*. *H. defensa* aids in defending the insect from parasitoids. *Sodalis glossinidius* is a secondary endosymbiont of tsetse flies that lives inter- and intracellularly in various host tissues, including the midgut and hemolymph. Phylogenetic studies have not indicated a correlation between evolution of *Sodalis* and tsetse. Unlike tsetse's P-symbiont *Wigglesworthia*, though, *Sodalis* has been cultured *in vitro*.

Viral Endosymbionts and Endogenous Retrovirus

During pregnancy in viviparous mammals, endogenous retroviruses (ERVs) are activated and produced in high quantities during the implantation of the embryo. On one hand, they are hypothesized to act as immunosuppressors, perhaps protecting the embryo from the immune system of the mother; on the other hand viral fusion proteins cause the formation of the placental syncytium in order to limit the exchange of migratory cells between the developing embryo and the body of the mother, where an epithelium will not be adequate because certain blood cells are specialized to be able to insert themselves between adjacent epithelial cells. The ERV is an endogenized form of what was once an infectious retrovirus. The immunodepressive action was important for the infection of the original virus. The fusion proteins may have been a way to spread the infection to other cells by simply merging them with the infected one. It is believed that the ancestors of modern viviparous mammals evolved after an infection of an ancestor with this virus, perhaps improving the ability of the fetus to survive the immune system of the mother.

The human genome project found several thousand ERVs, which are organized into 24 families.

Microbiota

A microbiota is "the ecological community of commensal, symbiotic and pathogenic microorganisms that literally share our body space". Joshua Lederberg coined the term, emphasising the importance of microorganisms inhabiting the human body in health and disease. Many scientific articles distinguish microbiome and microbiota to describe either the collective genomes of the microorganisms that reside in an environmental niche or the microorganisms themselves, respectively. However, by the original definitions, these terms are largely synonymous.

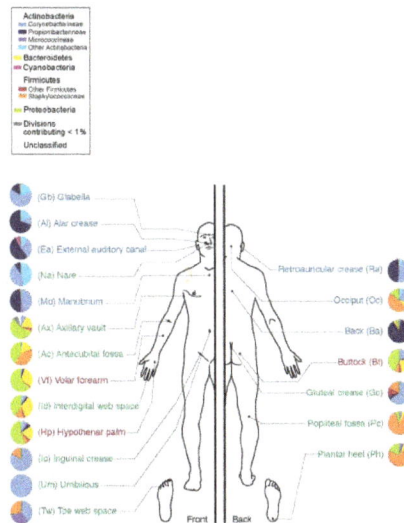

Depiction of the human skin and bacteria that predominate

The microbes being discussed generally do not cause disease unless they grow abnormally non-pathogenic organisms; they exist in harmony and symbiotically with their hosts. The microbiome and host may have emerged as a unit by the process of integration.

Introduction

All plants and animals, from protists to humans, live in close association with microbial organisms. Up until relatively recently, however, biologists have defined the interactions of plants and animals with the microbial world mostly in the context of disease states and of a relatively small number of symbiotic case studies. Organisms do not live in isolation, but have evolved in the context of complex communities. A number of advances have driven a change in the perception of microbiomes, including:

- the ability to perform genomic and gene expression analyses of single cells and even of entire microbial communities in the new disciplines of metagenomics and metatranscriptomics

- massive databases making this information accessible to researchers across multiple disciplines

- methods of mathematical analysis that help researchers to make sense of complex data sets

Increasingly, biologists have come to appreciate that microbes make up an important part of an organism's phenotype, far beyond the occasional symbiotic case study.

Pierre-Joseph van Beneden(1809-1894), a Belgian professor at the University of Louvain, developed the concept of commensalism during the nineteenth century. In his 1875 publication *Animal*

Parasites and Messmates, Van Beneden presented 264 examples of commensalism. His conception was widely accepted by his contemporaries and commensalism has continued to be used as a concept right up to the present day: microbiome is clearly linked to commensalism.

Microbiota by Host

There is a strengthening consensus among evolutionary biologists that one should not separate an organism's genes from the context of its resident microbes.

Humans

The human microbiota includes bacteria, fungi, and archaea. Micro-animals which live on the human body are excluded. The human microbiome refers to their genomes.

Humans are colonized by many microorganisms; the traditional estimate was that humans live with ten times more non-human cells than human cells; more recent estimates have lowered that to 3:1 and even to approximately the same number; all the numbers are estimates. Regardless of the exact number, the microbiota that colonize humans have not merely a commensal (a non-harmful coexistence), but rather a mutualistic relationship with their human hosts. Some of these organisms perform tasks that are known to be useful for the human host; for most, the role is not well understood. Those that are expected to be present, and that under normal circumstances do not cause disease, are deemed *normal flora* or *normal microbiota*.

The Human Microbiome Project took on the project of sequencing the genome of the human microbiota, focusing particularly on the microbiota that normally inhabit the skin, mouth, nose, digestive tract, and vagina. It reached a milestone in 2012 when it published initial results.

Non-human Animals

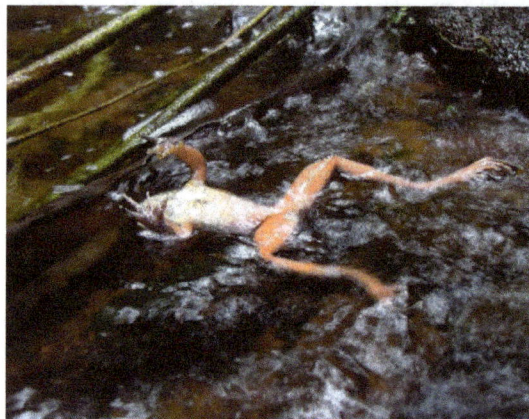

A chytrid-infected frog

- A massive, worldwide decline in amphibian populations has been well-publicised. Habitat loss and over-exploitation account for part of the problem, but many other processes seem to be at work. The spread of the virulent fungal disease chytridiomycosis represents an enigma. The ability of some species to coexist with the causative agent *Batrachochytrium dendrobatidis* appears to be due to the expression of antimicrobial skin peptides along with the presence of symbiotic microbes that benefit the host by resisting pathogen colonization or inhibiting their growth while being themselves resistant to high concentrations of antimicrobial skin peptides.

- The bovine rumen harbors a complex microbiome that converts plant cell wall biomass into proteins, short chain fatty acids, and gases. Multiple species are involved in this conversion. Traditional methods of characterizing the microbial population, based on culture analysis, missed many of the participants in this process. Comparative metagenomic studies yielded the surprising result that individual steer had markedly different community structures, predicted phenotype, and metabolic potentials, even though they were fed identical diets, were housed together, and were apparently functionally identical in their utilization of plant cell wall resources.

- Leaf-cutter ants form huge underground colonies with millions of workers, each colony harvesting hundreds of kilograms of leaves each year. Unable to digest the cellulose in the leaves directly, they maintain fungus gardens that are the colony's primary food source. The fungus itself does not digest cellulose. Instead, a microbial community containing a diversity of bacteria is responsible for cellulose digestion. Analysis of the microbial population's genomic content by community metagenome sequencing methods revealed the presence of many genes with a role in cellulose digestion. This microbiome's predicted carbohydrate-degrading enzyme profile is similar to that of the bovine rumen, but the species composition is almost entirely different.

- Mice are the most used models for human disease. As more and more diseases are linked to dysfunctional microbiomes, mice have become the most studied organism in this regard. Mostly it is the gut microbiota that have been studied in relation to allergic airway disease, obesity, gastrointestinal diseases and diabetes. Intriguingly, recent work has shown that perinatal shifting of microbiota through administration of low dose antibiotics can have long-lasting effects on future susceptibility to allergic airway disease. These studies showed a remarkable link between the frequency of certain subsets of microbes and disease severity. In aggregate these studies suggest that the presence of specific microbes, early in postnatal life, play an instructive role in the development of future immune responses. Mechanistically, a recent study done on gnotobiotic mice described a method in which certain strains of gut bacteria were found to transmit a particular phenotype to recipient germ-free mice, identifying an unanticipated range of bacterial strains that promoted accumulation of colonic regulatory T cells, as well as strains that modulated mouse adiposity and cecal metabolite concentrations. Another study showed that when adult germ-free mice were colonized with the gut flora of obese mice, there was a dramatic weight increase and an observed increased metabolism of monosaccharides and short-chain fatty acids. Looking at the gut flora compositions between normal and obese mice, obese mice had less Bacteroidetes than Firmicutes in abundance in gut flora and it is hypothesized that the microbiota of obese mice are more efficient at extracting energy from food. This combinatorial approach enables a systems-level understanding of microbial contributions to human biology. But also other mucoide tissues as lung and vagina have been studied in relation to diseases such as asthma, allergy and vaginosis

Plants

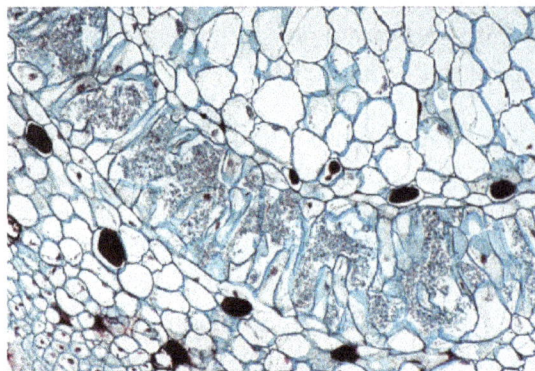

Light micrograph of a cross section of a coralloid root of a cycad, showing the layer that hosts
symbiotic cyanobacteria

- Plants exhibit a broad range of relationships with symbiotic microorganisms, ranging from parasitism, in which the association is disadvantageous to the host organism, to mutualism, in which the association is beneficial to both, to commensalism, in which the symbiont benefits while the host is not affected. Exchange of nutrients between symbiotic partners is an important part of the relationship: it may be bidirectional or unidirectional, and it may be context dependent. The strategies for nutrient exchange are highly diverse. Oomycetes and fungi have, through convergent evolution, developed similar morphology and occupy similar ecological niches. They develop hyphae, filamentous structures that penetrate the host cell. In those cases where the association is mutualistic, the plant often exchanges hexose sugars for inorganic phosphate from the fungal symbiont. It is speculated that such associations, which are very ancient, may have aided plants when they first colonized land.

- A huge range of bacterial symbionts colonize plants. Many of these are pathogenic, but others known as plant-growth promoting bacteria (PGPB) provide the host with essential services such as nitrogen fixation, solubilization of minerals such as phosphorus, synthesis of plant hormones, direct enhancement of mineral uptake, and protection from pathogens. PGPBs may protect plants from pathogens by competing with the pathogen for an ecological niche or a substrate, producing inhibitory allelochemicals, or inducing systemic resistance in host plants to the pathogen

- Plants are attractive hosts for microorganisms since they provide a variety of nutrients. Microorganisms on plants can be epiphytes (found on the plants) or endophytes (found inside plant tissue).

Immune System

The symbiotic relationship between a mammalian host and its microbiota has a significant impact on shaping the host's immune system. In many animals, the immune system and microbiota engage in "cross-talk", exchanging chemical signals. This allows the immune system to recognize the types of bacteria that are harmful to the host and combat them, while allowing the helpful bacteria to carry out their functions; in turn, the microbiota influence immune reactivity and targeting. Bacteria can be transferred from mother to child through direct contact and after birth, or through indirect contact through eggs, coprophagy, and several other pathways. As the infant microbiome is established, commensal bacteria quickly populate the gut, prompting a range of immune responses and "programming" the immune system with long-lasting effects. This early colonization helps to establish the symbiotic microbiome inside the animal host early in its life. The bacteria are also able to stimulate lymphoid tissue associated with the gut mucosa. This enables the tissue to produce antibodies for pathogens that may enter the gut.

It has been found that bacteria may also play a role in the activation of TLRs (toll-like receptors) in the intestines. TLRs are a type of PRR (pattern recognition receptor) used by host cells to help repair damage and recognize dangers to the host. This could be important in immune tolerance and autoimmune diseases. Pathogens could influence this symbiotic coexistence leading to immune dysregulation and susceptibility to diseases. This could provide new direction for managing immunological and metabolic diseases.

Co-evolution of Microbiota

Bleached branching coral (foreground) and normal branching coral (background). Keppel Islands, Great Barrier Reef

Organisms evolve within eco-systems so that the change of one organism affects the change of others. Co-evolution (also called "hologenome theory") proposes that an object of natural selection is not the individual organism, but the organism together with its associated organisms, includings its microbial communities.

Coral reefs. The hologenome theory originated in studies on coral reefs. Coral reefs are the largest structures created by living organisms, and contain abundant and highly complex microbial communities. Over the past several decades, major declines in coral populations have occurred. Climate change, water pollution and over-fishing are three stress factors that have been described as leading to disease susceptibility. Over twenty different coral diseases have been described, but of these, only a handful have had their causative agents isolated and characterized. Coral bleaching is the most serious of these diseases. In the Mediterranean Sea, the bleaching of *Oculina patagonica* was first described in 1994 and shortly determined to be due to infection by *Vibrio shiloi*. From 1994 to 2002, bacterial bleaching of *O. patagonica* occurred every summer in the eastern Mediterranean. Surprisingly, however, after 2003, *O. patagonica* in the eastern Mediterranean has been resistant to *V. shiloi* infection, although other diseases still cause bleaching. The surprise stems from the knowledge that corals are long lived, with lifespans on the order of decades, and do not have adaptive immune systems. Their innate immune systems do not produce antibodies, and they should seemingly not be able to respond to new challenges except over evolutionary time scales. The puzzle of how corals managed to acquire resistance to a specific pathogen led Eugene Rosenberg and Ilana Zilber-Rosenberg to propose the Coral Probiotic Hypothesis. This hypothesis proposes that a dynamic relationship exists between corals and their symbiotic microbial communities. By altering its composition, this holobiont can adapt to changing environmental conditions far more rapidly than by genetic mutation and selection alone. Extrapolating this hypothesis of adaptation and evolution to other organisms, including higher plants and animals, led to the proposal of the Hologenome Theory of Evolution.

The hologenome theory is still being debated. A major criticism has been the claim that *V. shiloi* was misidentified as the causative agent of coral bleaching, and that its presence in bleached *O. patagonica* was simply that of opportunistic colonization. If this is true, the basic observation leading to the theory would be invalid. Nevertheless, the theory has gained significant popularity as a way of explaining rapid changes in adaptation that cannot otherwise be explained by traditional mechanisms of natural selection. For those who accept the hologenome theory, the holobiont has become the principal unit of natural selection. On the other hand, it has been stated that the holobiont is the result of other step of integration that it is also observed at the cell (symbiogenesis, endosymbiosis) and genomic levels.

Research Methods

Targeted Amplicon Sequencing

Targeted amplicon sequencing relies on having some expectations about the composition of the community that is being studied. In target amplicon sequencing a phylogenetically informative marker is targeted for sequencing. Such a marker should be present in ideally all the expected organisms. It should also evolve in such a way that it is conserved enough that primers can target genes from a wide range of organisms while evolving quickly enough to allow for finer resolution at the taxonomic level. A common marker for human microbiome studies is the gene for bacterial 16S rRNA (*i.e.* "16S rDNA", the sequence of DNA which encodes the ribosomal RNA molecule). Since ribosomes are present in all living organisms, using 16S rDNA allows for DNA to be amplified from many more organisms than if another marker were used. The 16S rDNA gene contains both slowly evolving regions and fast evolving regions; the former can be used to design broad primers while the latter allow for finer taxonomic distinction. However, species-level resolution is not typically possible using the 16S rDNA. Primer selection is an important step, as anything that cannot be targeted by the primer will not be amplified and thus will not be detected. Different sets of primers have been shown to amplify different taxonomic groups due to sequence variation.

Targeted studies of eukaryotic and viral communities are limited and subject to the challenge of excluding host DNA from amplification and the reduced eukaryotic and viral biomass in the human microbiome.

After the amplicons are sequenced, molecular phylogenetic methods are used to infer the composition of the microbial community. This is done by clustering the amplicons into operational taxonomic units (OTUs) and inferring phylogenetic relationships between the sequences. Due to the complexity of the data, distance measures such as UniFrac distances are usually defined between microbiome samples, and downstream multivariate methods are carried out on the distance matrices. An important point is that the scale of data is extensive, and further approaches must be taken to identify patterns from the available information. Tools used to analyze the data include VAMPS, QIIME and mothur.

Metagenomic Sequencing

Metagenomics is also used extensively for studying microbial communities. In metagenomic sequencing, DNA is recovered directly from environmental samples in an untargeted manner with the goal of obtaining an unbiased sample from all genes of all members of the community. Recent studies use shotgun Sanger sequencing or pyrosequencing to recover the sequences of the reads. The reads can then be assembled into contigs. To determine the phylogenetic identity of a sequence, it is compared to available full genome sequences using methods such as BLAST. One drawback of this approach is that many members of microbial communities do not have a representative sequenced genome.

Despite the fact that metagenomics is limited by the availability of reference sequences, one significant advantage of metagenomics over targeted amplicon sequencing is that metagenomics data can elucidate the functional potential of the community DNA. Targeted gene surveys cannot do this as they only reveal the phylogenetic relationship between the same gene from different organisms. Functional analysis is done by comparing the recovered sequences to databases of metagenomic annotations such as KEGG. The metabolic pathways that these genes are involved in can then be predicted with tools such as MG-RAST, CAMERA and IMG/M.

RNA and Protein-based Approaches

Metatranscriptomics studies have been performed to study the gene expression of microbial communities through methods such as the pyrosequencing of extracted RNA. Structure based studies have also identified non-coding RNAs (ncRNAs) such as ribozymes from microbiota. Metaproteomics is a new approach that studies the proteins expressed by microbiota, giving insight into its functional potential.

Projects

The Human Microbiome Project (HMP) was a United States National Institutes of Health initiative with the goal of identifying and characterizing the microorganisms which are found in association with both healthy and diseased humans (their microbial flora). Launched in 2008, it was a five-year project, best characterized as a feasibility study, with a total budget of $115 million. The ultimate goal of this and similar NIH-sponsored microbiome projects is to test how changes in the human microbiome are associated with human health or disease.

The Earth Microbiome Project (EMP) is an initiative to collect natural samples and analyze the microbial community around the globe. Microbes are highly abundant, diverse and have an important role in the ecological system. Yet as of 2010, it was estimated that the total global environmental DNA sequencing effort had produced less than 1 percent of the total DNA found in a liter of seawater or a gram of soil, and the specific interactions between microbes are largely unknown. The EMP aims to process as many as 200,000 samples in different biomes, generating a complete database of microbes on earth to characterize environments and ecosystems by microbial composition and interaction. Using these data, new ecological and evolutionary theories can be proposed and tested.

The Brazilian Microbiome Project (BMP) aims to assemble a Brazilian Microbiome Consortium/Database. At present, many metagenomic projects underway in Brazil are widely known. Our goal is to co-ordinate and standardize these, together with future projects. This is the first attempt to collect and collate information about Brazilian microbial genetic and functional diversity in a systematic and holistic manner. New sequence data have been generated from samples collected in all Brazilian regions, however the success of the BMP depends on a massive collaborative effort of both the Brazilian and international scientific communities. Therefore, we invite all colleagues to participate in this project. There is no prioritization of specific taxonomic groups, studies could include any ecosystem, and all proposals and any help will be very welcome.

Privacy

The DNA of the microbes that inhabit a person's human body can uniquely identify the person. A risk to violating a person's privacy may exist, if the person anonymously donated microbe DNA data, and the data could be used to identify the person and their medical condition, and if the person's identity were revealed.

Human Microbiota

The human microbiota is the aggregate of microorganisms, a microbiome that resides on or within a number of tissues and biofluids, including the skin, mammary glands, placenta, seminal fluid, uterus, ovarian follicles, lung, saliva, oral mucosa, conjunctiva, and gastrointestinal tracts. They include bacteria, fungi, and archaea. Micro-animals which live on the human body are excluded. The human microbiome refers to their genomes.

Humans are colonized by many microorganisms; the traditional estimate was that humans live with ten times more non-human cells than human cells; more recent estimates have lowered that to 3:1 and even to approximately the same number; all the numbers are estimates. Some microbiota that colonize humans have not merely a commensal (a non-harmful coexistence), but rather a mutualistic relationship with their human hosts. Conversely, some non-pathogenic microbiota can harm human hosts via the metabolites that they produce, like trimethylamine. Certain microbiota perform tasks that are known to be useful for the human host; for most, the role is not well understood. Those that are expected to be present, and that under normal circumstances do not cause disease, are deemed *normal flora* or *normal microbiota*.

The Human Microbiome Project took on the project of sequencing the genome of the human microbiota, focusing particularly on the microbiota that normally inhabit the skin, mouth, nose, digestive tract, and vagina. It reached a milestone in 2012 when it published initial results.

Terminology

Though widely known as *flora or microflora*, this is a misnomer in technical terms, since the word root *flora* pertains to plants, and *biota* refers to the total collection of organisms in a particular ecosystem. Recently, the more appropriate term *microbiota* is applied, though its use has not eclipsed the entrenched use and recognition of *flora* with regard to bacteria and other microorganisms. Both terms are being used in different literature.

Relative Numbers

As of 2014, it was often reported in popular media and in the scientific literature that there are about 10 times as many microbial cells in the human body than there are human cells; this figure was based on estimates that the human microbiome includes around 100 trillion bacterial cells and an adult human typically has around 10 trillion human cells. In 2014 the American Academy of Microbiology published an FAQ that emphasized that the number of microbial cells and the number of human cells are both estimates, and noted that recent research had arrived at a new estimate of the number of human cells at around 37 trillion cells, meaning that the ratio of microbial to human cells is probably about 3:1. In 2016 another group published a new estimate of ratio as being roughly 1:1 (1.3:1, with "an uncertainty of 25% and a variation of 53% over the population of standard 70 kg males").

Study

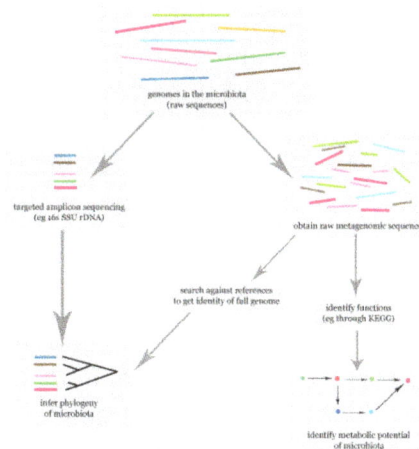

Flowchart illustrating how the human microbiome is studied on the DNA level.

The problem of elucidating the human microbiome is essentially identifying the members of a microbial community which includes bacteria, eukaryotes, and viruses. This is done primarily using DNA-based studies, though RNA, protein and metabolite based studies are also performed. DNA-based microbiome studies typically can be categorized as either targeted amplicon studies or more recently shotgun metagenomic studies. The former focuses on specific known marker genes and is primarily informative taxonomically, while the latter is an entire metagenomic approach which can also be used to study the functional potential of the community. One of the challenges that is present in human microbiome studies but not in other metagenomic studies is to avoid including the host DNA in the study.

Aside from simply elucidating the composition of the human microbiome, one of the major questions involving the human microbiome is whether there is a "core", that is, whether there is a subset of the community that is shared among most humans. If there is a core, then it would be possible to associate certain community compositions with disease states, which is one of the goals of the Human Microbiome Project. It is known that the human microbiome is highly variable both within a single subject and between different individuals. For example, the gut microbiota of humans is markedly dissimilar between individuals, a phenomenon which is also observed in mice.

On 13 June 2012, a major milestone of the Human Microbiome Project (HMP) was announced by the NIH director Francis Collins. The announcement was accompanied with a series of coordinated articles published in Nature and several journals in the Public Library of Science (PLoS) on the same day. By mapping the normal microbial make-up of healthy humans using genome sequencing techniques, the researchers of the HMP have created a reference database and the boundaries of normal microbial variation in humans. From 242 healthy U.S. volunteers, more than 5,000 samples were collected from tissues from 15 (men) to 18 (women) body sites such as mouth, nose, skin, lower intestine (stool), and vagina. All the DNA, human and microbial, were analyzed with DNA sequencing machines. The microbial genome data were extracted by identifying the bacterial specific ribosomal RNA, 16S rRNA. The researchers calculated that more than 10,000 microbial species occupy the human ecosystem and they have identified 81 – 99% of the genera.

Types

Bacteria

Populations of microbes (such as bacteria and yeasts) inhabit the skin and mucosal surfaces in various parts of the body. Their role forms part of normal, healthy human physiology, however if microbe numbers grow beyond their typical ranges (often due to a compromised immune system) or if microbes populate (such as through poor hygiene or injury) areas of the body normally not colonized or sterile (such as the blood, or the lower respiratory tract, or the abdominal cavity), disease can result (causing, respectively, bacteremia/sepsis, pneumonia, and peritonitis).

The Human Microbiome Project found that individuals host thousands of bacterial types, different body sites having their own distinctive communities. Skin and vaginal sites showed smaller diversity than the mouth and gut, these showing the greatest richness. The bacterial makeup for a given site on a body varies from person to person, not only in type, but also in abundance. Bacteria of the same species found throughout the mouth are of multiple subtypes, preferring to inhabit distinctly different locations in the mouth. Even the enterotypes in the human gut, previously thought to be well-understood, are from a broad spectrum of communities with blurred taxon boundaries.

It is estimated that 500 to 1,000 species of bacteria live in the human gut but belong to just a few phyla: Firmicutes and Bacteroidetes dominate but there are also Proteobacteria, Verrumicrobia, Actinobacteria, Fusobacteria and Cyanobacteria.

A number of types of bacteria, such as *Actinomyces viscosus* and *A. naeslundii*, live in the mouth, where they are part of a sticky substance called plaque. If this is not removed by brushing, it hardens into calculus (also called tartar). The same bacteria also secrete acids that dissolve tooth enamel, causing tooth decay.

The vaginal microflora consist mostly of various lactobacillus species. It was long thought that the most common of these species was *Lactobacillus acidophilus*, but it has later been shown that the most common one is *L. iners* followed by *L. crispatus*. Other lactobacilli found in the vagina are *L. jensenii, L. delbruekii* and *L. gasseri*. Disturbance of the vaginal flora can lead to infections such as bacterial vaginosis or candidiasis ("yeast infection").

Archaea

Archaea are present in the human gut, but, in contrast to the enormous variety of bacteria in this organ, the numbers of archaeal species are much more limited. The dominant group are the methanogens, particularly *Methanobrevibacter smithii* and *Methanosphaera stadtmanae*. However, colonization by methanogens is variable, and only about 50% of humans have easily detectable populations of these organisms.

As of 2007, no clear examples of archaeal pathogens were known, although a relationship has been proposed between the presence of some methanogens and human periodontal disease.

Fungi

Fungi, in particular yeasts, are present in the human gut. The best-studied of these are *Candida* species due to their ability to become pathogenic in immunocompromised and even in healthy hosts. Yeasts are also present on the skin, such as *Malassezia* species, where they consume oils secreted from the sebaceous glands.

Viruses

Viruses, especially bacterial viruses (bacteriophages), colonize various body sites. These colonized sites include the skin, gut, lungs, and oral cavity. Virus communities have been associated with some diseases, and do not simply reflect the bacterial communities.

Anatomical Areas

Skin

A study of twenty skin sites on each of ten healthy humans found 205 identified genera in nineteen bacterial phyla, with most sequences assigned to four phyla: Actinobacteria (51.8%), Firmicutes (24.4%), Proteobacteria (16.5%), and Bacteroidetes (6.3%). A large number of fungal genera are present on healthy human skin, with some variability by region of the body; however, during pathological conditions, certain genera tend to dominate in the affected region. For example, *Malassezia* is dominant in atopic dermatitis and *Acremonium* is dominant on dandruff-afflicted scalps.

The skin acts as a barrier to deter the invasion of pathogenic microbes. The human skin contains microbes that reside either in or on the skin and can be residential or transient. Resident microorganism types vary in relation to skin type on the human body. A majority of microbes reside on superficial cells on the skin or prefer to associate with glands. These glands such as oil or sweat glands provide the microbes with water, amino acids, and fatty acids. In addition, resident bacteria that associated with oil glands are often Gram-positive and can be pathogenic.

Conjunctiva

A small number of bacteria and fungi are normally present in the conjunctiva. Classes of bacteria include Gram-positive cocci (e.g., *Staphylococcus* and Streptococcus) and Gram-negative rods and cocci (e.g., *Haemophilus* and *Neisseria*) are present. Fungal genera include *Candida*, *Aspergillus*, and *Penicillium*. The lachrymal glands continuously secrete, keeping the conjunctiva moist, while intermittent blinking lubricates the conjunctiva and washes away foreign material. Tears contain bactericides such as lysozyme, so that microorganisms have difficulty in surviving the lysozyme and settling on the epithelial surfaces.

Gut

Tryptophan Metabolism by Human Gastrointestinal Microbiota

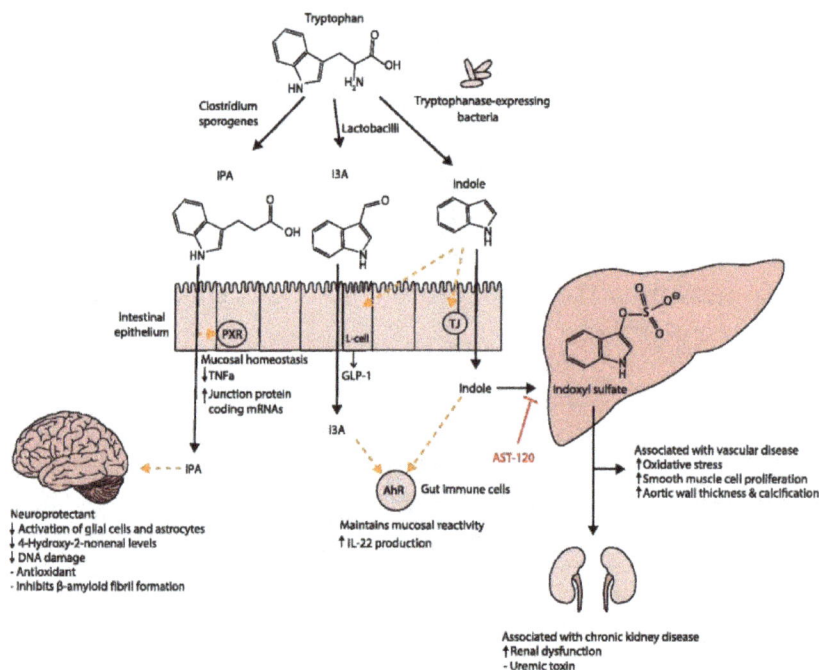

This diagram shows the biosynthesis of bioactive compounds (indole and certain derivatives) from tryptophan by bacteria in the gut. Indole is produced from tryptophan by bacteria that express tryptophanase. *Clostridium sporogenes* metabolizes indole into 3-indolepropionic acid (IPA), a highly potent neuroprotective antioxidant that scavenges hydroxyl radicals. In the intestine, IPA binds to pregnane X receptors (PXR) in intestinal cells, thereby facilitating mucosal homeostasis and barrier function. Following absorption from the intestine and distribution to the brain, IPA confers a neuroprotective effect against cerebral ischemia and Alzheimer's disease. *Lactobacillus* species metabolize tryptophan into indole-3-aldehyde (I3A) which acts on the aryl hydrocarbon receptor (AhR) in intestinal immune cells, in turn increasing interleukin-22 (IL-22) production. Indole itself acts as a glucagon-like peptide-1 (GLP-1) secretagogue in intestinal L cells and as a ligand for AhR. Indole can also be metabolized by the liver into indoxyl sulfate, a compound that is toxic in high concentrations and associated with vascular disease and renal dysfunction. AST-120 (activated charcoal), an intestinal sorbent that is taken by mouth, adsorbs indole, in turn decreasing the concentration of indoxyl sulfate in blood plasma.

The gut flora has the largest numbers of bacteria and the greatest number of species compared to other areas of the body. In humans the gut flora is established at one to two years after birth, and by that time the intestinal epithelium and the intestinal mucosal barrier that it secretes have co-developed in a way that is tolerant to, and even supportive of, the gut flora and that also provides a barrier to pathogenic organisms.

The relationship between some gut flora and humans is not merely commensal (a non-harmful coexistence), but rather a mutualistic relationship. Some human gut microorganisms benefit the host by fermentating dietary fiber into short-chain fatty acids (SCFAs), such as acetic acid and butyric acid, which are then absorbed by the host. Intestinal bacteria also play a role in synthesizing vitamin B and vitamin K as well as metabolizing bile acids, sterols, and xenobiotics. The systemic importance of the SCFAs and other compounds they produce are like hormones and the gut flora itself appears to function like an endocrine organ, and dysregulation of the gut flora has been correlated with a host of inflammatory and autoimmune conditions.

The composition of human gut flora changes over time, when the diet changes, and as overall health changes. A systematic review of 15 human randomized controlled trials from July 2016 found that certain commercially available strains of probiotic bacteria from the *Bifidobacterium* and *Lactobacillus* genera (*B. longum, B. breve, B. infantis, L. helveticus, L. rhamnosus, L. plantarum,* and *L. casei*), when taken by mouth in daily doses of 10^9–10^{10} colony forming units (CFU) for 1–2 months, possess treatment efficacy (i.e., improved behavioral outcomes) in certain central nervous system disorders – including anxiety, depression, autism spectrum disorder, and obsessive–compulsive disorder – and improved certain aspects of memory.

Vagina

Vaginal microbiota refers to those species and genera that colonize the lower reproductive tract of women. These organisms play an important role in protecting against infections and maintaining vaginal health. The most abundant vaginal microorganisms found in premenopausal women are from the genus *Lactobacillus*, which suppress pathogens by producing hydrogen peroxide and lactic acid. Bacterial species composition and ratios vary depending on the stage of the menstrual cycle. Ethnicity also influences vaginal flora. The occurrence of hydrogen peroxide-producing lactobacilli is lower in African American women and vaginal pH is higher. Other influential factors such as sexual intercourse and antibiotics have been linked to the loss of lactobacilli. Moreover, studies have found that sexual intercourse with a condom does appear to change lactobacilli levels, and does increase the level of *Escherichia coli* within the vaginal flora. Changes in the normal, healthy vaginal microbiota is an indication of infections, such as candidiasis or bacterial vaginosis. *Candida albicans* inhibits the growth of *Lactobacillus* species, while *Lactobacillus* species which produce hydrogen peroxide inhibit the growth and virulence of *Candida albicans* in both the vagina and the gut.

Fungal genera that have been detected in the vagina include *Candida, Pichia, Eurotium, Alternaria, Rhodotorula,* and *Cladosporium,* among others.

Placenta

Until recently the placenta was considered to be a sterile organ but commensal, nonpathogenic bacterial species and genera have been identified that reside in the placental tissue.

Uterus

Until recently, the upper reproductive tract of women was considered to be a sterile environment. A variety of microorganisms inhabit the uterus of healthy, asymptomatic women of reproductive age. The microbiome of the uterus differs significantly from that of the vagina and gastrointestinal tract.

Oral Cavity

The environment present in the human mouth allows the growth of characteristic microorganisms found there. It provides a source of water and nutrients, as well as a moderate temperature. Resident microbes of the mouth adhere to the teeth and gums to resist mechanical flushing from the mouth to stomach where acid-sensitive microbes are destroyed by hydrochloric acid.

Anaerobic bacteria in the oral cavity include: *Actinomyces, Arachnia, Bacteroides, Bifidobacterium, Eubacterium, Fusobacterium, Lactobacillus, Leptotrichia, Peptococcus, Peptostreptococcus, Propionibacterium, Selenomonas, Treponema,* and *Veillonella.*Genera of fungi that are frequently found in the mouth include *Candida, Cladosporium, Aspergillus, Fusarium, Glomus, Alternaria, Penicillium,* and *Cryptococcus,* among others. Bacteria accumulate on both the hard and soft oral tissues in biofilms. Bacterial adhesion is particularly important for oral bacteria.

Oral bacteria have evolved mechanisms to sense their environment and evade or modify the host. Bacteria occupy the ecological niche provided by both the tooth surface and gingival epithelium. However, a highly efficient innate host defense system constantly monitors the bacterial colonization and prevents bacterial invasion of local tissues. A dynamic equilibrium exists between dental plaque bacteria and the innate host defense system. Of particular interest is the role of oral microorganisms in the two major dental diseases: dental caries and periodontal disease. Additionally, research has correlated poor oral heath and the resulting ability of the oral microbiota to invade the body to affect cardiac health as well as cognitive function.

Lung

Much like the oral cavity, the upper and lower respiratory system possess mechanical deterrents to remove microbes. Goblet cells produce mucous which traps microbes and moves them out of the respiratory system via continuously moving ciliated epithelial cells. In addition, a bactericidal effect is generated by nasal mucus which contains the enzyme lysozyme.

Nonetheless, the upper and lower respiratory tract appears to have their own set of microbiota. Pulmonary bacterial microbiota belong to 9 major bacterial genera: *Prevotella, Sphingomonas, Pseudomonas, Acinetobacter, Fusobacterium, Megasphaera, Veillonella, Staphylococcus,* and *Streptococcus.* Some of the bacteria considered "normal biota" in the respiratory tract can cause serious disease especially in immunocompromised individuals; these include *Streptococcus pyogenes, Haemophilus influenzae, Streptococcus pneumoniae, Neisseria meningitidis,* and *Staphylococcus aureus.*

Fungal genera that compose the pulmonary mycobiome include *Candida, Malassezia, Neosartorya, Saccharomyces,* and *Aspergillus,* among others.

Unusual distributions of bacterial and fungal genera in the respiratory tract is observed in people with cystic fibrosis. Their bacterial flora often contains antibiotic-resistant and slow-growing bacteria, and the frequency of these pathogens changes in relation to age.

Disease

Communities of microflora have been shown to change their behavior in diseased individuals.

Cancer

Although cancer is generally a disease of host genetics and environmental factors, microorganisms are implicated in ~20% of human malignancies. Mucosal microbes can become part of the tumor microenvironment (TME) of aerodigestive tract malignancies. Intratumoral microbes can affect cancer growth and spread. Gut microbiota also detoxify dietary components, reducing inflammation and balancing host cell growth and proliferation. Coley's toxins were one of the earliest forms of cancer bacteriotherapy. Synthetic biology employs designer microbes and microbiota transplants against tumors.

Microbes and the microbiota affect carcinogenesis in three broad ways: (i) altering the balance of tumor cell proliferation and death, (ii) regulating immune system function and (iii) influencing metabolism of host-produced factors, foods and pharmaceuticals.

Modes of Action

Ten microbes are designated by the International Agency for Research on Cancer (IARC) as human carcinogens. Most of these microbes colonize large percentages of the human population, although only genetically susceptible individuals develop cancer. Tumors arising at boundary surfaces, such as the skin, oropharynx and respiratory, digestive and urogenital tracts, harbor a microbiota, which complicates cancer-microbe causality. Substantial microbe presence at a tumor site does not establish association or causal links. Instead, microbes may find the tumor's oxygen tension or nutrient profile supportive. Decreased populations of specific microbes may also increase risks.

Human oncoviruses can drive carcinogenesis by integrating oncogenes into host genomes. Human papillomaviruses (HPV) express oncoproteins such as E6 and E7. Viral integration selectively amplifies host genes in pathways with established cancer roles.

Microbes affect genomic stability, resistance to cell death and proliferative signaling. Many bacteria can damage DNA, to kill competitors/survive. These defense factors can lead to mutational events that contribute to carcinogenesis. Examples include colibactin encoded by the *pks* locus (expressed by B2 group *Escherichia coli* as well as by other Enterobacteriaceae), *Bacteroides fragilis* toxin (Bft) produced by enterotoxigenic *B. fragilis* and cytolethal distending toxin (CDT) produced by several ε- and γ-proteobacteria. Colibactin is of interest in colorectal carcinogenesis, given the detection of pks⁺ *E. coli* in human colorectal cancers and the ability of colibactin-expressing *E. coli* to potentiate intestinal tumorigenesis in mice. Data also support a role for enterotoxigenic *B. fragilis* in both human and animal models of colon tumors. Both colibactin and CDT can cause double-stranded DNA damage in mammalian cells. In contrast, Bft acts indirectly by eliciting high levels of reactive oxygen species (ROS), which in turn damage host DNA. Chronically high ROS levels can outpace DNA repair mechanisms, leading to DNA damage and mutations.

β-catenin

Several microbes possess proteins that engage host pathways involved in carcinogenesis. The Wnt/β-catenin signaling pathway, which regulates cells' polarity, growth and differentiation, is one example and is altered in many malignancies. Multiple cancer-associated bacteria can influence β-catenin signaling. Oncogenic type 1 strains of *Helicobacter pylori* express CagA, which is injected directly into the cytoplasm of host cells and modulates β-catenin to drive gastric cancer. This modulation leads to up-regulation of cellular proliferation, survival and migration genes, as well as angiogenesis—all processes central to carcinogenesis. Oral microbiota *Fusobacterium nucleatum* is associated with human colorectal adenomas and adenocarcinomas and amplified intestinal tumorigenesis in mice. *F. nucleatum* expresses FadA, a bacterial cell surface adhesion component that binds host E-cadherin, activating β-catenin. Enterotoxigenic *B. fragilis*, which is enriched in some human colorectal cancers, can stimulate E-cadherin cleavage via Btf, leading to β-catenin activation. *Salmonella* Typhi strains that maintain chronic infections secrete AvrA, which can activate epithelial β-catenin signaling and are associated with hepatobiliary cancers.

Several of these bacteria are normal microbiota constituents. The presence of these cancer-potentiating microbes and their access to E-cadherin in evolving tumors demonstrate that a loss of appropriate boundaries and barrier maintenance between host and microbe is a critical step in the development of some tumors.

Inflammation

Mucosal surface barriers are subject to environmental insult and must rapidly repair to maintain homeostasis. Compromised host or microbiota resiliency also reduce resistance to malignancy. Cancer and inflammatory disorders can then arise. Once barriers are breached, microbes can elicit proinflammatory or immunosuppressive programs.

Inflammation, whether high-grade as in inflammatory disorders or low-grade as in malignancies and obesity, drive a tumor-permissive milieu. Pro-inflammatory factors such as reactive oxygen and nitrogen species, cytokines and chemokines can also drive tumor growth and spread. Tumors can up-regulate and activate pattern recognition receptors (e.g. toll-like receptors), driving feed-forward loops of activation of cancer-associated inflammation regulator NF-κB. Cancer-associated microbes appear to activate NF-κB signaling within the TME. The activation of NF-κB by *F. nucleatum* may be the result of pattern recognition receptor engagement or FadA engagement of E-cadherin. Other pattern recognition receptors, such as nucleotide-binding oligomerization domain–like receptor (NLR) family members *NOD-2*, *NLRP3*, *NLRP6* and *NLRP12*, may play a role in mediating colorectal cancer.

Immune system TME engagement is not restricted to the innate immune system. Once the innate immune system is activated, adaptive immune responses ensue, often with tumor progression. The interleukin-23 (IL-23)–IL-17 axis, tumor necrosis factor–α (TNF-α)–TNF receptor signaling, IL-6–IL-6 family member signaling, and STAT3 activation all represent innate and adaptive pathways contributing to tumor progression and growth.

The microbiota adapts to host changes such as inflammation. Adaptation shift microbiota to a vulnerable tissue site. Genotoxin azoxymethane and colon barrier–disrupting agent dextran sodium sulfate independently result in colon tumors in susceptible mouse strains; combining them accelerates tumorigenesis.

Perturbations to a host immune system coupled with inflammatory stimulus may enrich bacterial clades that attach to host surfaces, invade host tissue, or trigger host inflammatory mediators. Fecal microbiota from *NOD2-* or *NLRP6*-deficient mice acquire features that enhance the susceptibility of wild-type mice to caCRC. In mice, gut microbiota modulate colon tumorigenesis, independent of genetic deficiencies. Germ-free mice developed more tumors when colonized from donors with caCRC, once followed by treatments that induced caCRC.

Research

Diabetic foot ulcers develop their own, distinctive microbiota. Investigations into characterizing and identifying the phyla, genera and species of bacteria or other microorganisms populating these ulcers may help identify one group of microbiota that promotes healing.

Environmental Health

Studies in 2009 questioned whether the decline in biota (including microfauna) as a result of human intervention might impede human health.

Biological Agent

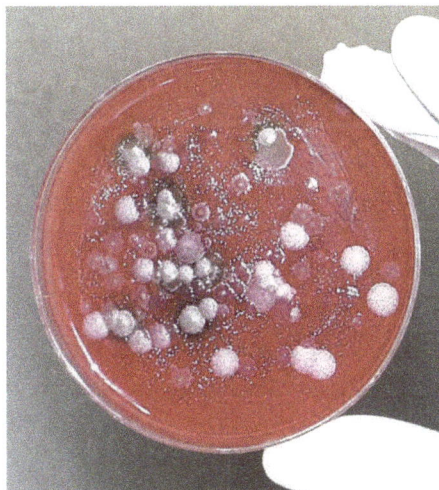

A culture of *Bacillus anthracis*, the anthrax bacillus.

A biological agent—also called bio-agent, biological threat agent, biological warfare agent, biological weapon, or bioweapon—is a bacterium, virus, protozoan, parasite, or fungus that can be used purposefully as a weapon in bioterrorism or biological warfare (BW). In addition to these living and/ or replicating pathogens, biological toxins are also included among the bio-agents. More than 1,200 different kinds of potentially weaponizable bio-agents have been described and studied to date.

Biological agents have the ability to adversely affect human health in a variety of ways, ranging from relatively mild allergic reactions to serious medical conditions, including death. Many of these organisms are ubiquitous in the natural environment where they are found in water, soil, plants, or animals. Bio-agents may be amenable to "weaponization" to render them easier to deploy or disseminate. Genetic modification may enhance their incapacitating or lethal properties, or render them impervious to conventional treatments or preventives. Since many bio-agents reproduce rapidly and require minimal resources for propagation, they are also a potential danger in a wide variety of occupational settings.

The Biological Weapons Convention (1972) is an international treaty banning the use or stockpiling of bio-agents; as of February 2015 there were 171 state signatories. Bio-agents are, however, widely studied for defensive purposes under various biosafety levels and within biocontainment facilities throughout the world. In 2008, according to a U.S. Congressional Research Service report, China, Cuba, Egypt, Iran, Israel, North Korea, Russia, Syria, and Taiwan were considered, with varying degrees of certainty, to be maintaining bio-agents in an offensive BW program capacity.

Classifications

Operational

The former US biological warfare program (1943-1969) categorized its weaponized anti-personnel bio-agents as either Lethal Agents (*Bacillus anthracis*, *Francisella tularensis*, Botulinum toxin) or Incapacitating Agents (*Brucella suis*, *Coxiella burnetii*, Venezuelan equine encephalitis virus, Staphylococcal enterotoxin B). In the next section ("List of biological and toxin agents of military importance") the military symbols for various weaponized agents are given; this nomenclature system began in the US and UK BW programs of World War II and continued as NATO nomenclature through the Cold War.

Legal

Since 1997, United States law has declared a list of bio-agents designated by the U.S. Department of Health and Human Services or the U.S. Department of Agriculture that have the "potential to pose a severe threat to public health and safety" to be officially defined as "select agents" and possession or transportation of them are tightly controlled as such. Select agents are divided into "HHS select agents and toxins", "USDA select agents and toxins" and "Overlap select agents and toxins".

Regulatory

The U.S. Centers for Disease Control and Prevention (CDC) breaks biological agents into three categories: Category A, Category B, and Category C. Category A agents pose the greatest threat to the U.S. Criteria for being a Category A agent include high rates of morbidity and mortality; ease of dissemination and communicability; ability to cause public panic; and special action required by public health officials to respond. Category A agents include anthrax, botulism, plague, smallpox, tularemia, and viral hemorrhagic fevers.

List of Bio-agents of Military Importance

The following pathogens and toxins were weaponized by one nation or another at some time. NATO abbreviations are included where applicable.

Bacterial Bio-agents

Disease	Causative Agent (Military Symbol)	Comments
Anthrax	Bacillus anthracis (N or TR)	
Brucellosis (bovine)	Brucella abortus	
Brucellosis (caprine)	Brucella melitensis (AM or BX)	
Brucellosis (porcine)	Brucella suis (US, AB or NX)	
Cholera	Vibrio cholerae (HO)	
Diphtheria	Corynebacterium diphtheriae (DK)	
Dysentery (bacterial)	Shigella dysenteriae, some species of Escherichia coli (Y)	
Glanders	Burkholderia mallei (LA)	
Listeriosis	Listeria monocytogenes (TQ)	
Melioidosis	Burkholderia pseudomallei (HI)	
Plague	Yersinia pestis (LE)	
Tularemia	Francisella tularensis (SR or JT)	

Chlamydial Bio-agents

Disease	Causative Agent (Military Symbol)	Comments
Psittacosis	Chlamydophila psittaci (SI)	

Rickettsial Bio-agents

Disease	Causative Agent (Military Symbol)	Comments
Q fever	Coxiella burnetii (OU)	
Rocky Mountain spotted fever	Rickettsia rickettsii (RI or UY)	
Typhus (human)	Rickettsia prowazekii (YE)	
Typhus (murine)	Rickettsia typhi (AV)	

Viral Bio-agents

Disease	Causative Agent (Military Symbol)	Comments
Equine Encephalitis (Eastern)	Eastern equine encephalitis virus (ZX)	
Equine Encephalitis (Venezuelan)	Venezuelan Equine Encephalomyelitis virus (FX)	
Equine Encephalitis (Western)	Western equine encephalitis virus (EV)	
Japanese B encephalitis	Japanese encephalitis virus (AN)	
Rift Valley fever	Rift Valley fever virus (FA)	
Smallpox	Variola virus (ZL)	
Yellow fever	Yellow fever virus (OJ or LU)	

Additionally, the Soviet Union is known to have weaponized Marburg virus (MARV) in the 1970s and '80s.

Mycotic Bio-agents

Disease	Causative Agent (Military Symbol)	Comments
Coccidiomycosis	Coccidioides immitis (OC)	

Biological Toxins

Toxin	Source of Toxin (Military Symbol)	Comments
Abrin	Rosary pea (*Abrus precatorius*)	
Botulinum toxins (A through G)	*Clostridium botulinum* bacteria or spores, and several other Clostridial species. (X or XR)	
Ricin	Castor bean (*Ricinus communis*) (W or WA)	
Saxitoxin	Various marine and brackish cyanobacteria, such as *Anabaena, Aphanizomenon, Lyngbya*, and *Cylindrospermopsis* (TZ)	
Staphyloccocal enterotoxin B	*Staphylococcus aureus* (UC or PG)	
Tetrodotoxin	Various marine bacteria, including *Vibrio alginolyticus, Pseudoalteromonas tetraodonis* (PP)	
Trichothecene mycotoxins	Various species of fungi, including *Fusarium, Trichoderma*, and *Stachybotrys*	

Biological Vectors

Vector (Military Symbol)	Disease	Comments
Mosquito (Aedes aegypti) (AP)	Malaria, Dengue fever, Chikungunya, Yellow fever, other Arboviruses	
Oriental rat flea (Xenopsylla cheopis)	Plague, Murine typhus	

Simulants

- MR - molasis residium

- BG - *Bacillus globigii*

- BS - *Bacillus globigii*

- U - *Bacillus globigii*

- SM - *Serratia marcescens*

- P - *Serratia marcescens*

- AF - *Aspergillus fumigatus* mutant C-2

- EC - *E. coli*

- BT - *Bacillus thuringiensis*

- EH - *Erwinia hebicola*

- FP - fluorescent particle

In popular Culture

Microbial Cyst

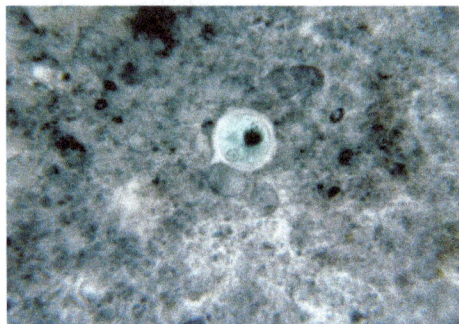

Cyst stage of *Entamoeba histolytica*

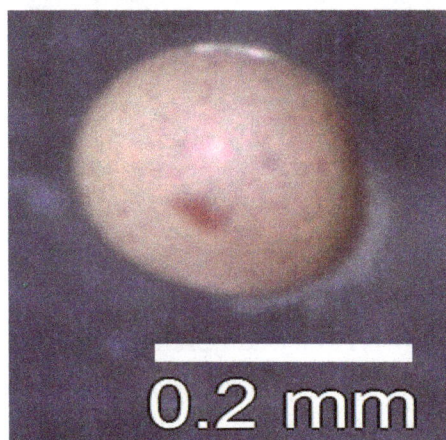

Cyst of *Artemia salina*

A microbial cyst is a resting or dormant stage of a microorganism, usually a bacterium or a protist or rarely an invertebrate animal, that helps the organism to survive in unfavorable environmental conditions. It can be thought of as a state of suspended animation in which the metabolic processes of the cell are slowed down and the cell ceases all activities like feeding and locomotion. Encystment also helps the microbe to disperse easily, from one host to another or to a more favorable environment. When the encysted microbe reaches an environment favorable to its growth and survival, the cyst wall breaks down by a process known as *excystation*.

Unfavorable environmental conditions such as lack of nutrients or oxygen, extreme temperatures, lack of moisture and presence of toxic chemicals, which are not conducive for the growth of the microbe trigger the formation of a cyst.

Cyst Formation Across Species

In Bacteria

In bacteria (for instance, *Azotobacter sp.*), encystment occurs by changes in the cell wall; the cytoplasm contracts and the cell wall thickens. Bacterial cysts differ from endospores in the way they are formed and also the degree of resistance to unfavorable conditions. Endospores are much more resistant than cysts.

In Protists

Protists, especially protozoan parasites, are often exposed to very harsh conditions at various stages in their life cycle. For example, *Entamoeba histolytica*, a common intestinal parasite that causes dysentery, has to endure the highly acidic environment of the stomach before it reaches the intestine and various unpredictable conditions like desiccation and lack of nutrients while it is outside the host. An encysted form is well suited to survive such extreme conditions, although protozoan cysts are less resistant to adverse conditions compared to bacterial cysts. In addition to survival, the chemical composition of certain protozoan cyst walls may play a role in their dispersal. The sialyl groups present in the cyst wall of *Entamoeba histolytica* confer a net negative charge to the cyst which prevents its attachment to the intestinal wall thus causing its elimination in the feces. Other protozoan intestinal parasites like *Giardia lamblia* and *Cryptosporidium* also produce cysts as part of their life cycle. In some protozoans, the unicellular organism multiplies during or after encystment and releases multiple trophozoites upon excystation.

In Nematodes

Some soil-dwelling plant parasitic nematodes, such as the soybean cyst nematode, or the potato cyst nematode form cysts as a normal part of their lifecycle.

Composition of the Cyst Wall

The composition of the cyst wall is variable in different organisms. The cyst walls of bacteria are formed by the thickening of the normal cell wall with added peptidoglycan layers whereas the walls of protozoan cysts are made of chitin, a type of glycopolymer. Nematode cyst walls are composed of chitin reinforced by collagen.

Colony-forming Unit

In microbiology, a colony-forming unit (CFU, cfu, Cfu) is a unit used to estimate the number of viable bacteria or fungal cells in a sample. Viable is defined as the ability to multiply via binary fission under the controlled conditions. Counting with colony-forming units requires culturing the microbes and counts only viable cells, in contrast with microscopic examination which counts all cells, living or dead. The visual appearance of a colony in a cell culture requires significant growth, and when counting colonies it is uncertain if the colony arose from one cell or a group of cells. Expressing results as colony-forming units reflects this uncertainty.

Theory

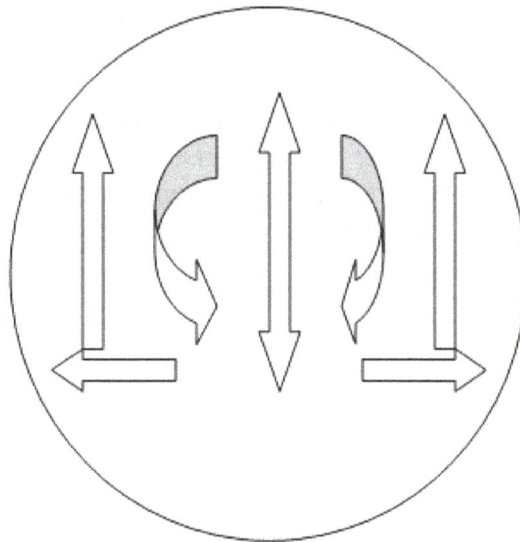

A dilution made with bacteria and peptoned water is placed in an Agar plate (Agar plate count for food samples or Trypticase soy agar for clinic samples) and spread over the plate by tipping in the pattern shown.

The purpose of plate counting is to estimate the number of cells present based on their ability to give rise to colonies under specific conditions of nutrient medium, temperature and time. Theoretically, one viable cell can give rise to a colony through replication. However, solitary cells are the exception in nature, and most likely the progenitor of the colony was a mass of cells deposited together. In addition, many bacteria grow in chains (e.g. Streptococcus) or clumps (e.g. Staphylococcus). Estimation of microbial numbers by CFU will, in most cases, undercount the number of living cells present in a sample for these reasons. This is because the counting of CFU assumes that every colony is separate and founded by a single viable microbial cell.

The plate count is linear for *E. coli* over the range of 30 - 300 CFU on a standard sized Petri dish. Therefore, to ensure that a sample will yield CFU in this range requires dilution of the sample and plating of several dilutions. Typically ten-fold dilutions are used, and the dilution series is plated in replicates of 2 or 3 over the chosen range of dilutions. The CFU/plate is read from a plate in the linear range, and then the CFU/g (or CFU/mL) of the original is deduced mathematically, factoring in the amount plated and its dilution factor.

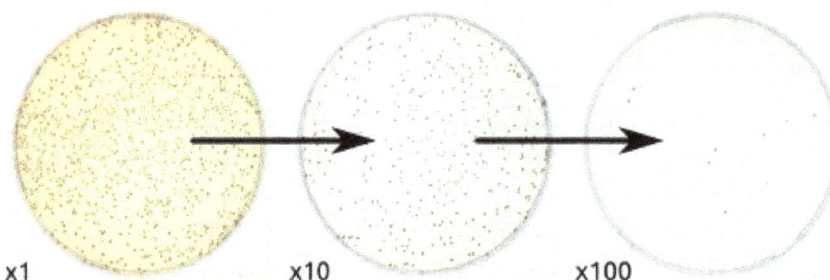

x1   x10   x100

A solution of bacteria at an unknown concentration is often serially diluted in order to obtain at least one plate with a countable number of bacteria. In this figure, the "x10" plate is suitable for counting.

An advantage to this method is that different microbial species may give rise to colonies that are clearly different from each other, both microscopically and macroscopically. The colony morphology can be of great use in the identification of the microorganism present.

A prior understanding of the microscopic anatomy of the organism can give a better understanding of how the observed CFU/mL relates to the number of viable cells per milliliter. Alternatively it is possible to decrease the average number of cells per CFU in some cases by vortexing the sample before conducting the dilution. However many microorganisms are delicate and would suffer a decrease in the proportion of cells that are viable when placed in a vortex.

Log Notation

Concentrations of colony-forming units can be expressed using logarithmic notation, where the value shown is the base 10 logarithm of the concentration.

Uses

Colony-forming units are used to quantify results in many microbiological plating and counting methods, including:

- The Pour Plate method wherein the sample is suspended in a petri dish using molten agar cooled to approximately 40-45 °C (just above the point of solidification to minimize heat-induced cell death). After the nutrient agar solidifies the plate is incubated.

- The Spread Plate method wherein the sample (in a small volume) is spread across the surface of a nutrient agar plate and allowed to dry before incubation for counting.

- The Membrane Filter method wherein the sample is filtered through a membrane filter, then the filter placed on the surface of a nutrient agar plate (bacteria side up). During incubation nutrients leach up through the filter to support the growing cells. As the surface area of most filters is less than that of a standard petri dish, the linear range of the plate count will be less.

- The Miles and Misra Methods or drop-plate method wherein a very small aliquot (usually about 10 microliters) of sample from each dilution in series is dropped onto a petri dish. The drop dish must be read while the colonies are very small to prevent the loss of CFU as they grow together.

However, with the techniques that require the use of an agar plate, no fluid solution can be used because the purity of the specimen cannot be unidentified and it is not possible to count the cells one by one in the liquid.

Tools for Counting Colonies

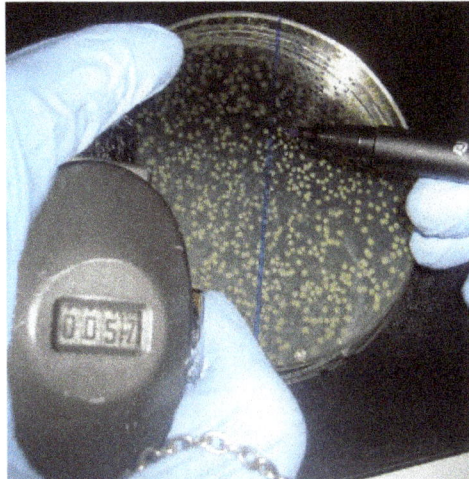

The traditional way of enumerating CFUs with a "click-counter" and a pen. When the colonies are too numerous, it is frequent to count CFUs only on a fraction of the dish.

Counting colonies is traditionally performed manually using a pen and a click-counter. This is generally a straightforward task, but can become very laborious and time-consuming when many plates have to be enumerated. Alternatively semi-automatic (software) and automatic (hardware + software) solutions can be used.

Software for Counting CFUs

Colonies can be enumerated from pictures of plates using software tools. The experimenters would generally take a picture of each plate they need to count and then analyse all the pictures (this can be done with a simple digital camera or even a webcam). Since it takes less than 10 seconds to take a single picture, as opposed to several minutes to count CFU manually, this approach generally saves a lot of time. In addition, it is more objective and allows extraction of other variables such as the size and colour of the colonies.

- OpenCFU is a free and open-source program designed to optimise user friendliness, speed and robustness. It offers a wide range of filters and control as well as a modern user interface. OpenCFU is written in C++ and uses OpenCV for image analysis.

- NICE is a program written in MATLAB providing an easy way to count colonies from images.

- ImageJ and CellProfiler: Some ImageJ macros and plugins and some CellProfiler pipelines can be used to count colonies. This often requires the user to change the code in order to achieve an efficient work-flow, but can prove useful and flexible. One main issue is the absence of specific GUI which can make the interaction with the processing algorithms tedious.

Automated Systems

Many of the automated systems are used to counteract human error as many of the research techniques done by humans counting individual cells have a high chance of error involved. Due to the fact that researchers regularly manually count the cells with the assistance of a transmitted light, this error prone technique can have a significant effect on the calculated concentration in the main liquid medium when the cells are in low numbers.

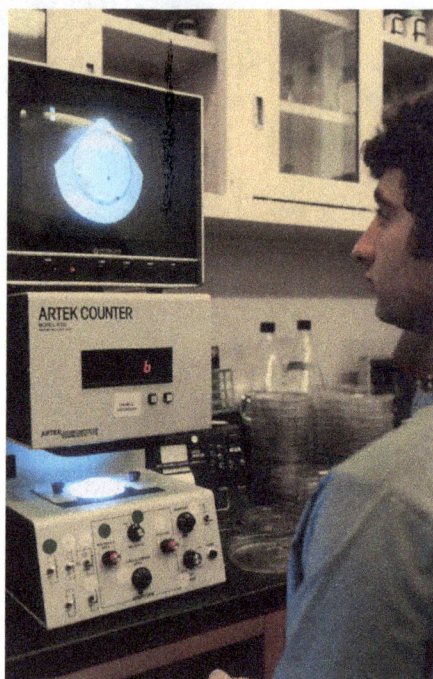

An automated colony counter using image processing.

Completely automated systems are also available from some biotechnology manufacturers. They are generally expensive and not as flexible as standalone software since the hardware and software are designed to work together for a specific set-up. Alternatively, some automatic systems use the spiral plating paradigm.

Some of the automated systems such as the systems from MATLAB allow the cells to be counted without having to stain them. This lets the colonies to be reused for other experiments without the risk of killing the microorganisms with stains. However, a disadvantage to these automated systems is that it is extremely difficult to differentiate between the microorganisms with dust or scratches on blood agar plates because both the dust and scratches can create a highly diverse combination of shapes and appearances.

Alternative Units

Instead of colony-forming units, the parameters Most Probable Number (MPN) and Modified Fishman Units (MFU) can be used. The Most Probable Number method counts viable cells and is useful when enumerating low concentrations of cells or enumerating microbes in products where particulates make plate counting impractical. Modified Fishman Units take into account bacteria which are viable, but non-culturable.

Viral Eukaryogenesis

Viral eukaryogenesis is the hypothesis that the cell nucleus of eukaryotic life forms evolved from a large DNA virus in a form of endosymbiosis within a methanogenic archaeon. The virus later evolved into the eukaryotic nucleus by acquiring genes from the host genome and eventually usurping its role. The hypothesis was proposed by Philip Bell in 2001 and gained support as large, complex DNA viruses (such as Mimivirus) capable of protein biosynthesis were discovered.

While viruses remain one of the more poorly understood aspects of biology, recent genomic research and the discovery of complex DNA viruses indicate that they may have played a role in the development of eukaryotic nuclei. This research further opens a hotly debated question: Are viruses living organisms? Most virologists took the stand that viruses were not alive; currently, viruses are hypothesized by some researchers to have been ancestors of modern eukaryotic cells—most importantly with respect to DNA, the shared genetic code of all eukaryotes and prokaryotes alive today.

Hypothesis

Like viruses, a eukaryotic nucleus contains linear chromosomes with specialized end sequences (in contrast to bacterial genomes, which have a circular topology); it uses mRNA capping, and separates transcription from translation. Eukaryotic nuclei are also capable of cytoplasmic replication. Some large viruses have their own DNA-directed RNA polymerase. Transfers of "infectious" nuclei have been documented in many parasitic red algae. Complex eukaryotic DNA viruses could also have originated through nuclear viriogenesis.

The viral eukaryogenesis hypothesis posits that eukaryotes are composed of three ancestral elements: a viral component that preceded the modern nucleus; another eukaryotic cell that forms the cytoplasm of modern cells; and a bacterium that, by endocytosis, became the modern mitochondrion. Molecular homology suggests that all life shares a single common ancestor, and evidence, alongside the fossil record, can be found in the evolution of prokaryotes, and Archaea prior to the eukaryotes. The viral eukaryogenesis hypothesis points to the cell cycle of eukaryotes, particularly sex and meiosis, as evidence. Little is known about DNA or the origins of reproduction in prokaryotic or eukaryotic cells. It is thus possible that viruses were involved in the creation of Earth's first cells.

In 2006, researchers suggested that the transition from RNA to DNA genomes first occurred in the viral world. A DNA-based virus may have provided storage for an ancient host that had previously used RNA to store its genetic information. Viruses may initially have adopted DNA as a way to resist RNA-degrading enzymes in the host cells. Hence, the contribution from such a new component may have been as significant as the contribution from chloroplasts or mitochondria. Following this hypothesis, archaea, Bacteria, and eukaryotes each obtained their DNA informational system from a different virus. The RNA cell at the origin of eukaryotes was probably more complex, featuring RNA processing. It has also been suggested that telomerase and telomeres, key aspects of eukaryotic cell replication, have viral origins. Further, the viral origins of the modern eukaryotic nucleus may have relied on multiple infections of archaeal cells carrying bacterial mitochondrial precursors with lysogenic viruses.

The viral eukaryogenesis hypothesis depicts a model of eukaryotic evolution in which a virus, similar to a modern pox virus, evolved into a nucleus via gene acquisition from existing bacterial and archaeal species. The lysogenic virus then became the information storage center for the cell, while the cell retained its capacities for gene translation and general function despite the viral genome's entry. Similarly, the bacterial species involved in this eukaryogenesis retained its capacity to produce energy in the form of ATP while also passing much of its genetic information into this new virus-nucleus organelle. It is hypothesized that the modern cell cycle, whereby mitosis, meiosis, and sex occur in all eukaryotes, evolved because of the balances struck by viruses, which characteristically follow a pattern of tradeoff between infecting as many hosts as possible and killing an individual host through viral proliferation. Hypothetically, viral replication cycles may mirror those of plasmids and viral lysogens. However, this theory is controversial, and additional experimentation involving archaeal viruses is necessary, as they are probably the most evolutionarily similar to modern eukaryotic nuclei.

Implications

A number of precepts in the theory are possible. For instance, a helical virus with a bilipid envelope bears a distinct resemblance to a highly simplified cellular nucleus (i.e., a DNA chromosome encapsulated within a lipid membrane). To consider the concept logically, a large DNA virus would take control of a bacterial or archaeal cell. Instead of replicating and destroying the host cell, it would remain within the cell, thus overcoming the tradeoff paradox typically faced by viruses. With the virus in control of the host cell's molecular machinery, it would effectively become a functional nucleus. Through the processes of mitosis and cytokinesis, the virus would thus hijack the entire cell—an extremely favorable way to ensure its survival.

References

- Brown, Amy Christian (2007). Understanding Food: Principles and Preparation (3 ed.). Cengage Learning. p. 546. ISBN 978-0-495-10745-3.

- Margulis, Lynn; Chapman, Michael J. (2009). Kingdoms & domains an illustrated guide to the phyla of life on Earth (4th ed.). Amsterdam: Academic Press/Elsevier. p. 493. ISBN 978-0-08-092014-6. Retrieved 2 August 2016.

- Madigan, Michael T. (2012). Brock biology of microorganisms (13th ed.). San Francisco: Benjamin Cummings. ISBN 9780321649638.

- Sherwood, Linda; Willey, Joanne; Woolverton, Christopher (2013). Prescott's Microbiology (9th ed.). New York: McGraw Hill. pp. 713–721. ISBN 9780073402406. OCLC 886600661.

- Kloepper, J. W (1993). "Plant growth-promoting rhizobacteria as biological control agents". In Metting, F. B. Jr. Soil microbial ecology: applications in agricultural and environmental management. New York: Marcel Dekker Inc. pp. 255–274. ISBN 0-8247-8737-4.

- Marchesi, J. R. (2010). "Prokaryotic and Eukaryotic Diversity of the Human Gut". Advances in Applied Microbiology Volume 72. Advances in Applied Microbiology. 72. pp. 43–62. doi:10.1016/S0065-2164(10)72002-5. ISBN 9780123809896. PMID 20602987.

- Sherwood, Linda; Willey, Joanne; Woolverton, Christopher (2013). Prescott's Microbiology (9th ed.). New York: McGraw Hill. pp. 713–721. ISBN 9780073402406. OCLC 886600661.

- Schwiertz, Andreas (2016). Microbiota of the human body : implications in health and disease. Switzerland: Springer. p. 1. ISBN 978-3-319-31248-4.

- Eugene W. Nester, Denise G. Anderson, C. Evans Roberts Jr., Nancy N. Pearsall, Martha T. Nester; Microbiology: A Human Perspective, 2004, Fourth Edition, ISBN 0-07-291924-8

- Kenneth Alibek and S. Handelman. Biohazard: The Chilling True Story of the Largest Covert Biological Weapons Program in the World - Told from Inside by the Man Who Ran it. 1999. Delta (2000) ISBN 0-385-33496-6.

- Samuel Baron MD, Rhonda C. Peake, Deborah A. James, Mardelle Susman, Carol Ann Kennedy, Mary Jo Durson Singleton, Steve Schuenke; Medical Microbiology; Fourth Edition, ISBN 0-9631172-1-1 (hardcover)1996

- Goldman, Emanuel; Green, Lorrence H (24 August 2008). Practical Handbook of Microbiology, Second Edition (Google eBook) (Second ed.). USA: CRC Press, Taylor and Francis Group. p. 864. ISBN 978-0-8493-9365-5. Retrieved 2014-10-16

Theories of Microbiology

The theories of microbiology elucidated in this section are germ theory of disease, fermentation theory, hologenome theory of evolution and germ theory denialism. The germ theory of disease claims that some microorganisms cause diseases in organisms. The growth of these microorganisms in their hosts causes diseases. The following chapter will not only provide an overview, it will also delve into the topics related to it.

Germ Theory of Disease

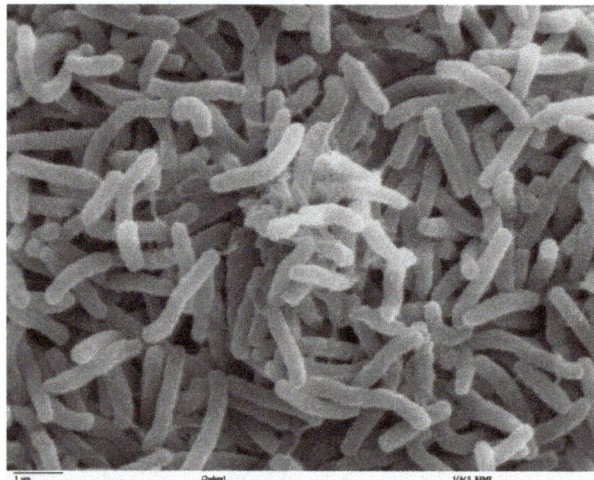

Scanning electron microscope image of *Vibrio cholerae*. This is the bacterium that causes cholera.

The germ theory of disease states that some diseases are caused by microorganisms. These small organisms, too small to see without magnification, invade humans, animals, and other living hosts. Their growth and reproduction within their hosts can cause a disease. "Germ" may refer to not just a bacterium but to any type of microorganisms, especially one which causes disease, such as protist, fungus, virus, prion, or viroid. Microorganisms that cause disease are called pathogens, and the diseases they cause are called infectious diseases. Even when a pathogen is the principal cause of a disease, environmental and hereditary factors often influence the severity of the disease, and whether a particular host individual becomes infected when exposed to the pathogen.

The germ theory was proposed by Girolamo Fracastoro in 1546, but scientific evidence in support of this accumulated slowly and Galen's miasma theory remained dominant among scientists and doctors. A transitional period began in the late 1850s as the work of Louis Pasteur and Robert Koch provided convincing evidence; by 1880, miasma theory was still competing with the germ theory of disease. Eventually, a "golden era" of bacteriology ensued, in which the theory quickly led to the identification of the actual organisms that cause many diseases. Viruses were discovered in the 1890s.

Miasma Theory

A representation of the cholera epidemic of the 19th century depicts the spread of the disease in the form of poisonous air.

The miasma theory of disease transmission held that diseases such as cholera, chlamydia or the Black Death were caused by a *miasma*, a noxious form of "bad air". The theory held that the origin of these epidemic diseases was a miasma, emanating from rotting organic matter. Miasma was considered to be a poisonous vapor or mist filled with particles from decomposed matter (miasmata) that caused illnesses. The miasmatic position was that diseases were the product of environmental factors such as contaminated water, foul air, and poor hygienic conditions. Such infection was not passed between individuals but would affect individuals within the locale that gave rise to such vapors. It was identifiable by its foul smell.

This was the predominant theory of disease transmission before the germ theory of disease took hold in the last decade of the 19th century.

Development

Pre-19th Century

There is some evidence that the Ancient Romans had a germ theory of disease that was later forgotten. Notably, Marcus Terentius Varro wrote, in his 'Rerum rusticarum libri III': "Precautions must also be taken in the neighborhood of swamps [...] because there are bred certain minute creatures which cannot be seen by the eyes, which float in the air and enter the body through the mouth and nose and there cause serious diseases."

Girolamo Fracastoro proposed in 1546 that epidemic diseases are caused by transferable seed-like entities that transmit infection by direct or indirect contact, or even without contact over long distances.

Italian physician Francesco Redi provided early evidence against spontaneous generation. He devised an experiment in 1668 in which he used three jars. He placed a meatloaf and egg in each of the three jars. He had one of the jars open, another one tightly sealed, and the last one covered with gauze. After a few days, he observed that the meatloaf in the open jar was covered by maggots, and the jar covered with gauze had maggots on the surface of the gauze. However, the tightly sealed jar had no maggots inside or outside it. He also noticed that the maggots were found only on surfaces that were accessible by flies. From this he concluded that spontaneous generation is not a plausible theory.

Microorganisms are said to be first directly observed in the 1670s by Anton van Leeuwenhoek, an early pioneer in microbiology. Yet Athanasius Kircher may have done so prior. "When Rome was struck by the bubonic plague in 1656, Kircher spent days on end caring for the sick. Searching for a cure, Kircher observed microorganisms under the microscope and invented the germ theory of disease, which he outlined in his *Scrutinium pestis physico-medicum* (Rome 1658)." Building on Leeuwenhoek's work, physician Nicolas Andry argued in 1700 that microorganisms he called "worms" were responsible for smallpox and other diseases.

In 1720, Richard Bradley theorised that plague and 'all pestilential distempers' were caused by 'poisonous insects', living creatures viewable only with the help of microscopes.

In a 1767 report to the College of Physicians in London, John Zephaniah Holwell mentions the practice of Smallpox vaccinations by Ayurvedic doctors and their explanations of the cause of the disease. According to his report, the Ayurvedic doctors explained to him that disease is caused by 'multitudes of imperceptible animalculae (microorganisms) floating in the atmosphere; that these are the cause of all epidemical diseases'. They further explained to him that they (microorganisms) circulate in and out of the bodies of all animals during respiration and that the severity of disease is proportional to the air charged with the animalculae, and the quantity of them received with the food' He was convinced by the physicians that when the disease of smallpox has run its course either naturally or after inoculations in its weak form, the patients were safe. The major difference between the two being that while natural infections are mostly fatal, the inoculations were merely an inconvenience. According to Henderson and Moss the practice of this method of inoculation has been prevalent in India from just before 1000 A.D.

Agostino Bassi

The Italian Agostino Bassi was the first person to prove that a disease was caused by a microorganism when he conducted a series of experiments between 1808 and 1813, demonstrating that a "vegetable parasite" caused a disease in silkworms known as calcinaccio. This disease was devastating the French silk industry at the time. The "vegetable parasite" is now known to be a fungus pathogenic to insects called *Beauveria bassiana* (named after Bassi).

Ignaz Semmelweis

Ignaz Semmelweis was a Hungarian obstetrician working at the Vienna General Hospital (Allgemeines Krankenhaus) in 1847, when he noticed the dramatically high incidence of death from puerperal fever among women who delivered at the hospital with the help of the doctors and medical students. Births attended by the midwives were relatively safe. Investigating further, Semmelweis made the connection between puerperal fever and examinations of delivering women by doctors, and further realized that these physicians had usually come directly from autopsies. Asserting that puerperal fever was a contagious disease and that matter from autopsies were implicated in its development, Semmelweis made doctors wash their hands with chlorinated lime water before examining pregnant women, thereby reducing mortality from childbirth from 18% to 2.2% at his hospital. Nevertheless, he and his theories were rejected by most of the contemporary medical establishment.

John Snow

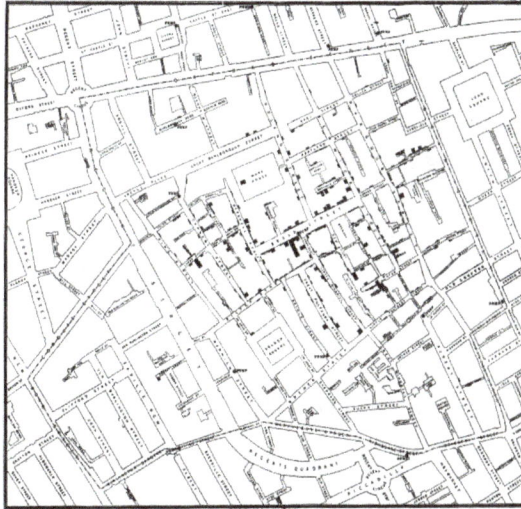

Original map by John Snow showing the clusters of cholera cases in the London epidemic of 1854

John Snow was a skeptic of the then-dominant miasma theory. Even though the germ theory of disease pioneered by Girolamo Fracastoro had not yet achieved full development or widespread currency, Snow demonstrated a clear understanding of germ theory in his writings. He first published his theory in an 1849 essay *On the Mode of Communication of Cholera,* in which he correctly suggested that the fecal-oral route was the mode of communication, and that the disease replicated itself in the lower intestines. He even proposed in his 1855 edition of the work, that the structure of cholera was that of a cell.

Having rejected effluvia and the poisoning of the blood in the first instance, and being led to the conclusion that the disease is something that acts directly on the alimentary canal, the excretions of the sick at once suggest themselves as containing some material which being accidentally swallowed might attach itself to the mucous membrane of the small intestines, and there multiply itself by appropriation of surrounding matter, in virtue of molecular changes going on within it, or capable of going on, as soon as it is placed in congenial circumstances.

—John Snow (1849)

Snow's 1849 recommendation that water be "filtered and boiled before it is used" is one of the first practical applications of germ theory in the area of public health and is the antecedent to the modern boil water advisory.

In 1855 he published a second edition of his article, documenting his more elaborate investigation of the effect of the water supply in the Soho, London epidemic of 1854.

By talking to local residents, he identified the source of the outbreak as the public water pump on Broad Street (now Broadwick Street). Although Snow's chemical and microscope examination of a water sample from the Broad Street pump did not conclusively prove its danger, his studies of the pattern of the disease were convincing enough to persuade the local council to disable the well pump by removing its handle. This action has been commonly credited as ending the outbreak, but Snow observed that the epidemic may have already been in rapid decline.

Snow later used a dot map to illustrate the cluster of cholera cases around the pump. He also used statistics to illustrate the connection between the quality of the water source and cholera cases. He showed that the Southwark and Vauxhall Waterworks Company was taking water from sewage-polluted sections of the Thames and delivering the water to homes, leading to an increased incidence of cholera. Snow's study was a major event in the history of public health and geography. It is regarded as one of the founding events of the science of epidemiology.

Later, researchers discovered that this public well had been dug only three feet from an old cesspit, which had begun to leak fecal bacteria.The diapers of a baby, who had contracted cholera from another source, had been washed into this cesspit. Its opening was originally under a nearby house, which had been rebuilt farther away after a fire. The city had widened the street and the cesspit was lost. It was common at the time to have a cesspit under most homes. Most families tried to have their raw sewage collected and dumped in the Thames to prevent their cesspit from filling faster than the sewage could decompose into the soil.

After the cholera epidemic had subsided, government officials replaced the handle on the Broad Street pump. They had responded only to the urgent threat posed to the population, and afterward they rejected Snow's theory. To accept his proposal would have meant indirectly accepting the fecal-oral method transmission of disease, which they dismissed as "too depressing."

Louis Pasteur

The more formal experiments on the relationship between germ and disease were conducted by Louis Pasteur between 1860 and 1864. He discovered the pathology of the puerperal fever and the pyogenic vibrio in the blood, and suggested using boric acid to kill these microorganisms before and after confinement.

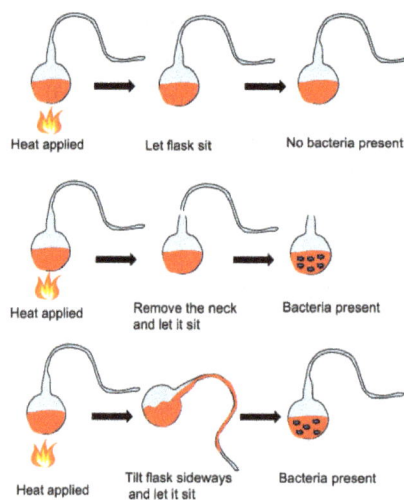

Louis Pasteur's pasteurization experiment illustrates the fact that the spoilage of liquid was caused by particles in the air rather than the air itself. These experiments were importantpieces of evidence supporting the idea of Germ Theory of Disease.

Pasteur further demonstrated between 1860 and 1864 that fermentation and the growth of microorganisms in nutrient broths did not proceed by spontaneous generation. He exposed freshly boiled broth to air in vessels that contained a filter to stop all particles passing through to the growth medium, and even with no filter at all, with air being admitted via a long tortuous tube that would not pass dust particles. Nothing grew in the broths: therefore the living organisms that grew in such broths came from outside, as spores on dust, rather than being generated within the broth.

Pasteur discovered that another serious disease of silkworms, pébrine, was caused by a small microscopic organism now known as Nosema *bombycis* (1870). Pasteur saved the silk industry in France by developing a method to screen silkworms eggs for those that are not infected, a method that is still used today to control this and other silkworm diseases.

Robert Koch

Robert Koch is known for developing four basic criteria (known as Koch's postulates) for demonstrating, in a scientifically sound manner, that a disease is caused by a particular organism. These postulates grew out of his seminal work with anthrax using purified cultures of the pathogen that had been isolated from diseased animals.

Koch's postulates were developed in the 19th century as general guidelines to identify pathogens that could be isolated with the techniques of the day. Even in Koch's time, it was recognized that some infectious agents were clearly responsible for disease even though they did not fulfill all of the postulates. Attempts to rigidly apply Koch's postulates to the diagnosis of viral diseases in the late 19th century, at a time when viruses could not be seen or isolated in culture, may have impeded the early development of the field of virology. Currently, a number of infectious agents are accepted as the cause of disease despite their not fulfilling all of Koch's postulates. Therefore, while Koch's postulates retain historical importance and continue to inform the approach to microbiologic diagnosis, fulfillment of all four postulates is not required to demonstrate causality.

Koch's postulates have also influenced scientists who examine microbial pathogenesis from a molecular point of view. In the 1980s, a molecular version of Koch's postulates was developed to guide the identification of microbial genes encoding virulence factors.

Koch's Postulates:

1. The microorganism must be found in abundance in all organisms suffering from the disease, but should not be found in healthy organisms.

2. The microorganism must be isolated from a diseased organism and grown in pure culture.

3. The cultured microorganism should cause disease when introduced into a healthy organism.

4. The microorganism must be reisolated from the inoculated, diseased experimental host and identified as being identical to the original specific causative agent.

However, Koch abandoned the universalist requirement of the first postulate altogether when he discovered asymptomatic carriers of cholera and, later, of typhoid fever. Asymptomatic or subclinical infection carriers are now known to be a common feature of many infectious diseases, especially viruses such as polio, herpes simplex, HIV, and hepatitis C. As a specific example, all doctors and virologists agree that poliovirus causes paralysis in just a few infected subjects, and the success of the polio vaccine in preventing disease supports the conviction that the poliovirus is the causative agent.

The third postulate specifies "should", not "must", because as Koch himself proved in regard to both tuberculosis and cholera, not all organisms exposed to an infectious agent will acquire the infection. Noninfection may be due to such factors as general health and proper immune functioning; acquired immunity from previous exposure or vaccination; or genetic immunity, as with the resistance to malaria conferred by possessing at least one sickle cell allele.

The second postulate may also be suspended for certain microorganisms or entities that cannot (at the present time) be grown in pure culture, such as prions responsible for Creutzfeldt–Jakob disease. In summary, a body of evidence that satisfies Koch's postulates is sufficient but not necessary to establish causation.

Sanitation

In the 1870s, Joseph Lister was instrumental in developing practical applications of the germ theory of disease with respect to sanitation in medical settings and aseptic surgical techniques—partly through the use of carbolic acid (phenol) as an antiseptic.

Fermentation Theory

Louis Pasteur, Archives Photographiques

The fermentation theory was studied in depth and brought to light first by Louis Pasteur. This theory states that it is the idea or concept of how fermentation is brought on by microbes and put to the concept of spontaneous generation to rest. Even though this theory is now outdated and has been replaced by the germ theory of disease, for a long time it held true, and Louis was on the forefront of explaining why it seemed organisms appeared out of nothing instead of claiming it was just a spontaneous act of God. Fermentation was a process that has been used for thousands of years, but no one could explain exactly what was happening and why. From Pasteur's discovery of why and how fermentation occurs, the process has been studied intensely and is now a mastered art used in everyday life with processes of making things such as alcoholic beverages, some foods like yogurt or even manufacturing some medications.

Fermentation

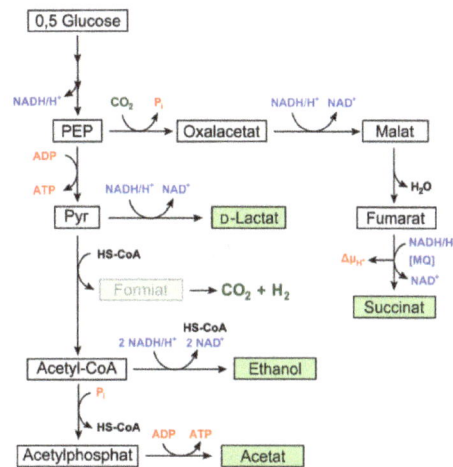

Process of Fermentation

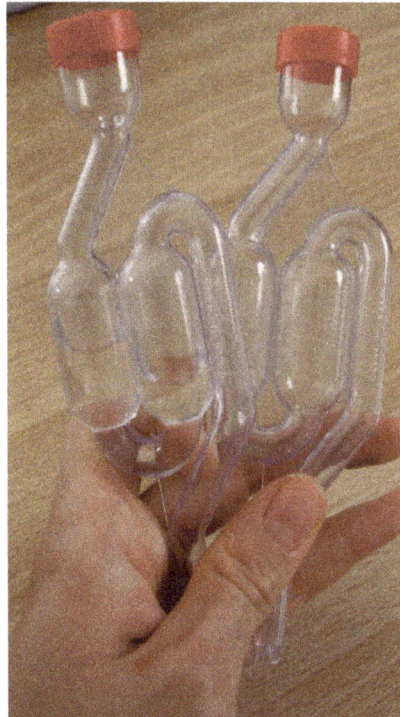

Example of a typical curved neck used today

Simply put, fermentation is the anaerobic metabolic process that converts sugar into acids, gases, or alcohols. This metabolic process is used in oxygen starved environments. Yeast and many other microbes commonly use this process in order to carry our their anaerobic respiration to survive. Even the human body carries out fermentation processes from time to time. When someone runs a long distance race, lactic acid will build up in their muscles over the course of the race. That lactic acid is the by-product of fermentation taking place in their body, which tries to produce ATP so the body can continue to run since they could not process the oxygen intake fast enough. Although fermentation will give a lower yield of ATP production than aerobic respiration does, it can occur at a much higher rate. Fermentation has been used by humans consciously since around 5000 BCE where there were jars recovered in the Iran Zagros Mountains area in which contained remnants of a microbes similar those present in the process of making grapes into wine.

What the Theory is

Before the 1870s, when Pasteur published his work on this theory, it was believe that microorganisms and even some small animals such as frogs would spontaneously appear, which was coined as spontaneous generation. Spontaneous generation was the explained theory that when elements of the Earth such as clay or mud would mix with water and sunlight in certain amounts, creatures would just appear out of that concoction. A common way that this idea was "proven" over and over again was by taking a piece of raw meat and placing it in open air, which would almost always produce maggots. This idea was accepted and believe to be true before Louis Pasteur shook the Earth with his new ideas that organisms actually came from traceable beginnings. Pasteur demonstrated that fermentation is caused by the growth of microorganisms, and the emergent growth of bacteria in nutrient broths is due to biogenesis rather than spontaneous generation. He exposed boiled broths to air in vessels that contained a filter to prevent all particles from passing through to the growth medium. Yet, when the vessels were open to the air surrounding it, the organisms appeared. It could be concluded that spontaneous generation could be disproven. The organisms did not just appear but were coming from the air, yet we were not able to see them at such a small level. His famous experiment was used with a curved neck placed on top of a beaker. This curved neck was the key to proving his findings because it showed that the germs and microbes had to fall into the broth inside. The curved neck did not allow this to happen.

The fermentation theory of disease is the (now obsolete) concept that many diseases, including the diseases which were "epidemic, endemic and contagious", owe their origin to the presence of a "morbific principle" in the system, acting in a manner analogous to, although not identical with, the process of fermentation. It was rendered obsolete by the germ theory of disease, which led to the new science of bacteriology.

Application

Fermenting a broth in a beaker has become more than just a way to prove that organisms don't just appear out of thin air. Today, the process of fermentation is used for a multitude of everyday applications. Some of those processes include medications which we ingest, beverages we consume, and even food we eat. Currently, companies like Genencor International uses the production of enzymes involved in fermentation to build a revenue of over $400 million a year. Many medications such as antibiotics are produced by the fermentation process. An example is the important drug cortisone, which can be prepared by the fermentation of a plant steroid known as diosgenin.

Burton Union fermentation system, Coors Visitor Centre

The enzymes used in the reaction are provided by the mold Rhizopus nigricans. Just as it is commonly known, alcohol of all types and brands are also produced by way of fermentation and distillation. Moonshine is a classic example of how this is carried out. Finally, foods such as yogurt are made by fermentation processes as well. Yogurt is a fermented milk product that contains the characteristic bacterial cultures Lactobacillus bulgaricus and Streptococcus thermopiles.

Hologenome Theory of Evolution

The hologenome theory of evolution, also known as the hologenome concept of evolution, recasts the individual animal or plant (and other multicellular organisms) as a community or a "holobiont" - the host plus all of its symbiotic microbes. Consequently, the collective genomes of the holobiont form a "hologenome". Holobionts and hologenomes are structural entities that replace misnomers in the context of host-microbiota symbioses such as superorganism (i.e., an integrated social unit composed of conspecifics), organ, and metagenome. Variation in the hologenome may encode phenotypic plasticity of the holobiont and can be subject to evolutionary changes caused by selection and drift, if portions of the hologenome are transmitted between generations with reasonable fidelity. One of the important outcomes of recasting the individual as a holobiont subject to evolutionary forces is that genetic variation in the hologenome can be brought about by changes in the host genome and also by changes in the microbiome, including new acquisitions of microbes, horizontal gene transfers, and changes in microbial abundance within hosts. Although there is a rich literature on binary host–microbe symbioses, the hologenome concept distinguishes itself by including the vast symbiotic complexity inherent in many multicellular hosts. For recent literature on holobionts and hologenomes published in an open access platform.

The Origin of the Hologenome Theory

In 1991, Lynn Margulis coined the term holobiont in a book chapter entitled Symbiogenesis and

Symbionticism in *Symbiosis as a Source of Evolutionary Innovation: Specation and Morphogenesis* (MIT Press).

In September 1994, Richard Jefferson first introduced the term hologenome at a presentation at Cold Spring Harbor Laboratory during a Symposium "A Decade of PCR", published by Cold Spring Harbor Laboratory Press as a video series. At the CSH Symposium and earlier, the unsettling number and diversity of microbes that were being discovered through the powerful tool of PCR-amplification of 16S ribosomal RNA genes was exciting, but confusing interpretations in diverse studies. A number of speakers referred to microbial contributions to mammalian or plant DNA samples as 'contamination'. In his lecture, Jefferson argued that these were likely not contamination, but rather essential components of the samples that reflected the actual genetic composition of the organism being studied, and that the logic of the organism's performance and capabilities would be embedded only in the hologenome. Observations on the ubiquity of microbes in plant and soil samples as well as laboratory work on molecular genetics of vertebrate-associated microbial enzymes informed this assertion.

In 2008, Eugene Rosenberg and Ilana Zilber-Rosenburg independently derived the term hologenome and developed the hologenome theory of evolution. This theory was originally based on the pair's observations of Vibrio shiloi-mediated bleaching of the coral Oculina patagonica; and since its first introduction, the theory has been promoted as a fusion of Lamarckism and Darwinism and expanded to all of evolution, not just that of corals. The history of the development of the hologenome theory and the logic undergirding its development was the focus of a cover article by Carrie Arnold in New Scientist in January, 2013. The most comprehensive treatment of the theory, including updates by the Rosenbergs on neutrality, pathogenesis and multi-level selection, can be found in their 2013 book.

In 2013, Robert Brucker and Seth Bordenstein re-invigorated the hologenome concept by showing that gut microbiomes, for the first time, are distinguishable between very closely related Nasonia wasp species and contribute to hybrid death. This publication has formed the basis of the recent resurgence and discussion in holobionts and hologenomes, namely by equating interactions between hosts and microbes as part of a conceptual continuum to interactions between genes in the same genome. In 2015, Seth R. Bordenstein and Kevin R. Theis outlined additional history and summarized a conceptual framework that aligns with pre-existing theories in biology. They discuss questions and critiques, show how neutral and selective processes are applicable to the intellectual framework, and discuss future research questions.

Observations from Vertebrate Biology Supporting Hologenome Theory

Multicellular life is made possible by the coordination of physically and temporally distinct processes, most prominently through hormones. Hormones mediate critical activities in vertebrates, including ontogeny, somatic and reproductive physiology, sexual development, performance and behaviour.

Many of these hormones - including most steroids and thyroxines - are secreted, typically as inactivated hormone conjugates to which a glucuronic acid or sulfate has been attached, through the endocrine and apocrine systems into epithelial corridors in which microbiota are widespread and diverse, including gut, urinary tract, lung and skin. There, the inactivated hormones can be re-activated by cleavage of the glucuronide or sulfate residue, allowing the newly re-activated hormone to be reabsorbed. Thus the concentration and bioavailability of many of the hormones is impacted by microbial cleavage of conjugated intermediaries, itself determined by a highly complex and diverse population with redundant enzymatic capabilities. Aspects of this phenomenon of entero-hepatic circulation have been known for decades, and represented in a very large and complex literature, but had previously been viewed as an ancillary effect of detoxification and excretion of metabolites and xenobiotics, including effects on lifetimes of pharmaceuticals, including birth control formulations.

The basic premise is that through the microbial population structure and activity and the diversity of enzyme specificities and kinetic parameters produced by these populations, a spectrum of hormones can be re-activated and resorbed from epithelia, potentially modulating effective time and dose relationships of many vertebrate hormones. The ability to alter and modulate, amplify and suppress, disseminate and recruit new capabilities as microbially-encoded 'traits' means that sampling, sensing and responding to the environment become intrinsic features and emergent capabilities of the holobiont, with mechanisms that can provide rapid, sensitive, nuanced and persistent performance changes.

Studies by Froebe et al. in 1990 indicating that essential mating pheromones, including androstenols, required activation by skin-associated microbial glucuronidases and sulfatases. In the absence of microbial populations in the skin, no detectable aromatic pheromone was released, as the pro-pheromone remained water-soluble and non-volatile. This effectively meant that the microbes in the skin were essential to produce a mating signal.

Observations from Coral Biology Supporting a Hologenome Theory

Subsequent re-articulation describing the hologenome theory by Rosenberg and Zilber-Rosenberg, published 13 years after Jefferson's definition of the theory, was based on their observations of corals, and the coral probiotic hypothesis.

Unbleached and bleached coral

Coral reefs are the largest structures created by living organisms, and contain abundant and highly complex microbial communities. A coral "head" is a colony of genetically identical polyps, which secrete an exoskeleton near the base. Depending on the species, the exoskeleton may be hard, based on calcium carbonate, or soft and proteinaceous. Over many generations, the colony creates a large skeleton that is characteristic of the species. Diverse forms of life take up residence in a coral colony, including photosynthetic algae such as *Symbiodinium*, as well as a wide range of bacteria including nitrogen fixers, and chitin decomposers, all of which form an important part of coral nutrition. The association between coral and its microbiota is species dependent, and different bacterial populations are found in mucus, skeleton and tissue from the same coral fragment.

Over the past several decades, major declines in coral populations have occurred. Climate change, water pollution and overfishing are three stress factors that have been described as leading to disease susceptibility. Over twenty different coral diseases have been described, but of these, only a handful have had their causative agents isolated and characterized.

Coral bleaching is the most serious of these diseases. In the Mediterranean Sea, the bleaching of *Oculina patagonica* was first described in 1994 and, through a rigorous application of Koch's Postulates, determined to be due to infection by *Vibrio shiloi*. From 1994 to 2002, bacterial bleaching of *O. patagonica* occurred every summer in the eastern Mediterranean. Surprisingly, however, after 2003, *O. patagonica* in the eastern Mediterranean has been resistant to *V. shiloi* infection, although other diseases still cause bleaching.

The surprise stems from the knowledge that corals are long lived, with lifespans on the order of decades, and do not have adaptive immune systems. Their innate immune systems do not produce antibodies, and they should seemingly not be able to respond to new challenges except over evolutionary time scales. Yet multiple researchers have documented variations in bleaching susceptibility that may be termed 'experience-mediated tolerance'. The puzzle of how corals managed to acquire resistance to a specific pathogen led Eugene Rosenberg and Ilana Zilber-Rosenberg to propose the Coral Probiotic Hypothesis. This hypothesis proposes that a dynamic relationship exists between corals and their symbiotic microbial communities. Beneficial mutations can arise and spread among the symbiotic microbes much faster than in the host corals. By altering its microbial composition, the "holobiont" can adapt to changing environmental conditions far more rapidly than by genetic mutation and selection in the host species alone.

Extrapolating the Coral Probiotic Hypothesis to other organisms, including higher plants and animals, led to the Rosenberg's support for and publications around the Hologenome Theory of Evolution.

Hologenome Theory

Definition

The framework of the hologenome theory of evolution is as follows (condensed from Rosenberg *et al.*, 2007):

- "All animals and plants establish symbiotic relationships with microorganisms."

- "Different host species contain different symbiont populations and individuals of the same species can also contain different symbiont populations."

- "The association between a host organism and its microbial community affect both the host and its microbiota."

- "The genetic information encoded by microorganisms can change under environmental demands more rapidly, and by more processes, than the genetic information encoded by the host organism."

- "... the genome of the host can act in consortium with the genomes of the associated symbiotic microorganisms to create a hologenome. This hologenome...can change more rapidly than the host genome alone, thereby conferring greater adaptive potential to the combined holobiont evolution."

- "Each of these points taken together [led Rosenberg *et al.* to propose that] the holobiont with its hologenome should be considered as the unit of natural selection in evolution."

Some authors supplement the above principles with an additional one. If a given holobiont is to be considered a unit of natural selection:

- The hologenome must be heritable from generation to generation.

Ten principles of holobionts and hologenomes were presented in PLOS Biology to clarify and append what these terms and concepts are and are not, to explain how they both support and extend existing theory in the life sciences, and to discuss their potential ramifications for the multifaceted approaches of zoology and botany

- I. Holobionts and hologenomes are units of biological organization

- II. Holobionts and hologenomes are not organ systems, superorganisms, or metagenomes

- III. The hologenome is a comprehensive gene system

- IV. The hologenome concept reboots elements of Lamarckian evolution

- V. Hologenomic variation integrates all mechanisms of mutation

- VI. Hologenomic evolution is most easily understood by equating a gene in the nuclear genome to a microbe in the microbiome

- VII. The hologenome concept fits squarely into genetics and accommodates multilevel selection theory

- VIII. The hologenome is shaped by selection and neutrality

- IX. Hologenomic speciation blends genetics and symbiosis

- X. Holobionts and their hologenomes do not change the rules of evolutionary biology

Horizontally Versus Vertically Transmitted Symbionts

Many case studies clearly demonstrate the importance of an organism's associated microbiota to its existence. However, horizontal *versus* vertical transmission of endosymbionts must be distinguished. Endosymbionts whose transmission is predominantly vertical may be considered as contributing to the heritable genetic variation present in a host species.

In the case of colonial organisms such as corals, the microbial associations of the colony persist even though individual members of the colony, reproducing asexually, live and die. Corals also have sexual mode of reproduction, resulting in planktonic larva; it is less clear whether microbial associations persist through this stage of growth. Also, the bacterial community of a colony may change with the seasons.

Many insects maintain heritable obligate symbiosis relationships with bacterial partners. For example, normal development of female wasps of the species *Asobara tabida* is dependent on *Wolbachia* infection. If "cured" of the infection, their ovaries degenerate. Transmission of the infection is vertical through the egg cytoplasm.

In contrast, many obligate symbiosis relationships have been described in the literature where transmission of the symbionts is via horizontal transfer. A well-studied example is the nocturnally feeding squid *Euprymna scolopes*, which camouflages its outline against the moonlit ocean surface by emitting light from its underside with the aid of the symbiotic bacterium *Vibrio fischeri*. The Rosenbergs cite this example within the context of the hologenome theory of evolution. Squid and bacterium maintain a highly co-evolved relationship. The newly hatched squid collects its bacteria from the sea water, and lateral transfer of symbionts between hosts permits faster transfer of beneficial mutations within a host species than are possible with mutations within the host genome.

Primary Versus Secondary Symbionts

Green Vegetable Bug (Nezara viridula) in Fronton, France

Another traditional distinction between endosymbionts has been between primary and secondary symbionts. Primary endosymbionts reside in specialized host cells that may be organized into larger, organ-like structures (in insects, the bacteriome). Associations between hosts and primary endosymbionts are usually ancient, with an estimated age of tens to hundreds of millions of years. According to endosymbiotic theory, extreme cases of primary endosymbionts include mitochondria, plastids (including chloroplasts), and possibly other organelles of eukaryotic cells. Primary endosymbionts are usually transmitted exclusively vertically, and the relationship is always mutualistic and generally obligate for both partners. Primary endosymbiosis is surprisingly common. An estimated 15% of insect species, for example, harbor this type of endosymbiont. In contrast, secondary endosymbiosis is often facultative, at least from the host point of view, and the associations are less ancient. Secondary endosymbionts do not reside in specialized host tissues, but may dwell in the body cavity dispersed in fat, muscle, or nervous tissue, or may grow within the gut. Transmission may be via vertical, horizontal, or both vertical and horizontal transfer. The relationship between host and secondary endosymbiont is not necessarily beneficial to the host; indeed, the relationship may be parasitic.

The distinction between vertical and horizontal transfer, and between primary and secondary endosymbiosis is not absolute, but follows a continuum, and may be subject to environmental influences. For example, in the stink bug *Nezara viridula*, the vertical transmission rate of symbionts, which females provide to offspring by smearing the eggs with gastric caeca, was 100% at 20 °C, but decreased to 8% at 30 °C. Likewise, in aphids, the vertical transmission of bacteriocytes containing the primary endosymbiont *Buchnera* is drastically reduced at high temperature. In like manner, the distinction between commensal, mutualistic, and parasitic relationships is also not absolute. An example is the relationship between legumes and rhizobial species: N_2 uptake is energetically more costly than the uptake of fixed nitrogen from the soil, so soil N is preferred if not limiting. During the early stages of nodule formation, the plant-rhizobial relationship actually resembles a pathogenesis more than it does a mutualistic association.

Neo-Lamarckism Within a Darwinian Context

Lamarckism, the concept that an organism can pass on characteristics that it acquired during its lifetime to its offspring (also known as inheritance of acquired characteristics or soft inheritance) incorporated two common ideas of its time:

- Use and disuse – individuals lose characteristics they do not require (or use) and develop characteristics that are useful.

- Inheritance of acquired traits – individuals inherit the traits of their ancestors.

Although Lamarckian theory was rejected by the neo-Darwinism of the modern evolutionary synthesis in which evolution occurs through random variations being subject to natural selection, the hologenome theory has aspects that harken back to Lamarckian concepts. In addition to the traditionally recognized modes of variation (*i.e.* sexual recombination, chromosomal rearrangement, mutation), the holobiont allows for two additional mechanisms of variation that are specific to the hologenome theory: (1) changes in the relative population of existing microorganisms (*i.e.* amplification and reduction) and (2) acquisition of novel strains from the environment, which may be passed on to offspring.

Changes in the relative population of existing microorganisms corresponds to Lamarckian "use and disuse", while the ability to acquire novel strains from the environment, which may be passed on to offspring, corresponds to Lamarckian "inheritance of acquired traits". The hologenome theory, therefore, is said by its proponents to incorporate Lamarckian aspects within a Darwinian framework.

Additional Case Studies

Pea aphids extracting sap from the stem and leaves of garden peas

The pea aphid *Acyrthosiphon pisum* maintains an obligate symbiotic relationship with the bacterium *Buchnera aphidicola*, which is transmitted maternally to the embryos that develop within the mother's ovarioles. Pea aphids live on sap, which is rich in sugars but deficient in amino acids. They rely on their *Buchnera* endosymbiotic population for essential amino acids, supplying in exchange nutrients as well as a protected intracellular environment that allows *Buchnera* to grow and reproduce. The relationship is actually more complicated than mutual nutrition; some strains of *Buchnera* increases host thermotolerance, while other strains do not. Both strains are present in field populations, suggesting that under some conditions, increased heat tolerance is advantageous to the host, while under other conditions, decreased heat tolerance but increased cold tolerance may be advantageous. One can consider the variant *Buchnera* genomes as alleles for the larger hologenome. The association between *Buchnera* and aphids began about 200 million years ago, with host and symbiont co-evolving since that time; in particular, it has been discovered that genome size in various *Buchnera* species has become extremely reduced, in some cases down to 450 kb, which is far smaller even than the 580 kb genome of *Mycoplasma genitalium*.

Development of mating preferences, *i.e.* sexual selection, is considered to be an early event in speciation. In 1989, Dodd reported mating preferences in *Drosophila* that were induced by diet. It has recently been demonstrated that when otherwise identical populations of *Drosophila* were switched in diet between molasses medium and starch medium, that the "molasses flies" preferred to mate with other molasses flies, while the "starch flies" preferred to mate with other starch flies. This mating preference appeared after only one generation and was maintained for at least 37 generations. The origin of these differences were changes in the flies' populations of a particular bacterial symbiont, *Lactobacillus plantarum*. Antibiotic treatment abolished the induced mating preferences. It appears that the symbiotic bacteria changed the levels of cuticular hydrocarbon sex pheromones.

Zilber-Rosenberg and Rosenberg (2008) have tabulated many of the ways in which symbionts are transmitted and their contributions to the fitness of the holobiont, beginning with mitochondria found in all eukaryotes, chloroplast in plants, and then various associations described in specific systems. The microbial contributions to host fitness included provision of specific amino acids, growth at high temperatures, provision of nutritional needs from cellulose, nitrogen metabolism, recognition signals, more efficient food utilization, protection of eggs and embryos against metabolism, camouflage against predators, photosynthesis, breakdown of complex polymers, stimulation of the immune system, angiogenesis, vitamin synthesis, fiber breakdown, fat storage, supply of minerals from the soil, supply of organics, acceleration of mineralization, carbon cycling, and salt tolerance.

Criticism

The hologenome theory is debated. A major criticism by Ainsworth *et al.* has been their claim that *V. shiloi* was misidentified as the causative agent of coral bleaching, and that its presence in bleached *O. patagonica* was simply that of opportunistic colonization.

If this is true, the original observation that led to Rosenberg's later articulation of the theory would be invalid. On the other hand, Ainsworth *et al.* performed their samplings in 2005, two years after the Rosenberg group discovered *O. patagonica* no longer to be susceptible to *V. shiloi* infection; therefore their finding that bacteria are not the primary cause of present-day bleaching in Mediterranean coral *O. patagonica* should not be considered surprising. The rigorous satisfaction of Koch's postulates, as employed in Kushmaro *et al.* (1997), is generally accepted as providing a definitive identification of infectious disease agents.

Baird *et al.* (2009) have questioned basic assumptions made by Reshef *et al.* (2006) in presuming that (1) coral generation times are too slow to adjust to novel stresses over the observed time scales, and that (2) the scale of dispersal of coral larvae is too large to allow for adaptation to local environments. They may simply have underestimated the potential rapidity of conventional means of natural selection. In cases of severe stress, multiple cases have been documented of ecologically significant evolutionary change occurring over a handful of generations. Novel adaptive mechanisms such as switching symbionts might not be necessary for corals to adjust to rapid climate change or novel stressors.

Organisms in symbiotic relationships evolve to accommodate each other, and the symbiotic relationship increases the overall fitness of the participant species. Although the hologenome theory is still being debated, it has gained a significant degree of popularity within the scientific community as a way of explaining rapid adaptive changes that are difficult to accommodate within a traditional Darwinian framework.

Definitions and uses of the words holobiont and hologenome also differ between proponents and skeptics, and the misuse of the terms has led to confusions over what comprises evidence related to the hologenome. Ongoing discourse is attempting to clear this confusion. Theis et al. clarify that "critiquing the hologenome concept is not synonymous with critiquing coevolution, and arguing that an entity is not a primary unit of selection dismisses the fact that the hologenome concept has always embraced multilevel selection."

For instance, Chandler and Turelli (2014) criticize the conclusions of Brucker and Bordenstein (2013), noting that their observations are also consistent with an alternative explanation. Brucker and Bordenstein (2014) responded to these criticisms, claiming they were unfounded because of factual inaccuracies and altered arguments and definitions that were not advanced by Brucker and Bordenstein (2013).

Recently, Forest L Rohwer and colleagues developed a novel statistical test to examine the potential for the hologenome theory of evolution in coral species. They found that coral species do not inherit microbial communities, and are instead colonized by a core group of microbes that associate with a diversity of species. The authors conclude: "Identification of these two symbiont communities supports the holobiont model and calls into question the hologenome theory of evolution." However, other studies in coral adhere to the original and pluralistic definitions of holobionts and hologenomes. David Bourne, Kathleen Morrow and Nicole Webster clarify that "The combined genomes of this coral holobiont form a coral hologenome, and genomic interactions within the hologenome ultimately define the coral phenotype."

Germ Theory Denialism

Germ theory denialism is the belief that germs do not cause infectious disease, and that the germ theory of disease is wrong. It usually involves arguing that Louis Pasteur's model of infectious disease was wrong, and that Antoine Béchamp's was right. One of the first movements to deny the germ theory was the Sanitary Movement, which was nevertheless central in developing America's public health infrastructure. One well-known advocate of this form of denialism is Bill Maher, who has claimed that Pasteur recanted germ theory on his deathbed. Shikha Dalmia, writing in *The Washington Examiner*, referred to Maher as a "germ theory denier" after he made these comments on *Real Time with Bill Maher* on March 4, 2007. However, in response to criticism of his views, Maher said, on the October 16, 2009 episode of his show, that he accepted microorganisms as the cause of some disease, but expressed skepticism about other topics in medicine, such as vaccination. Similarly, the following month, Maher wrote that he "understand[s] germ theory," but that he still thought that "Western medicine ignores too much [sic] the fact that the terrain in which bacteria can thrive is crucial and often controllable, which shouldn't even be controversial." Harriet Hall published an article in *Skeptic* where she describes her experience arguing with germ theory denialists.

An opposite kind of pseudoscience is the belief that germs cause *all* diseases, including ones like cancer, which the preponderance of evidence indicates are not caused by microorganisms. That belief was promoted by Hulda Regehr Clark in her book *The Cure for all Cancers*.

References

- Chapelle, Frank (2005). "Ch. 5: Hidden Life, Hidden Death". Wellsprings. New Brunswick NJ: Rutgers University Press. p. 82. ISBN 978-0-8135-3614-9.

- Brock TD (1999). Robert Koch: a life in medicine and bacteriology. Washington DC: American Society of Microbiology Press. ISBN 1-55581-143-4.

- Margulis, Lynn; Fester, René (1991-01-01). Symbiosis as a Source of Evolutionary Innovation: Speciation and Morphogenesis. MIT Press. ISBN 9780262132695.

- Number 6 in a series of 7 VHS recordings, 'A Decade of PCR: Celebrating 10 Years of Amplification,' released by Cold Spring Harbor Laboratory Press, 1994. ISBN 0-87969-473-4. http://www.cshlpress.com/default.tpl?..

- Kushmaro A, Kramarsky-Winter E (2004). "Bacteria as a source of coral nutrition". In Rosenberg E, Loya Y. Coral Health and Disease. Berlin Heidelberg: Springer-Verlag. pp. 231–241. ISBN 3-540-20772-4.

- Pérez-Giménez J, Quelas JI, Lodeiro AR (2011). "Competition for Nodulation". In El-Shemy HA. Soybean Physiology and Biochemistry. InTech. ISBN 978-953-307-534-1.

- Rosenberg, Eugene; Zilber-Rosenberg, Ilana (2016-05-04). "Microbes Drive Evolution of Animals and Plants: the Hologenome Concept". mBio. 7 (2): e01395–15. doi:10.1128/mBio.01395-15. ISSN 2150-7511. PMC 4817260 . PMID 27034283.

- Dalmia, Shikha (10 May 2013). "Shikha Dalmia: 'Scientific' liberals should accept results of science". The Washington Examiner. Retrieved 22 March 2014.

- Maher, Bill (15 November 2009). "Vaccination: A Conversation Worth Having". Huffington Post. Retrieved 22 March 2014.

- "[[List of Real Time with Bill Maher episodes (2008–2012)|]]". Real Time with Bill Maher. Season 7. Episode 172. October 16, 2009. 24 minutes in. I am not a germ theory denier. I understand that germs and viruses cause diseases.

- Feldhaar H (2011). "Bacterial symbionts as mediators of ecologically important traits of insect hosts". Ecological entomology. 5 (5): 533–543. doi:10.1111/j.1365-2311.2011.01318.x.

- Rosenberg E, Zilber-Rosenberg I (2011). "Symbiosis and development: The hologenome concept". Birth Defects Research Part C: Embryo Today: Reviews. 93 (1): 56–66. doi:10.1002/bdrc.20196.

5

Methods and Techniques of Microbiology

Methods and techniques are important components of any field of study. Bacteriological water analysis is a method that is usually used to analyze water in order to estimate the number of bacteria present in the water. Microbiological culture is the method applied on organisms to multiply them and then they are used as a tool to determine the cause of an infectious disease. This text helps the reader in developing an understanding of the methods and techniques that are used in microbiology.

Microbiological Culture

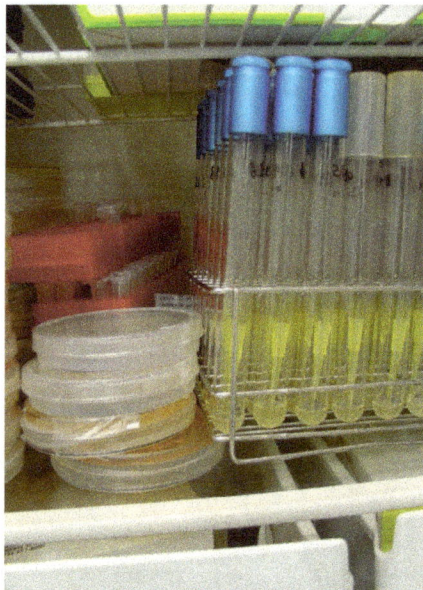

Solid and liquid microbial cultures

A microbiological culture, or microbial culture, is a method of multiplying microbial organisms by letting them reproduce in predetermined culture media under controlled laboratory conditions. Microbial cultures are used to determine the type of organism, its abundance in the sample being tested, or both. It is one of the primary diagnostic methods of microbiology and used as a tool to determine the cause of infectious disease by letting the agent multiply in a predetermined medium. For example, a throat culture is taken by scraping the lining of tissue in the back of the throat and blotting the sample into a medium to be able to screen for harmful microorganisms, such as *Streptococcus pyogenes*, the causative agent of strep throat. Furthermore, the term culture is more generally used informally to refer to "selectively growing" a specific kind of microorganism in the lab.

Microbial cultures are foundational and basic diagnostic methods used extensively as a research tool in molecular biology. It is often essential to isolate a pure culture of microorganisms. A pure (or *axenic*) culture is a population of cells or multicellular organisms growing in the absence of other species or types. A pure culture may originate from a single cell or single organism, in which case the cells are genetic clones of one another.

For the purpose of gelling the microbial culture, the medium of agarose gel (agar) is used. Agar is a gelatinous substance derived from seaweed. A cheap substitute for agar is guar gum, which can be used for the isolation and maintenance of thermophiles.

Bacterial Culture

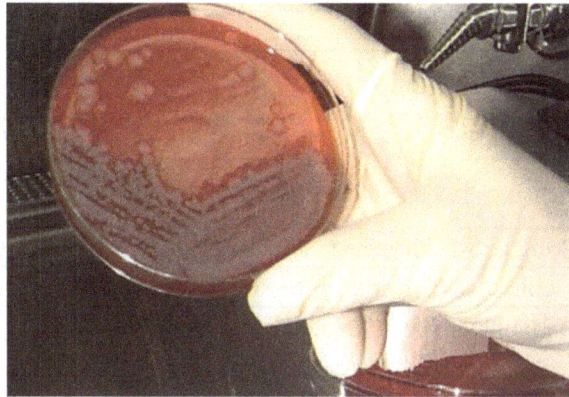

A culture of *Bacillus anthracis*

Microbiological cultures can be grown in petri dishes of differing sizes that have a thin layer of agar-based growth medium. Once the growth medium in the petri dish is inoculated with the desired bacteria, the plates are incubated at the best temperature for the growing of the selected bacteria (for example, usually at 37 degrees Celsius for cultures from humans or animals, or lower for environmental cultures).

Another method of bacterial culture is liquid culture, in which the desired bacteria are suspended in liquid broth, a nutrient medium. These are ideal for preparation of an antimicrobial assay. The experimenter would inoculate liquid broth with bacteria and let it grow overnight (they may use a shaker for uniform growth). Then they would take aliquots of the sample to test for the antimicrobial activity of a specific drug or protein (antimicrobial peptides).

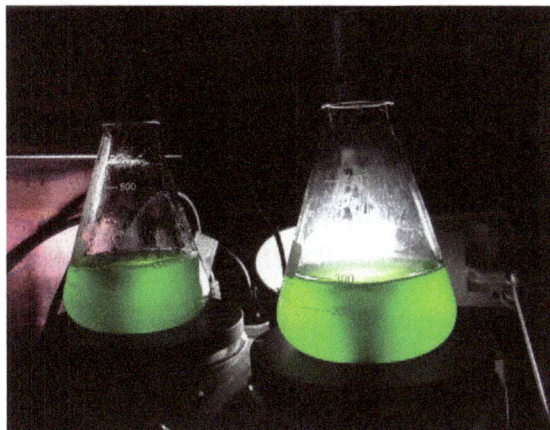

Liquid cultures of cyanobacterium *Synechococcus* PCC 7002

As an alternative, the microbiologist may decide to use static liquid cultures. These cultures are not shaken and they provide the microbes with an oxygen gradient.

Culture Collections

Microbial culture collections focus on the acquisition, authentication, production, preservation, catalogueing and distribution of viable cultures of standard reference microorganisms, cell lines and other materials for research in microbial systematics. Culture collection are also repositories of type strains.

Major national culture collections.		
Collection Acronym	**Name**	**Location**
ATCC	American Type Culture Collection	Manassas, Virginia
NCTC	National Collection of Type Cultures	Public Health England, London, United Kingdom
BCCM	Belgium Coordinated Collection of Microorganism	Ghent, Belgium
CIP	Collection d'Institut Pasteur	Paris, France
DSMZ	Deutsche Sammlung von Mikroorganismen und Zellkulturen	Braunschweig, Germany
JCM	Japan Collection of Microorganisms	Tsukuba, Ibaraki, Japan
NCCB	Netherlands Culture Collection of Bacteria	Utrecht, Netherlands
NCIMB	National Collection of Industrial, Food and Marine Bacteria	Aberdeen, Scotland
STCC	Spanish Type Culture Collection, Valencia University	Valencia, Spain

Virus and Phage Culture

Virus or phage cultures require host cells in which the virus or phage multiply. For bacteriophages, cultures are grown by infecting bacterial cells. The phage can then be isolated from the resulting plaques in a lawn of bacteria on a plate. Virus cultures are obtained from their appropriate eukaryotic host cells.

Eukaryotic Cell Culture

Isolation of Pure Cultures

For single-celled eukaryotes, such as yeast, the isolation of pure cultures uses the same techniques as for bacterial cultures. Pure cultures of multicellular organisms are often more easily isolated by simply picking out a single individual to initiate a culture. This is a useful technique for pure culture of fungi, multicellular algae, and small metazoa, for example.

Developing pure culture techniques is crucial to the observation of the specimen in question. The most common method to isolate individual cells and produce a pure culture is to prepare a streak plate. The streak plate method is a way to physically separate the microbial population, and is done by spreading the inoculate back and forth with an inoculating loop over the solid agar plate. Upon incubation, colonies will arise and single cells will have been isolated from the biomass.

Bacteriological Water Analysis

Bacteriological water analysis is a method of analysing water to estimate the numbers of bacteria present and, if needed, to find out what sort of bacteria they are. It represents one aspect of water quality. It is a microbiological analytical procedure which uses samples of water and from these samples determines the concentration of bacteria. It is then possible to draw inferences about the suitability of the water for use from these concentrations. This process is used, for example, to routinely confirm that water is safe for human consumption or that bathing and recreational waters are safe to use.

The interpretation and the action trigger levels for different waters vary depending on the use made of the water. Very stringent levels applying to drinking water whilst more relaxed levels apply to marine bathing waters, where much lower volumes of water are expected to be ingested by users.

Approach

The common feature of all these routine screening procedures is that the primary analysis is for indicator organisms rather than the pathogens that might cause concern. Indicator organisms are bacteria such as non-specific coliforms, *Escherichia coli* and *Pseudomonas aeruginosa* that are very commonly found in the human or animal gut and which, if detected, may suggest the presence of sewage. Indicator organisms are used because even when a person is infected with a more pathogenic bacteria, they will still be excreting many millions times more indicator organisms than pathogens. It is therefore reasonable to surmise that if indicator organism levels are low, then pathogen levels will be very much lower or absent. Judgements as to suitability of water for use are based on very extensive precedents and relate to the probability of any sample population of bacteria being able to be infective at a reasonable statistical level of confidence.

Analysis is usually performed using culture, biochemical and sometimes optical methods. When indicator organisms levels exceed pre-set triggers, specific analysis for pathogens may then be undertaken and these can be quickly detected (where suspected) using specific culture methods or molecular biology.

Methodologies

Because the analysis is always based on a very small sample taken from a very large volume of water, all methods rely on statistical principles.

Multiple Tube Method

One of the oldest methods is called the multiple tube method. In this method a measured sub-sample (perhaps 10 ml) is diluted with 100 ml of sterile growth medium and an aliquot of 10 ml is then decanted into each of ten tubes. The remaining 10 ml is then diluted again and the process repeated. At the end of 5 dilutions this produces 50 tubes covering the dilution range of 1:10 through to 1:10000.

The tubes are then incubated at a pre-set temperature for a specified time and at the end of the process the number of tubes with growth in is counted for each dilution. Statistical tables are then used to derive the concentration of organisms in the original sample. This method can be enhanced by using indicator medium which changes colour when acid forming species are present and by including a tiny inverted tube called a Durham tube in each sample tube. The Durham inverted tube catches any gas produced. The production of gas at 37 degrees Celsius is a strong indication of the presence of *Escherichia coli*.

ATP Testing

An ATP test is the process of rapidly measuring active microorganisms in water through detection adenosine triphosphate (ATP). ATP is a molecule found only in and around living cells, and as such it gives a direct measure of biological concentration and health. ATP is quantified by measuring the light produced through its reaction with the naturally-occurring enzyme firefly luciferase using a luminometer. The amount of light produced is directly proportional to the amount of biological energy present in the sample.

Second generation ATP tests are specifically designed for water, wastewater and industrial applications where, for the most part, samples contain a variety of components that can interfere with the ATP assay.

Plate Count

The plate count method relies on bacteria growing a colony on a nutrient medium so that the colony becomes visible to the naked eye and the number of colonies on a plate can be counted. To be effective, the dilution of the original sample must be arranged so that on average between 30 and 300 colonies of the target bacterium are grown. Fewer than 30 colonies makes the interpretation statistically unsound whilst greater than 300 colonies often results in overlapping colonies and imprecision in the count. To ensure that an appropriate number of colonies will be generated several dilutions are normally cultured. This approach is widely utilised for the evaluation of the effectiveness of water treatment by the inactivation of representative microbial contaminants such as "E. coli" following ASTM D5465.

The laboratory procedure involves making serial dilutions of the sample (1:10, 1:100, 1:1000, etc.) in sterile water and cultivating these on nutrient agar in a dish that is sealed and incubated. Typical media include plate count agar for a general count or MacConkey agar to count Gram-negative bacteria such as *E. coli*. Typically one set of plates is incubated at 22°C and for 24 hours and a second set at 37°C for 24 hours. The composition of the nutrient usually includes reagents that resist the growth of non-target organisms and make the target organism easily identified, often by a colour change in the medium. Some recent methods include a fluorescent agent so that counting of the colonies can be automated. At the end of the incubation period the colonies are counted by eye, a procedure that takes a few moments and does not require a microscope as the colonies are typically a few millimetres across.

Membrane Filtration

Most modern laboratories use a refinement of total plate count in which serial dilutions of the sample are vacuum filtered through purpose made membrane filters and these filters are themselves laid on nutrient medium within sealed plates. The methodology is otherwise similar to conventional total plate counts. Membranes have a printed millimetre grid printed on and can be reliably used to count the number of colonies under a binocular microscope.

Pour Plate Method

When the analysis is looking for bacterial species that grow poorly in air, the initial analysis is done by mixing serial dilutions of the sample in liquid nutrient agar which is then poured into bottles which are then sealed and laid on their sides to produce a sloping agar surface. Colonies that develop in the body of the medium can be counted by eye after incubation.

The total number of colonies is referred to as the Total Viable Count (TVC). The unit of measurement is cfu/ml (or colony forming units per millilitre) and relates to the original sample. Calculation of this is a multiple of the counted number of colonies multiplied by the dilution used.

Pathogen Analysis

When samples show elevated levels of indicator bacteria, further analysis is often undertaken to look for specific pathogenic bacteria. Species commonly investigated in the temperate zone include *Salmonella typhi* and *Salmonella Typhimurium*. Depending on the likely source of contamination investigation may also extend to organisms such as *Cryptosporidium spp*. In tropical areas analysis of *Vibrio cholerae* is also routinely undertaken.

Types of Nutrient Media Used in Analysis

MacConkey agar is culture medium designed to grow Gram-negative bacteria and stain them for lactose fermentation. It contains bile salts (to inhibit most Gram-positive bacteria), crystal violet dye (which also inhibits certain Gram-positive bacteria), neutral red dye (which stains microbes fermenting lactose), lactose and peptone. Alfred Theodore MacConkey developed it while working as a bacteriologist for the Royal Commission on Sewage Disposal in the United Kingdom.

Endo agar contains peptone, lactose, dipotassium phosphate, agar, sodium sulfite, basic fuchsin and was originally developed for the isolation of *Salmonella typhi*, but is now commonly used in water analysis. As in MacConkey agar, coliform organisms ferment the lactose, and the colonies become red. Non-lactose-fermenting organisms produce clear, colourless colonies against the faint pink background of the medium.

mFC medium is used in membrane filtration and contains selective and differential agents. These include Rosolic acid to inhibit bacterial growth in general, except for faecal coliforms, Bile salts inhibit non-enteric bacteria and Aniline blue indicates the ability of faecal coliforms to ferment lactose to acid that causes a pH change in the medium.

TYEA medium contains tryptone, yeast extract, common salt and L-arabinose per liter of glass distilled water and is a non selective medium usually cultivated at two temperatures (22 and 36°C) to determine a general level of contamination (a.k.a. colony count).

Impedance Microbiology

Impedance microbiology is a rapid microbiological technique used to measure the microbial concentration (mainly bacteria but also yeasts) of a sample by monitoring the electrical parameters of the growth medium. The ability of microbial metabolism to change the electrical conductivity of the growth medium was discovered by Stewart and further studied by other scientists such as Oker-Blom, Parson and Allison in the first half of 20th century. However, it was only in the late 1970s that, thanks to computer-controlled systems used to monitor impedance, the technique showed its full potential, as discussed in the works of Fistenberg-Eden & Eden, Ur & Brown and Cady.

Principle of Operation

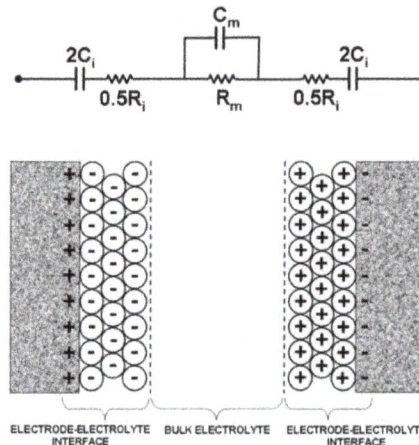

Figure : Equivalent electrical circuit to model a couple of electrodes in direct contact with a liquid medium

When a pair of electrodes are immersed in the growth medium, the system composed of electrodes and electrolyte can be modeled with the electrical circuit of Fig. 1, where R_m and C_m are the resistance and capacitance of the bulk medium, while R_i and C_i are the resistance and capacitance of the electrode-electrolyte interface. However, when frequency of the sinusoidal test signal applied to the electrodes is relatively low (lower than 1 MHz) the bulk capacitance C_m can be neglected and the system can be modeled with a simpler circuit consisting only of a resistance R_s and a capacitance C_s in series. The resistance R_s accounts for the electrical conductivity of the bulk medium while the capacitance C_s is due to the capacitive double-layer at the electrode-electrolyte interface. During the growth phase, bacterial metabolism transforms uncharged or weakly charged compounds of the bulk medium in highly charged compounds that change the electrical properties of the medium. This results in a decrease of resistance R_s and an increase of capacitance C_s.

Impedance Microbiology technique works this way:

1. The sample with the initial unknown bacterial concentration (C_0) is placed at a temperature favoring bacterial growth (in the range 37 to 42 °C if mesophilic microbial population is the target) and the electrical parameters R_s and C_s are measured at regular time intervals of few minutes by means of a couple of electrodes in direct contact with the sample.

2. Until the bacterial concentration is lower than a critical threshold C_{TH} the electrical parameters R_s and C_s remain essentially constant (at their baseline values). C_{TH} depends on various parameters such as electrode geometry, bacterial strain, chemical composition of the growth medium etc., but it is always in the range 10^6 to 10^7 cfu/ml.

3. When the bacterial concentration increases over C_{TH}, the electrical parameters deviate from their baseline values (generally in the case of bacteria there is a decrease of R_s and an increase of C_s, the opposite happens in the case of yeasts).

4. The time needed for the electrical parameters R_s and C_s to deviate from their baseline value is referred as Detect Time (DT) and is the parameter used to estimate the initial unknown bacterial concentration C_0.

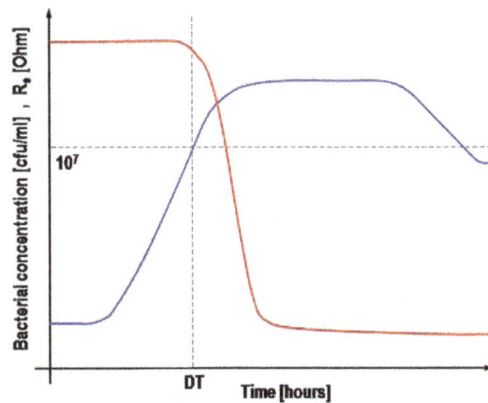

Figure: R_s curve and bacterial concentration curve as function of time

In Fig. a typical curve for R_s as well as the corresponding bacterial concentration are plotted vs. time. Fig. shows typical R_s curves vs time for samples characterized by different bacterial concentration. Since DT is the time needed for the bacterial concentration to grow from the initial value C_0 to C_{TH}, highly contaminated samples are characterized by lower values of DT than samples with low bacterial concentration. Given C_1, C_2 and C_3 the bacterial concentration of three samples with $C_1 > C_2 > C_3$, it is $DT_1 < DT_2 < DT_3$. Data from literature show how DT is a linear function of the logarithm of C_0:

$$DT = A \cdot \log_{10}(C_0) + B$$

where the parameters A and B are dependent on the particular type of samples under test, the bacterial strains, the type of enriching medium used and so on. These parameters can be calculated by calibrating the system using a set of samples whose bacterial concentration is known and calculating the linear regression line that will be used to estimate the bacterial concentration from the measured DT.

Impedance microbiology has different advantages on the standard Plate Count technique to measure bacterial concentration:

1. It is characterized by faster response time. In the case of mesophilic bacteria, the response time range from 2 – 3 hours for highly contaminated samples (10^5 - 10^6 cfu/ml) to over 10 hours for samples with very low bacterial concentration (less than 10 cfu/ml). As a comparison, for the same bacterial strains the Plate Count technique is characterized by response times from 48 to 72 hours.

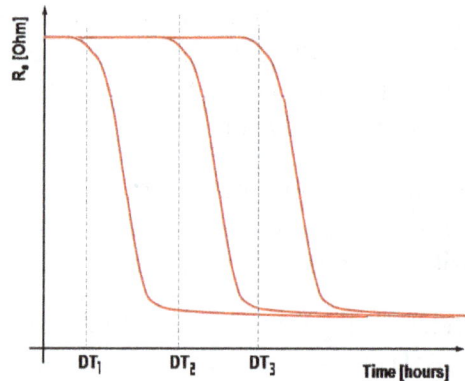

Figure 3: R$_s$ curves for samples featuring different bacterial concentration as function of time

2. Impedance microbiology is a method that can be easily automated, i.e. implemented as part of an industrial machine or realized as an embedded portable sensor, while Plate Count is a manual method that needs to be carried out in a laboratory by long trained personnel.

Instrumentation

Over the past decades different instruments (either laboratory built or commercially available) to measure bacterial concentration using impedance microbiology have been built. One of the best selling and well accepted instruments in the industry is the Bactometer by Biomerieux. The original instrument of 1984 features a multi-incubator system capable of monitoring up to 512 samples simultaneously with the ability to set 8 different incubation temperatures. Other instruments with performance comparable to the Bactometer are Malthus by Malthus Instruments Ltd (Bury, UK), RABIT by Don Whitley Scientific (Shipley, UK) and Bac Trac by Sy-Lab (Purkensdorf, Austria). A portable embedded system for microbial concentration measurement in liquid and semi-liquid media using impedance microbiology has been recently proposed. The system is composed of a thermoregulated incubation chamber where the sample under test is stored and a controller for thermoregulation and impedance measurements.

Applications

Impedance microbiology has been extensively used in the past decades to measure the concentration of bacteria and yeasts in different type of samples, mainly for quality assurance in the food industry. Some applications are, the determination of the shelf life of pasteurized milk and the measure of total bacterial concentration in raw-milk, frozen vegetables, grain products, meat products and beer. The technique has been also used in environmental monitoring to detect the coliform concentration in water samples, in the pharmaceutical industry to test the efficiency of novel antibacterial agents and the testing of final products.

Streaking (Microbiology)

In microbiology, streaking is a technique used to isolate a pure strain from a single species of microorganism, often bacteria. Samples can then be taken from the resulting colonies and a microbiological culture can be grown on a new plate so that the organism can be identified, studied, or tested.

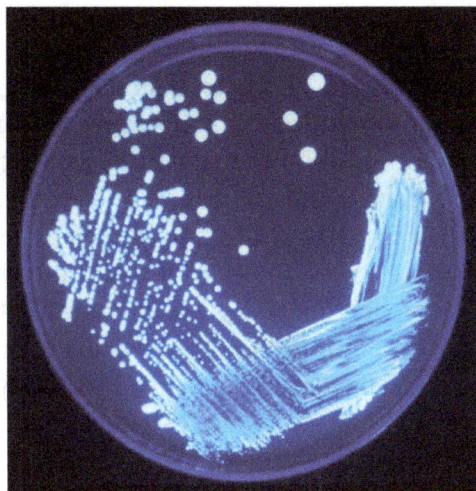

A plate which has been streaked showing the colonies thinning as the streaking moves clockwise.

The modern streak plate method has progressed from the efforts of Robert Koch and other micro-biologists to obtain microbiological cultures of bacteria in order to study them. The dilution or iso-lation by streaking method was first developed by Loeffler and Gaffky in Koch's laboratory, which involves the dilution of bacteria by systematically streaking them over the exterior of the agar in a petri dish to obtain isolated colonies which will then grow into quantity of cells, or isolated colo-nies. If the agar surface grows microorganisms which are all genetically same, the culture is then considered as a microbiological culture.

Technique

Streaking is rapid and ideally a simple process of isolation dilution. The technique is done by di-luting a comparatively large concentration of bacteria to a smaller concentration. The decrease of bacteria should show that colonies are sufficiently spread apart to effect the separation of the different types of microbes. Streaking is done using a sterile tool, such as a cotton swab or com-monly an inoculation loop. Aseptic techniques are used to maintain microbiological cultures and to prevent contamination of the growth medium.There are many different types of methods used to streak a plate. Picking a technique is a matter of individual preference and can also depend on how large the number of microbes the sample contains.

The three-phase streaking pattern, known as the T-Streak, is recommended for beginners.The streaking is done using a sterile tool, such as a cotton swab or commonly an inoculation loop. The inoculation loop is first sterilized by passing it through a flame. When the loop is cool, it is dipped into an inoculum such as a broth or patient specimen containing many species of bacteria. The inoculation loop is then dragged across the surface of the agar back and forth in a zigzag motion until approximately 30% of the plate has been covered. The loop then is re-sterilized and the plate is turned 90 degrees. Starting in the previously streaked section, the loop is dragged through it two to three times continuing the zigzag pattern. The procedure is then repeated once more being cautious to not touch the previously streaked sectors. Each time the loop gathers fewer and fewer bacteria until it gathers just single bacterial cells that can grow into a colony.The plate should show the heaviest growth in the first section. The second section will have less growth and a few isolated colonies, while the final section will have the least amount of growth and many isolated colonies.

Growth Medium

The sample is spread across one quadrant of a petri dish containing a growth medium. Bacteria need different nutrients to grow. This includes water, a source of energy, sources of carbon, sulfur, nitrogen, phosphorus, certain minerals, and other vitamins and growth factors. A very common type of media used in microbiology labs is known as agar, a gelatinous substance derived from seaweed. The nutrient agar has a lot of ingredients with unknown amounts of nutrients in them. On one hand, this can be a very selective media to use because as mentioned bacteria are particular. If there is a certain nutrient in the media the bacteria could most certainly not grow and could die. On the other hand, this media is very complex. Complex media is important because it allows for a wide range of microbial growth. The bacteria growth can be supported by this media greatly due in part to the numerous amounts of nutrients. Choice of which growth medium is used depends on which microorganism is being cultured, or selected for.

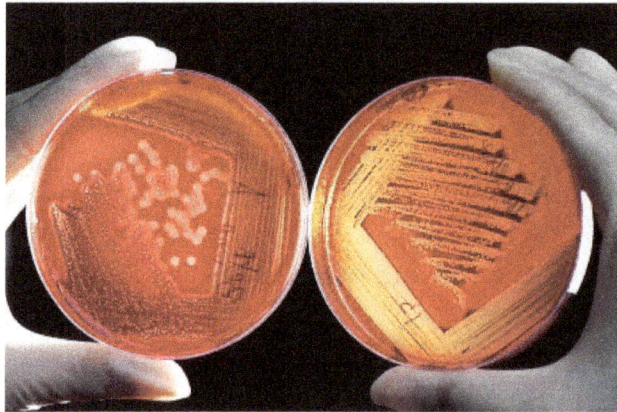

Different labs have different standards as to the direction and style of the streaking.

Incubation

Dependent on the strain, the plate may then be incubated, usually for 24 to 36 hours, to allow the bacteria to reproduce. At the end of incubation there should be enough bacteria to form visible colonies in the areas touched by the inoculation loop. From these mixed colonies, single bacterial or fungal species can be identified based on their morphological (size/shape/colour) differences, and then sub-cultured to a new media plate to yield a pure culture for further analysis.

Automated equipment is used at industrial level for streak plating the solid media in order to achieve better sterilization and consistency of streaking and for reliably faster work. While streaking manually it is important to avoid scratching the solid medium as subsequent streak lines will be damaged and non-uniform deposition of inoculum at damaged sites on the medium yield clustered growth of microbes which may extend into nearby streak lines.

Importance

Bacteria exist in water, soil and food, on skin, and intestinal tract normal flora. The assortment of microbes that exist in the environment and on human bodies is enormous. The human body has billions of bacteria which creates the normal flora fighting against the invading pathogens. Bacteria frequently occur in mixed populations. It is very rare to find a single occurring species of bacteria. To be able to study the cultural, morphological, and physiological characteristics of an individual species, it is vital that the bacteria be divided from the other species that generally originate in the environment. This is important in determining a bacterium in a clinical sample. When the bacteria is streaked and isolated, the causative agent of a bacterial disease can be identified.

Thermal Shift Assay

A thermal shift assay quantifies the change in thermal denaturation temperature of a protein under varying conditions. The differing conditions that can be examined are very diverse, e.g. pH, salts, additives, drugs, drug leads, oxidation/reduction, or mutations. The binding of low molecular weight ligands can increase the thermal stability of a protein, as described by Koshland (1958) and Linderstrom-Lang and Schellman (1959). Almost half of enzymes require a metal ion co-factor. Thermostable proteins are often more useful than their non-thermostable counterparts, e.g. DNA polymerase in the polymerase chain reaction, so protein engineering often includes adding mutations to increase thermal stability. Protein crystallisation is more successful for proteins with a higher melting point and adding buffer components that stabilise proteins improve the likelihood of protein crystals forming. If examining pH then the possible effects of the buffer molecule on thermal stability should be taken into account along with the fact that pKa of each buffer molecule changes uniquely with temperature. Additionally, any time a charged species is examined the effects of the counterion should be accounted for.

Thermal stability of proteins has traditionally been investigated using biochemical assays, circular dichroism, or differential scanning calorimetry. Biochemical assays require a catalytic activity of the protein in question as well as a specific assay. Circular dichroism and differential scanning calorimetry both consume large amounts of protein and are low-throughput methods. The thermofluor assay was the first high-throughput thermal shift assay and its utility and limitations has spurred the invention of a plethora of alternate methods. Each method has its strengths and weaknesses but they all struggle with intrinsically disordered proteins without any clearly defined tertiary structure as the essence of a thermal shift assay is measuring the temperature at which a protein goes from well-defined structure to disorder.

Thermofluor

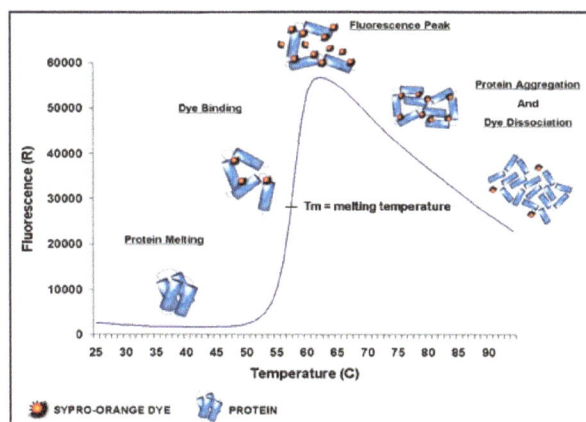

Diagram

The technique was first described by Semisotnov et al. (1991) using 1,8-ANS and quartz cuvettes. Millennium Pharmaceuticals were the first to describe a high-throughput version using a plate reader and Wyeth Research published a variation of the method with SYPRO Orange instead of 1,8-ANS. SYPRO Orange has an excitation/emission wavelength profile compatible with qPCR machines which are almost ubiquitous in institutions that perform molecular biology research. The name Differential Scanning Fluorimetry (DSF) was introduced later but thermofluor is preferable as thermofluor is no longer trademarked and differential scanning fluorimetry is easily confused with differential scanning calorimetry.

SYPRO Orange binds nonspecifically to hydrophobic surfaces, and water strongly quenches its fluorescence. When the protein unfolds, the exposed hydrophobic surfaces bind the dye, resulting in an increase in fluorescence by excluding water. Detergent micelles will also bind the dye and increase background noise dramatically - this effect is lessened by switching to the dye ANS but that requires UV excitation. The stability curve and its midpoint value (melting temperature, T_m also known as the temperature of hydrophobic exposure, T_h) are obtained by gradually increasing the temperature to unfold the protein and measuring the fluorescence at each point. Curves are measured for protein only and protein + ligand, and ΔTm is calculated. The method might not work very well for protein-protein interactions if one of the interaction partners contains large hydrophobic patches as it is highly non-trivial to dissect prevention of aggregation, stabilisation of a native folds and steric hindrance of dye access to hydrophobic sites. In addition, partly aggregated protein can also limit the relative fluorescence increase upon heating; in extreme cases there will be no fluorescence increase at all because all protein is already in aggregates before heating. Knowing this effect can be very useful as a high relative fluorescence increase suggests a significant fraction of folded protein in the starting material.

This assay allows high-throughput screening of ligands to the target protein and it is widely used in the early stages of drug discovery in the pharmaceutical industry, structural genomics efforts, and high-throughput protein engineering.

A Typical Assay

1. Materials: A fluorometer equipped with temperature control or similar instrumentation (RT-PCR machines); suitable fluorescent dye; a suitable assay plate, such as 96 well RT-PCR plate.

2. Compound solutions: Test ligands are prepared at a 50- to 100-fold concentrated solution, generally in the 10-100 mM range. For titration, a typical experimental protocol employs a set of 12 well, comprising 11 different concentrations of a test compound with a single negative control well.

3. Protein solution: Typically, target protein is diluted from a concentrated stock to a working concentration of ~0.5-5 µM protein with dye into a suitable assay buffer. The exact concentrations of protein and dye are defined by experimental assay development studies.

4. Centrifugation and oil dispense: A brief centrifugation (~1000 × g-force, 1 min) of the assay plate to mix compounds into the protein solution, 1-2 µl silicone oil to prevent the

evaporation during heating is overlaid onto the solution (some systems use plastic seals instead), followed by an additional centrifugation step (~1000 × g-force, 1 min).

5. Instrumental set up: A typical temperature ramp rates range from 0.1-10 °C/min but generally in the range of 1 °C/min. The fluorescence in each well is measured at regular intervals, 0.2-1 °C/image, over a temperature range spanning the typical protein unfolding temperatures of 25-95 °C.

CPM, Thiol-specific Dyes

Alexandrov et al. (2008) published a variation on the thermofluor assay where SYPRO Orange was replaced by N-[4-(7-diethylamino-4-methyl-3-coumarinyl)phenyl]maleimide (CPM), a compound that only fluoresces after reacting with a nucleophile. CPM has a high preference for thiols over other typical biological nucleophiles and therefore will react with cysteine side chains before others. Cysteines are typically buried in the interior of a folded protein as they are hydrophobic. When a protein denatures cysteine thiols become available and a fluorescent signal can be read from reacted CPM. The exitation and emission wavelengths for reacted CPM are 387 nm/ 463 nm so a fluorescence plate reader or a qPCR machine with specialised filters is required. Alexandrov et al. used the technique successfully on the membrane proteins Apelin GPCR and FAAH as well as β-lactoglobin which fibrillates on heating rather than going to a molten globule.

DCVJ, Rigidity Sensitive Dyes

4-(dicyanovinyl)julolidine (DCVJ) is a molecular rotor probe with fluorescence that is strongly dependent on the rigidity of its environment. When protein denatures DCVJ increases in fluorescence. It has been reported to work with 40 mg/ml of antibody.

Intrinsic Tryptophan Fluorescence Lifetime

The lifetime of tryptophan fluorescence differs between folded and unfolded protein and by measuring the lifetime of UV-excited fluorescence at temperature intervals you can get a measurement of the melting temperature of your protein.

A prominent advantage is that no reporter dyes are required to be added as tryptophan is an intrinsic part of the protein. This can also be a disadvantage as not all proteins contain tryptophan. Intrinsic fluorescence lifetime works with membrane proteins and detergent micelles but a powerful UV fluorescer in the buffer could drown out the signal and few articles are published using the technique for thermal shift assays.

Intrinsic Tryptophan Fluorescence Wavelength

The excitation and emission wavelengths of tryptophan are dependent on the immediate environment and therefore differs between folded and unfolded protein, just as the fluorescence lifetime. Currently there are at least two machines on the market that can read this shift in wavelength in a high-throughput manner while heating the samples.

The advantages and disadvantages are the same as for fluorescence lifetime except that there are more examples in the scientific literature of use.

Static Light Scattering

Static light scattering allows monitoring of the sizes of the species in solution. Since proteins typically aggregate upon denaturation (or form fibrils) the detected species size will go up.

This is label-free and independent of specific residues in the protein or buffer composition. The only requirement is that the protein actually aggregates/fibrillates after denaturation and that the protein of interest has been purified.

FastPP

In fast parallel proteolysis the researcher adds a thermostable protease (thermolysin) and takes out samples in parallel upon heating in a thermal gradient cycler. Optionally, for instance for lowly expressed proteins, a Western blot is then run to determine at what temperature a protein becomes degraded. For pure or highly enriched proteins, direct SDS-PAGE detection is possible facilitating Commassie-fluorescence based direct quantification. FastPP exploits that proteins become increasingly susceptible to proteolysis when unfolded and that thermolysin cleaves at hydrophobic residues which are typically found in the core of proteins.

To reduce the workload, Western Blots could be replaced by SDS-PAGE gel polyhistidine-tag staining, provided that the protein has such a tag and is expressed in adequate amounts.

FastPP can be used on unpurified, complex mixtures of proteins and proteins fused with other proteins, such as GST or GFP, as long as the sequence that is the target of the Western blot, e.g. His-tag, is directly linked to the protein of interest. However, commercially available thermolysin is dependent on calcium ions for activity and denatures itself just above 85 degrees Celsius. So calcium must be present and calcium chelators absent in the buffer - other compounds that interfere with the function (such as high concentrations of detegents) of the protease could also be problematic.

FASTpp has also been used to monitor binding-coupled folding of intrinsically disordered proteins (IDPs).

CETSA

CETSA stands for cellular thermal shift assay. Very similar to fastPP but instead of adding a protease samples are centrifuged after heating and the supernatant is run on Western blots or two target-directed antibodies are added and their relative proximity is detected through FRET. When a protein denatures it aggregates and becomes part of the pellet and can therefore not be found in the supernatant. An intrinsic limitation of this assay is that not all proteins aggregate upon unfolding but might instead populate highly soluble, relatively compact molten-globule (like) conformations. In addition, it is possible that proteins will co-precipitate with their less stable protein interaction partners and therefore show lower apparent stability than their actual thermodynamic stability.

Just as for FastPP the Western blot could be replaced by staining for a polyhistidine-tag in the SDS-PAGE gel. Additionally, instead of using two antibodies the FRET method could be performed with a single fluorescent antibody and a fluorescently labelled peptide recognised by the antibody.

A technique that has been shown to work on both complex mixtures and whole cells but unfortunately requires antibodies. The protein of interest should not be fused to other proteins such as MBP or GFP as they could keep a denatured protein in solution.

A CETSA assay for membrane proteins was described recently. This method not only allows to screen ligands but also thermostabilizing conditions.

ThermoFAD

Thermofluor variant specific for flavin-binding proteins. Analogous to thermofluor binding assays, a small volume of protein solution is heated up and the fluorescence increase is followed as function of temperature. In contrast to Thermofluor, no external fluorescent dye is needed because the flavin cofactor is already present in the flavin-binding protein and its fluorescence properties change upon unfolding.

SEC-TS

Size exclusion chromatography can be used directly to access protein stability in the presence or absence of ligands. Samples of purified protein are heated in a water bath or thermocycler, cooled, centrifuged to remove aggregated proteins, and run on an analytical HPLC. As the melting temperature is reached and protein precipitates or aggregates, peak height decreases and void peak height increases. This can be used to identify ligands and inhibitors, and optimize purification conditions.

While of lower through-put than FSEC-TS, requiring large amounts of purified protein, SEC-TS avoids any influence of the fluorescent tag on apparent protein stability.

FSEC-TS

In fluorescence-detection size exclusion chromatography the protein of interest is fluorescently tagged, e.g. with GFP, and run through a gel filtration column on an FPLC system equipped with a fluorescence detector. The resulting chromatogram allows allows the researcher to estimate the monodispersity and expression level of the tagged protein in the current buffer. Since only fluorescence is measured, only the tagged protein is seen in the chromatogram. FSEC is typically used to compare membrane protein orthologs or screen detergents to solubilise specific membrane proteins in.

For fluorescence-detection size-exclusion chromatography-based thermostability assay (FSEC-TS) the samples are heated in the same manner as in FastPP and CETSA and after centrifugation to clear away precipitate the supernatant is treated in the same manner as FSEC. Larger aggregates are seen in the void volume while the peak height for the protein of interest decreases when the unfolding temperature is reached.

GFP has a T_m of ~76 C so the technique is limited to temperature below ~70 C. Of all the techniques mentioned it may have the lowest throughput but it is free of antibody requirements and works on complex mixtures of proteins.

Radioligand Binding Thermostability Assay

GPCRs are pharmacologically important transmembrane proteins. Their x-ray crystal structures were revealed long after other transmembrane proteins of lesser interest. The difficulty in obtaining protein crystals of GPCRs was likely due to their high flexibility. Less flexible versions were obtained by truncating, mutating, and inserting T4 lysozyme in the recombinant sequence. One of the methods researchers used to guide these alterations was radioligand binding thermostability assay.

The assay is performed by incubating the protein with a radiolabelled ligand of the protein for 30 minutes at a given temperature, then quench on ice, run through a gel filtration mini column, and quantify the radiation levels of the protein that comes off the column. The radioligand concentration is high enough to saturate the protein. Denatured protein is unable to bind the radioligand and the protein and radioligand will be separated in the gel filtration mini column. When screening mutants selection will be for thermal stability in the specific conformation, i.e. if the radioligand is an agonist selection will be for the agonist binding conformation and if it is an antagonist then the screening is for stability in the antagonist binding conformation.

Radioassays have the advantage of working with minute amounts of protein. But it is work with radioactive substances and large amount of manual labour is involved. A high-affinity ligand has to be known for the protein of interest and the buffer must not interfere with the binding of the radioligand. Other thermal shift assays can also select for specific conformations if a ligand of the appropriate type is added to the experiment.

ATP Test

The ATP test is a process of rapidly measuring actively growing microorganisms through detection of adenosine triphosphate, or ATP.

ATP Testing Method

ATP is a molecule found in and around living cells, and as such it gives a direct measure of biological concentration and health. ATP is quantified by measuring the light produced through its reaction with the naturally occurring firefly enzyme luciferase using a luminometer. The amount of light produced is directly proportional to the amount of ATP present in the sample.

ATP Tests Can be Used to:

- Control biological treatment reactors
- Guide biocide dosing programs
- Determine drinking water cleanliness
- Manage fermentation processes
- Assess soil activity
- Determine corrosion / deposit process type
- Measure equipment or product sanitation

1st Generation Testing Vs. 2nd Generation Testing

1st generation ATP tests are derived from hygiene monitoring uses where samples are relatively free of interferences. 2nd Generation tests are specifically designed for water, wastewater and industrial applications where, for the most part, samples contain a variety of components that can interfere with the ATP assay.

How ATP is Measured

ATP is a molecule found only in and around living cells, and as such it gives a direct measure of biological concentration and health. ATP is quantified by measuring the light produced through its reaction with the naturally-occurring firefly enzyme Luciferase using a Luminometer. The amount of light produced is directly proportional to the amount of biological energy present in the sample.

Within a water sample containing microorganisms, there are two types of ATP:

- Intracellular ATP – ATP contained within living biological cells.

- Extracellular ATP – ATP located outside of biological cells that has been released from dead or stressed organisms.

Accurate measurement of these two types of ATP is critical to utilizing ATP-based measurements. Being able to accurately measure these different types of ATP offers the ability to assess biological health and activity, and subsequently control water and wastewater processes.

Phenotypic Testing of Mycobacteria

In microbiology, the phenotypic testing of mycobacteria uses a number of methods. The most-commonly used phenotypic tests to identify and distinguish *Mycobacterium* strains and species from each other are described below.

Tests

Acetamide as sole C and N sources

Media: KH_2PO_4 (0.5 g), MgSO>$_4$*$7H_2$0 (0.5 g), purified agar (20 g), distilled water (1000 ml). The medium is supplemented with acetamide to a final concentration of 0.02M, adjusted to a pH of 7.0 and sterilized by autoclaving at 115°C for 30 minutes. After sloping, the medium is inoculated with one loop of the cultures and incubated. Growth is read after incubation for two weeks (rapid growers) or four weeks (slow growers).

Arylsulfatase test

Arylsulfatase enzyme is present in most mycobacteria. The rate by which arylsulfatase enzyme breaks down phenolphthalein disulfate into phenolphthalein (which forms a red color in the presence of sodium bicarbonate) and other salts is used to differentiate certain strains of Mycobacteria. 3 day arylsulfatase test is used to identify potentially pathogenic rapid growers such as M. fortuitum and M. chelonae. Slow growing M. marinum and M. szulgai are positive in the 14-day arylsulfatase test.

Catalase, semiquantitative activity

Most mycobacteria produce the enzyme catalase, but they vary in the quantity produced. Also, some forms of catalase are inactivated by heating at 68°C for 20 minutes (others are stable). Organisms producing the enzyme catalase have the ability to decompose hydrogen peroxide into water and free oxygen. The test differs from that used to detect catalase in other types of bacteria by using 30% hydrogen peroxide in a strong detergent solution (10% polysorbate 80).

Citrate

Sole carbon source

Egg medium

Growth on Löwenstein–Jensen medium (LJ medium)

L-Glutamate

Sole carbon and nitrogen source

Growth rate

The growth rate is the length of time required to form mature colonies visible without magnification on solid media. Mycobacteria forming colonies visible to the naked eye within seven days on subculture are known as rapid growers, while those requiring longer periods are termed slow growers.

Iron uptake

The ability to take up iron from an inorganic iron containing reagent helps differentiate some species of mycobacteria.

Lebek medium

Lebek is a semisolid medium used to test the oxygen preferences of mycobacterial isolates. Aerophilic growth is indicated by growth on (and above) the surface of the glass wall of the tube; microaerophilic growth is indicated by growth below the surface.

MacConkey agar without crystal violet

Niacin accumulation (paper strip method)

Niacin is formed as a metabolic byproduct by all mycobacteria, but some species possess an enzyme that converts free niacin to niacin ribonucleotide. M. tuberculosis (and some other species) lack this enzyme, and accumulate niacin as a water-soluble byproduct in the culture medium.

Nitrate reduction

Mycobacteria containing nitroreductase catalyze the reduction from nitrate to nitrite. The presence of nitrite in the test medium is detected by addition of sulfanilamide and n-naphthylethylendiamine. If nitrate is present, red diazonium dye is formed.

Photoreactivity of mycobacteria;

Some mycobacteria produce carotenoid pigments without light; others require photoactivation for pigment production. Photochromogens produce non-pigmented colonies when grown in the dark, and pigmented colonies after exposure to light and re-incubation. Scotochromogens produce deep-yellow-to-orange colonies when grown in either light or darkness. Non-photochromogens are non-pigmented in light and darkness or have a pale-yellow, buff or tan pigment which does not intensify after light exposure.

Picrate tolerance

Grows on Sauton agar containing picric acid (0.2% w/v) after three weeks

Pigmentation

Some mycobacteria produce carotenoid pigments without light; others require photoactivation for pigment production.

Pyrazinamide sensitivity (PZA)

The deamidation of pyrazinamide to pyrazinoic acid (assumed to be the active component of the drug pyrazinamide) in four days is a useful physiologic characteristic by which *M. tuberculosis*-complex members can be distinguished.

Sodium chloride tolerance

Growth on LJ medium containing 5% NaCl

Thiophene-2carboxylic acid hydrazide (TCH) sensitivity

The growth of *M. bovis* and M. africanum subtype II is inhibited by thiophene-2carboxylic acid hydrazide; growth of *M. tuberculosis* and *M. africanum* subtype I is uninhibited.

Polysorbate 80 hydrolysis

A test for lipase using polysorbate 80 (polyoxyethylene sorbitan monooleate, a detergent). Certain mycobacteria possess a lipase that splits it into oleic acid and polyoxyethylated sorbitol. The test solution also contains phenol red, which is stabilised by the polysorbate 80; when the latter 80 is hydrolysed, the phenol red changes from yellow to pink.

Urease (adaptation to mycobacteria)

With an inoculation loop, several loopfuls of mycobacteria test colonies are transferred to 0.5 mL of urease substrate, mixed to emulsify and incubated at 35 °C for three days; a colour change (from amber-yellow to pink-red) is sought.

Electroporation

Electroporation, or electropermeabilization, is a microbiology technique in which an electrical field is applied to cells in order to increase the permeability of the cell membrane, allowing chemicals, drugs, or DNA to be introduced into the cell. In microbiology, the process of electroporation is often used to transform bacteria, yeast, or plant protoplasts by introducing new coding DNA. If bacteria and plasmids are mixed together, the plasmids can be transferred into the bacteria after electroporation, though depending on what is being transferred cell-penetrating peptides or CellSqueeze could also be used. Electroporation works by passing thousands of volts across a distance of one to two millimeters of suspended cells in an electroporation cuvette (1.8 - 2.5 kV, 12.5 - 18 kV/cm). Afterwards, the cells have to be handled carefully until they have had a chance to divide, producing new cells that contain reproduced plasmids. This process is approximately ten times more effective than chemical transformation.

Electroporation is also highly efficient for the introduction of foreign genes into tissue culture cells, especially mammalian cells. For example, it is used in the process of producing knockout mice, as well as in tumor treatment, gene therapy, and cell-based therapy. The process of introducing foreign DNA into eukaryotic cells is known as transfection. Electroporation is highly effective for transfecting cells in suspension using electroporation cuvettes. Electroporation has proven efficient for use on tissues in vivo, for in utero applications as well as in ovo transfection. Adherent cells can also be transfected using electroporation, providing researchers with an alternative to trypsinizing their cells prior to transfection.

Laboratory Practice

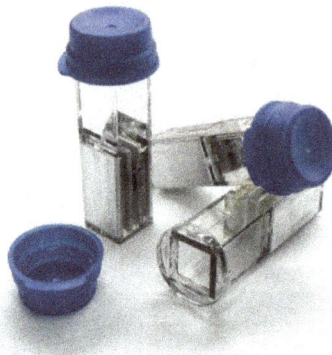

Cuvettes for electroporation. These are plastic with aluminium electrodes and a blue lid. They hold a maximum of 400 µl.

Electroporation is performed with electroporators, purpose-built appliances which create an electrostatic field in a cell solution. The cell suspension is pipetted into a glass or plastic cuvette which has two aluminum electrodes on its sides. For bacterial electroporation, typically a suspension of around 50 microliters is used. Prior to electroporation, this suspension of bacteria is mixed with the plasmid to be transformed. The mixture is pipetted into the cuvette, the voltage and capacitance are set, and the cuvette is inserted into the electroporator. The process requires direct contact between the electrodes and the suspension. Immediately after electroporation, one milliliter of liquid medium is added to the bacteria (in the cuvette or in an Eppendorf tube), and the tube is incubated at the bacteria's optimal temperature for an hour or more to allow recovery of the cells and expression of the plasmid, followed by bacterial culture on agar plates.

The success of the electroporation depends greatly on the purity of the plasmid solution, especially on its salt content. Solutions with high salt concentrations might cause an electrical discharge (known as arcing), which often reduces the viability of the bacteria. For a further detailed investigation of the process, more attention should be paid to the output impedance of the porator device and the input impedance of the cells suspension (e.g. salt content).

Since the cell membrane is not able to pass current (except in ion channels), it acts as an electrical capacitor. Subjecting membranes to a high-voltage electric field results in their temporary breakdown, resulting in pores that are large enough to allow macromolecules (such as DNA) to enter or leave the cell.

Medical Applications

The first research looking at how electroporation might be used on human cells was conducted by researchers at Eastern Virginia Medical School and Old Dominion University, and published in 2003.

The first successful treatment of malignant cutaneous tumors implanted in mice was completed in 2007 by a group of scientists who achieved complete tumor ablation in 12 out of 13 mice. They accomplished this by sending 80 pulses of 100 microseconds at 0.3 Hz with an electrical field magnitude of 2500 V/cm to treat the cutaneous tumors.

A higher voltage of electroporation was found in pigs to irreversibly destroy target cells within a narrow range while leaving neighboring cells unaffected, and thus represents a promising new treatment for cancer, heart disease and other disease states that require removal of tissue. Irreversible electroporation (IRE) has since proven effective in treating human cancer, with surgeons at Johns Hopkins and other institutions now using the technology to treat pancreatic cancer previously thought to be unresectable.

A recent technique called non-thermal irreversible electroporation (N-TIRE) has proven successful in treating many different types of tumors and other unwanted tissue. This procedure is done using small electrodes (about 1mm in diameter), placed either inside or surrounding the target tissue to apply short, repetitive bursts of electricity at a predetermined voltage and frequency. These bursts of electricity increase the resting transmembrane potential (TMP), so that nanopores form in the plasma membrane. When the electricity applied to the tissue is above the electric field threshold of the target tissue, the cells become permanently permeable from the formation of nanopores. As a result, the cells are unable to repair the damage and die due to a loss of homeostasis. N-TIRE is unique to other tumor ablation techniques in that it does not create thermal damage to the tissue around it.

Contrastingly, reversible electroporation occurs when the electricity applied with the electrodes is below the electric field threshold of the target tissue. Because the electricity applied is below the cells' threshold, it allows the cells to repair their phospholipid bilayer and continue on with their normal cell functions. Reversible electroporation is typically done with treatments that involve getting a drug or gene (or other molecule that is not normally permeable to the cell membrane) into the cell. Not all tissue has the same electric field threshold; therefore careful calculations need to be made prior to a treatment to ensure safety and efficacy.

One major advantage of using N-TIRE is that, when done correctly according to careful calculations, it only affects the target tissue. Proteins, the extracellular matrix, and critical structures such as blood vessels and nerves are all unaffected and left healthy by this treatment. This allows for a quicker recovery, and facilitates a more rapid replacement of dead tumor cells with healthy cells.

Before doing the procedure, scientists must carefully calculate exactly what needs to be done, and treat each patient on an individual case-by-case basis. To do this, imaging technology such as CT scans and MRI's are commonly used to create a 3D image of the tumor. From this information, they can approximate the volume of the tumor and decide on the best course of action including the insertion site of electrodes, the angle they are inserted in, the voltage needed, and more, using software technology. Often, a CT machine will be used to help with the placement of electrodes during the procedure, particularly when the electrodes are being used to treat tumors in the brain.

The entire procedure is very quick, typically taking about five minutes. The success rate of these procedures is high and is very promising for future treatment in humans. One disadvantage to using N-TIRE is that the electricity delivered from the electrodes can stimulate muscle cells to contract, which could have lethal consequences depending on the situation. Therefore, a paralytic agent must be used when performing the procedure. The paralytic agents that have been used in such research are successful; however, there is always some risk, albeit slight, when using anesthetics.

A more recent technique has been developed called high-frequency irreversible electroporation (H-FIRE). This technique uses electrodes to apply bipolar bursts of electricity at a high frequency, as opposed to unipolar bursts of electricity at a low frequency. This type of procedure has the same tumor ablation success as N-TIRE. However, it has one distinct advantage, H-FIRE does not cause muscle contraction in the patient and therefore there is no need for a paralytic agent.

Drug and Gene Delivery

Electroporation can also be used to help deliver drugs or genes into the cell by applying short and intense electric pulses that transiently permeabilize cell membrane, thus allowing transport of molecules otherwise not transported through a cellular membrane. This procedure is referred to as electrochemotherapy when the molecules to be transported are chemotherapeutic agents or gene electrotransfer when the molecule to be transported is DNA. Scientists from Karolinska Institutet and the University of Oxford use electroporation of exosomes to deliver siRNAs, antisense oligonucleotides, chemotherapeutic agents and proteins specifically to neurons after inject them systemically (in blood). Because these exosomes are able to cross the blood brain barrier this protocol could solve the issue of poor delivery of medications to the central nervous system and cure Alzheimer's, Parkinson's Disease and brain cancer among other diseases.

Physical Mechanism

Electroporation allows cellular introduction of large highly charged molecules such as DNA which would never passively diffuse across the hydrophobic bilayer core. This phenomenon indicates that the mechanism is the creation of nm-scale water-filled holes in the membrane. Although electroporation and dielectric breakdown both result from application of an electric field, the mechanisms involved are fundamentally different. In dielectric breakdown the barrier material is ionized, creating a conductive pathway. The material alteration is thus chemical in nature. In contrast, during electroporation the lipid molecules are not chemically altered but simply shift position, opening up a pore which acts as the conductive pathway through the bilayer as it is filled with water.

Electroporation is a dynamic phenomenon that depends on the local transmembrane voltage at each point on the cell membrane. It is generally accepted that for a given pulse duration and shape, a specific transmembrane voltage threshold exists for the manifestation of the electroporation phenomenon (from 0.5 V to 1 V). This leads to the definition of an electric field magnitude threshold for electroporation (E_{th}). That is, only the cells within areas where $E \geq E_{th}$ are electroporated. If a second threshold (E_{ir}) is reached or surpassed, electroporation will compromise the viability of the cells, *i.e.*, irreversible electroporation (IRE).

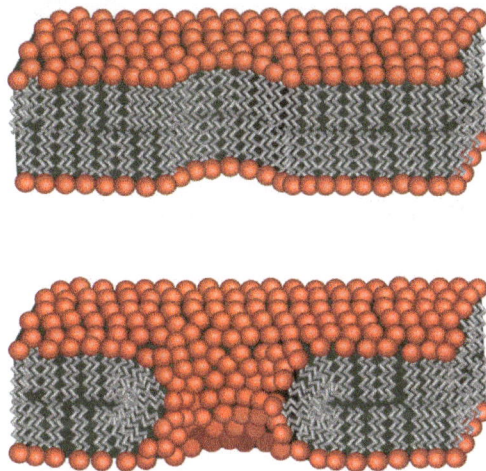

Schematic showing the theoretical arrangement of lipids in a hydrophobic pore (top)
and a hydrophilic pore (bottom).

Electroporation is a multi-step process with several distinct phases. First, a short electrical pulse must be applied. Typical parameters would be 300–400 mV for < 1 ms across the membrane (note- the voltages used in cell experiments are typically much larger because they are being applied across large distances to the bulk solution so the resulting field across the actual membrane is only a small fraction of the applied bias). Upon application of this potential the membrane charges like a capacitor through the migration of ions from the surrounding solution. Once the critical field is achieved there is a rapid localized rearrangement in lipid morphology. The resulting structure is believed to be a "pre-pore" since it is not electrically conductive but leads rapidly to the creation of a conductive pore. Evidence for the existence of such pre-pores comes mostly from the "flickering" of pores, which suggests a transition between conductive and insulating states. It has been suggested that these pre-pores are small (~3 Å) hydrophobic defects. If this theory is correct, then the transition to a conductive state could be explained by a rearrangement at the pore edge, in which the lipid heads fold over to create a hydrophilic interface. Finally, these conductive pores can either heal, resealing the bilayer or expand, eventually rupturing it. The resultant fate depends on whether the critical defect size was exceeded which in turn depends on the applied field, local mechanical stress and bilayer edge energy.

Phage Display

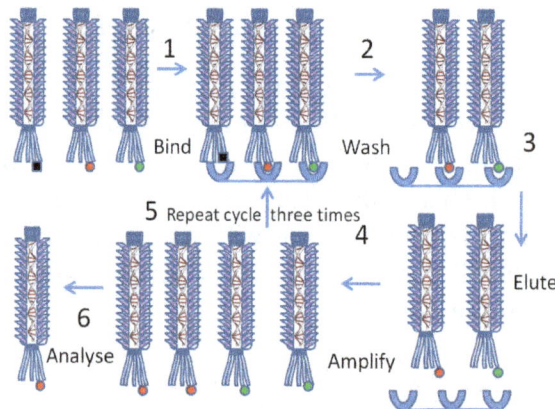

The sequence of events that are followed in phage display screening to identify polypeptides that bind with high affinity to desired target protein or DNA sequence.

Phage display is a laboratory technique for the study of protein–protein, protein–peptide, and protein–DNA interactions that uses bacteriophages (viruses that infect bacteria) to connect proteins with the genetic information that encodes them. In this technique, a gene encoding a protein of interest is inserted into a phage coat protein gene, causing the phage to "display" the protein on its outside while containing the gene for the protein on its inside, resulting in a connection between genotype and phenotype. These displaying phages can then be screened against other proteins, peptides or DNA sequences, in order to detect interaction between the displayed protein and those other molecules. In this way, large libraries of proteins can be screened and amplified in a process called *in vitro* selection, which is analogous to natural selection.

The most common bacteriophages used in phage display are M13 and fd filamentous phage, though T4, T7, and λ phage have also been used.

History

Phage display was first described by George P. Smith in 1985, when he demonstrated the display of peptides on filamentous phage by fusing the peptide of interest onto gene III of filamentous phage. A patent by George Pieczenik claiming priority from 1985 also describes the generation of phage display libraries. This technology was further developed and improved by groups at the Laboratory of Molecular Biology with Greg Winter and John McCafferty, The Scripps Research Institute with Lerner and Barbas and the German Cancer Research Center with Breitling and Dübel for display of proteins such as antibodies for therapeutic protein engineering.

Principle

Like the two-hybrid system, phage display is used for the high-throughput screening of protein interactions. In the case of M13 filamentous phage display, the DNA encoding the protein or peptide of interest is ligated into the pIII or pVIII gene, encoding either the minor or major coat protein, respectively. Multiple cloning sites are sometimes used to ensure that the fragments are inserted in all three possible reading frames so that the cDNA fragment is translated in the proper frame. The phage gene and insert DNA hybrid is then inserted (a process known as "transduction") into *Escherichia coli* (E. coli) bacterial cells such as TG1, SS320, ER2738, or XL1-Blue *E. coli*. If a "phagemid" vector is used (a simplified display construct vector) phage particles will not be released from the *E. coli* cells until they are infected with helper phage, which enables packaging of the phage DNA and assembly of the mature virions with the relevant protein fragment as part of their outer coat on either the minor (pIII) or major (pVIII) coat protein. By immobilizing a relevant DNA or protein target(s) to the surface of a microtiter plate well, a phage that displays a protein that binds to one of those targets on its surface will remain while others are removed by washing. Those that remain can be eluted, used to produce more phage (by bacterial infection with helper phage) and so produce a phage mixture that is enriched with relevant (i.e. binding) phage. The repeated cycling of these steps is referred to as 'panning', in reference to the enrichment of a sample of gold by removing undesirable materials. Phage eluted in the final step can be used to infect a suitable bacterial host, from which the phagemids can be collected and the relevant DNA sequence excised and sequenced to identify the relevant, interacting proteins or protein fragments.

The use of a helper phage can be eliminated by using 'bacterial packaging cell line' technology.

Elution can be done combining low-pH elution buffer with sonification, which, in addition to loosening the peptide-target interaction, also serves to detach the target molecule from the immobilization surface. This ultrasound-based method enables single-step selection of a high-affinity peptide.

Applications

Applications of phage display technology include determination of interaction partners of a protein (which would be used as the immobilised phage "bait" with a DNA library consisting of all coding sequences of a cell, tissue or organism) so that the function or the mechanism of the function of that protein may be determined. Phage display is also a widely used method for *in vitro* protein evolution (also called protein engineering). As such, phage display is a useful tool in drug discovery. It is used for finding new ligands (enzyme inhibitors, receptor agonists and antagonists) to target proteins. The technique is also used to determine tumour antigens (for use in diagnosis and therapeutic targeting) and in searching for protein-DNA interactions using specially-constructed DNA libraries with randomised segments.

Competing methods for *in vitro* protein evolution include yeast display, bacterial display, ribosome display, and mRNA display.

Antibody maturation in vitro

The invention of antibody phage display revolutionised antibody drug discovery. Initial work was done by laboratories at the MRC Laboratory of Molecular Biology (Greg Winter and John McCafferty), the Scripps Research Institute (Richard Lerner and Carlos F. Barbas) and the German Cancer Research Centre (Frank Breitling and Stefan Dübel). In 1991, The Scripps group reported the first display and selection of human antibodies on phage. This initial study described the rapid isolation of human antibody Fab that bound tetanus toxin and the method was then extended to rapidly clone human anti-HIV-1 antibodies for vaccine design and therapy.

Phage display of antibody libraries has become a powerful method for both studying the immune response as well as a method to rapidly select and evolve human antibodies for therapy. Antibody phage display was later used by Carlos F. Barbas at The Scripps Research Institute to create synthetic human antibody libraries, a principle first patented in 1990 by Breitling and coworkers (Patent CA 2035384), thereby allowing human antibodies to be created in vitro from synthetic diversity elements.

Antibody libraries displaying millions of different antibodies on phage are often used in the pharmaceutical industry to isolate highly specific therapeutic antibody leads, for development into antibody drugs primarily as anti-cancer or anti-inflammatory therapeutics. One of the most successful was HUMIRA (adalimumab), discovered by Cambridge Antibody Technology as D2E7 and developed and marketed by Abbott Laboratories. HUMIRA, an antibody to TNF alpha, was the world's first fully human antibody, which achieved annual sales exceeding $1bn.

General Protocol

Below is the sequence of events that are followed in phage display screening to identify polypeptides that bind with high affinity to desired target protein or DNA sequence:

1. Target proteins or DNA sequences are immobilised to the wells of a microtiter plate.

2. Many genetic sequences are expressed in a bacteriophage library in the form of fusions with the bacteriophage coat protein, so that they are displayed on the surface of the viral particle. The protein displayed corresponds to the genetic sequence within the phage.

3. This phage-display library is added to the dish and after allowing the phage time to bind, the dish is washed.

4. Phage-displaying proteins that interact with the target molecules remain attached to the dish, while all others are washed away.

5. Attached phage may be eluted and used to create more phage by infection of suitable bacterial hosts. The new phage constitutes an enriched mixture, containing considerably less irrelevant phage (i.e. non-binding) than were present in the initial mixture.

6. Steps 3 to 6 are optionally repeated one or more times, further enriching the phage library in binding proteins.

7. Following further bacterial-based amplification, the DNA within in the interacting phage is sequenced to identify the interacting proteins or protein fragments.

Selection of the Coat Protein

Filamentous Phages

pIII

pIII is the protein that determines the infectivity of the virion. pIII is composed of three domains (N1, N2 and CT) connected by glycine-rich linkers. The N2 domain binds to the F pilus during virion infection freeing the N1 domain which then interacts with a TolA protein on the surface of the bacterium. Insertions within this protein are usually added in position 249 (within a linker region between CT and N2), position 198 (within the N2 domain) and at the N-terminus (inserted between the N-terminal secretion sequence and the N-terminus of pIII). However, when using the BamHI site located at position 198 one must be careful of the unpaired Cysteine residue (C201) that could cause problems during phage display if one is using a non-truncated version of pIII.

An advantage of using pIII rather than pVIII is that pIII allows for monovalent display when using a phagemid (Ff-phage derived plasmid) combined with a helper phage. Moreover, pIII allows for the insertion of larger protein sequences (>100 amino acids) and is more tolerant to it than pVIII. However, using pIII as the fusion partner can lead to a decrease in phage infectivity leading to problems such as selection bias caused by difference in phage growth rate or even worse, the phage's inability to infect its host. Loss of phage infectivity can be avoided by using a phagemid plasmid and a helper phage so that the resultant phage contains both wild type and fusion pIII.

cDNA has also been analyzed using pIII via a two complementary leucine zippers system, Direct Interaction Rescue or by adding an 8-10 amino acid linker between the cDNA and pIII at the C-terminus.

pVIII

pVIII is the main coat protein of Ff phages. Peptides are usually fused to the N-terminus of pVIII. Usually peptides that can be fused to pVIII are 6-8 amino acids long. The size restriction seems to have less to do with structural impediment caused by the added section and more to do with the size exclusion caused by pIV during coat protein export. Since there are around 2700 copies of the protein on a typical phages, it is more likely that the protein of interest will be expressed polyvalently even if a phagemid is used. This makes the use of this protein unfavorable for the discovery of high affinity binding partners.

To overcome the size problem of pVIII, artificial coat proteins have been designed. An example is Weiss and Sidhu's inverted artificial coat protein (ACP) which allows the display of large proteins at the C-terminus. The ACP's could display a protein of 20kDa, however, only at low levels (mostly only monovalently).

pVI

pVI has been widely used for the display of cDNA libraries. The display of cDNA libraries via phage display is an attractive alternative to the yeast-2-hybrid method for the discovery of interacting proteins and peptides due to its high throughput capability. pVI has been used preferentially to pVIII and pIII for the expression of cDNA libraries because one can add the protein of interest to the C-terminus of pVI without greatly affecting pVI's role in phage assembly. This means that the stop codon in the cDNA is no longer an issue. However, phage display of cDNA is always limited by the inability of most prokaryotes in producing post-translational modifications present in eukaryotic cells or by the misfolding of multi-domain proteins.

While pVI has been useful for the analysis of cDNA libraries, pIII and pVIII remain the most utilized coat proteins for phage display.

pVII and pIX

In an experiment in 1995, display of Glutathione S-transferase was attempted on both pVII and pIX and failed. However, phage display of this protein was completed successfully after the addition of a periplasmic signal sequence (pelB or ompA) on the N-terminus. In a recent study, it has been shown that AviTag, FLAG and His could be displayed on pVII without the need of a signal sequence. Then the expression of single chain Fv's (scFv), and single chain T cell receptors (scTCR) were expressed both with and without the signal sequence.

PelB (an amino acid signal sequence that targets the protein to the periplasm where a signal peptidase then cleaves off PelB) improved the phage display level when compared to pVII and pIX fusions without the signal sequence. However, this led to the incorporation of more helper phage genomes rather than phagemid genomes. In all cases, phage display levels were lower than using pIII fusion. However, lower display might be more favorable for the selection of binders due to lower display being closer to true monovalent display. In five out of six occasions, pVII and pIX fusions without pelB was more efficient than pIII fusions in affinity selection assays. The paper even goes on to state that pVII and pIX display platforms may outperform pIII in the long run.

The use of pVII and pIX instead of pIII might also be an advantage because virion rescue may be undertaken without breaking the virion-antigen bond if the pIII used is wild type. Instead, one could cleave in a section between the bead and the antigen to elute. Since the pIII is intact it does not matter whether the antigen remains bound to the phage.

T7 phages

The issue of using Ff phages for phage display is that they require the protein of interest to be translocated across the bacterial inner membrane before they are assembled into the phage. Some proteins cannot undergo this process and therefore cannot be displayed on the surface of Ff phages. In these cases, T7 phage display is used instead. In T7 phage display, the protein to be displayed is attached to the C-terminus of the gene 10 capsid protein of T7.

The disadvantage of using T7 is that the size of the protein that can be expressed on the surface is limited to shorter peptides because large changes to the T7 genome cannot be accommodated like it is in M13 where the phage just makes its coat longer to fit the larger genome within it. However, it can be useful for the production of a large protein library for scFV selection where the scFV is expressed on an M13 phage and the antigens are expressed on the surface of the T7 phage.

Bioinformatics Resources and Tools

Databases and computational tools for mimotopes have been an important part of phage display study. Databases, programs and web servers have been widely used to exclude target-unrelated peptides, characterize small molecules-protein interactions and map protein-protein interactions. Users can use three dimensional structure of a protein and the peptides selected from phage display experiment to map conformational eptiopes. Some of the fast and efficient computational methods are available online (e.g. EpiSearch http://curie.utmb.edu/episearch.html).

References

- Scott JS, Barbas CF III, Burton, DA (2001). Phage Display: A Laboratory Manual. Plainview, N.Y: Cold Spring Harbor Laboratory Press. ISBN 0-87969-740-7.

- Lowman HB, Clackson T (2004). "1.3". Phage display: a practical approach. Oxford [Oxfordshire]: Oxford University Press. pp. 10–11. ISBN 0-19-963873-X.

- Madigan, Michael T. (2012). Brock biology of microorganisms (13th ed.). San Francisco: Benjamin Cummings. ISBN 9780321649638.

- Ashok, Y; Nanekar, R; Jaakola, VP (December 2015). "Defining thermostability of membrane proteins by western blotting.". Protein engineering, design & selection : PEDS. 28 (12): 539–42. doi:10.1093/protein/gzv049. PMID 26384510.

- ACHARYA, TANKESHWAR. "Key biochemical methods used to distinguish Mycobacterial group". Retrieved 21 November 2014.

- "A Potential Boon for Pancreatic Cancer Patients". Johns Hopkins Surgery: News From the Johns Hopkins Department of Surgery. 2014-06-23.

- Negi SS, Braun W (2009). "Automated Detection of Conformational Epitopes Using Phage Display Peptide Sequences". Bioinform Biol Insights. 3: 71–81. PMC 2808184 . PMID 20140073.

- Hanaor, Dorian A. H.; Sorrell, Charles C. (2014). "Sand Supported Mixed-Phase TiO2 Photocatalysts for Water Decontamination Applications". Advanced Engineering Materials. 16 (2): 248–254. doi:10.1002/adem.201300259.

- Kreisig, T; Prasse, AA; Zscharnack, K; Volke, D; Zuchner, T (8 July 2014). "His-tag protein monitoring by a fast mix-and-measure immunoassay.". Scientific Reports. 4: 5613. doi:10.1038/srep05613. PMID 25000910.

- Grossi, M.; Lazzarini, R.; Lanzoni, M.; Pompei, A.; Matteuzzi, D.; Riccò, B. (2013). "A portable sensor with disposable electrodes for water bacterial quality assessment". IEEE Sensors Journal. 13 (5): 1775–1781. doi:10.1109/JSEN.2013.2243142.

- Czyzewski, BK; Wang, DN (11 March 2012). "Identification and characterization of a bacterial hydrosulphide ion channel.". Nature. 483 (7390): 494–7. doi:10.1038/nature10881. PMID 22407320.

- Mancusso, R; Gregorio, GG; Liu, Q; Wang, DN (22 November 2012). "Structure and mechanism of a bacterial sodium-dependent dicarboxylate transporter.". Nature. 491 (7425): 622–6. doi:10.1038/nature11542. PMID 23086149.

6

Subdisciplines of Microbiology

The subdisciplines explained in this text are immunology, virology, microbial ecology, mycology, microbial genetics etc. The branch of biology that covers the analyses of immune systems is known as immunology whereas virology is the study of viruses. This section is a compilation of the various subdisciplines of microbiology that forms an integral part of the broader subject matter.

Immunology

Immunology is a branch of biology that covers the study of immune systems in all organisms. It was the Russian biologist Ilya Ilyich Mechnikov who boosted studies on immunology, and received the Nobel Prize in 1908 for his work. He jabbed the thorn of a rose on a starfish and noted that, 24 hours later, cells were surrounding the tip. It was an active response of the body, trying to maintain its integrity. It was Mechnikov who first observed the phenomenon of phagocytosis, in which the body defends itself against a foreign body, and coined the term. It charts, measures, and contextualizes the: physiological functioning of the immune system in states of both health and diseases; malfunctions of the immune system in immunological disorders (such as autoimmune diseases, hypersensitivities, immune deficiency, and transplant rejection); the physical, chemical and physiological characteristics of the components of the immune system *in vitro*, *in situ,* and *in vivo*. Immunology has applications in numerous disciplines of medicine, particularly in the fields of organ transplantation, oncology, virology, bacteriology, parasitology, psychiatry, and dermatology.

Prior to the designation of immunity from the etymological root *immunis*, which is Latin for "exempt"; early physicians characterized organs that would later be proven as essential components of the immune system. The important lymphoid organs of the immune system are the thymus and bone marrow, and chief lymphatic tissues such as spleen, tonsils, lymph vessels, lymph nodes, adenoids, and liver. When health conditions worsen to emergency status, portions of immune system organs including the thymus, spleen, bone marrow, lymph nodes and other lymphatic tissues can be surgically excised for examination while patients are still alive.

Many components of the immune system are typically cellular in nature and not associated with any specific organ; but rather are embedded or circulating in various tissues located throughout the body.

Classical Immunology

Classical immunology ties in with the fields of epidemiology and medicine. It studies the relationship between the body systems, pathogens, and immunity. The earliest written mention of immunity can be traced back to the plague of Athens in 430 BCE. Thucydides noted that people who had recovered from a previous bout of the disease could nurse the sick without contracting the illness a second time. Many other ancient societies have references to this phenomenon, but it was not until the 19th and 20th centuries before the concept developed into scientific theory.

The study of the molecular and cellular components that comprise the immune system, including their function and interaction, is the central science of immunology. The immune system has been divided into a more primitive innate immune system and, in vertebrates, an acquired or adaptive immune system. The latter is further divided into humoral (or antibody) and cell-mediated components.

The humoral (antibody) response is defined as the interaction between antibodies and antigens. Antibodies are specific proteins released from a certain class of immune cells known as B lymphocytes, while antigens are defined as anything that elicits the generation of antibodies ("anti"body "gen"erators). Immunology rests on an understanding of the properties of these two biological entities and the cellular response to both.

Immunological research continues to become more specialized, pursuing non-classical models of immunity and functions of cells, organs and systems not previously associated with the immune system (Yemeserach 2010).

Clinical Immunology

Clinical immunology is the study of diseases caused by disorders of the immune system (failure, aberrant action, and malignant growth of the cellular elements of the system). It also involves diseases of other systems, where immune reactions play a part in the pathology and clinical features.

The diseases caused by disorders of the immune system fall into two broad categories:

- immunodeficiency, in which parts of the immune system fail to provide an adequate response (examples include chronic granulomatous disease and primary immune diseases);

- autoimmunity, in which the immune system attacks its own host's body (examples include systemic lupus erythematosus, rheumatoid arthritis, Hashimoto's disease and myasthenia gravis).

Other immune system disorders include various hypersensitivities (such as in asthma and other allergies) that respond inappropriately to otherwise harmless compounds.

The most well-known disease that affects the immune system itself is AIDS, an immunodeficiency characterized by the suppression of CD4+ ("helper") T cells, dendritic cells and macrophages by the Human Immunodeficiency Virus (HIV).

Clinical immunologists also study ways to prevent the immune system's attempts to destroy allografts (transplant rejection).

Developmental Immunology

The body's capability to react to antigen depends on a person's age, antigen type, maternal factors and the area where the antigen is presented. Neonates are said to be in a state of physiological immunodeficiency, because both their innate and adaptive immunological responses are greatly suppressed. Once born, a child's immune system responds favorably to protein antigens while not as well to glycoproteins and polysaccharides. In fact, many of the infections acquired by neonates are caused by low virulence organisms like *Staphylococcus* and *Pseudomonas*. In neonates, opsonic activity and the ability to activate the complement cascade is very limited. For example, the mean level of C3 in a newborn is approximately 65% of that found in the adult. Phagocytic activity is also greatly impaired in newborns. This is due to lower opsonic activity, as well as diminished up-regulation of integrin and selectin receptors, which limit the ability of neutrophils to interact with adhesion molecules in the endothelium. Their monocytes are slow and have a reduced ATP production, which also limits the newborn's phagocytic activity. Although, the number of total lymphocytes is significantly higher than in adults, the cellular and humoral immunity is also impaired. Antigen-presenting cells in newborns have a reduced capability to activate T cells. Also, T cells of a newborn proliferate poorly and produce very small amounts of cytokines like IL-2, IL-4, IL-5, IL-12, and IFN-g which limits their capacity to activate the humoral response as well as the phagocitic activity of macrophage. B cells develop early during gestation but are not fully active.

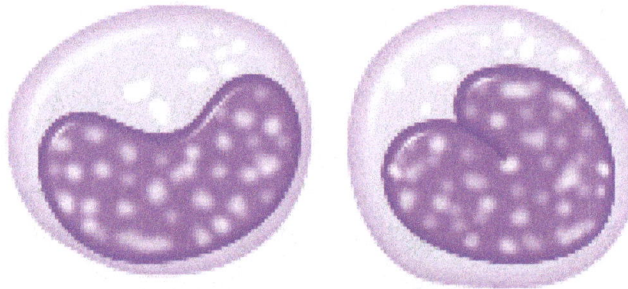

Artist's impression of monocytes.

Maternal factors also play a role in the body's immune response. At birth, most of the immunoglobulin present is maternal IgG. Because IgM, IgD, IgE and IgA don't cross the placenta, they are almost undetectable at birth. Some IgA is provided by breast milk. These passively-acquired antibodies can protect the newborn for up to 18 months, but their response is usually short-lived and of low affinity. These antibodies can also produce a negative response. If a child is exposed to the antibody for a particular antigen before being exposed to the antigen itself then the child will produce a dampened response. Passively acquired maternal antibodies can suppress the antibody response to active immunization. Similarly the response of T-cells to vaccination differs in children compared to adults, and vaccines that induce Th1 responses in adults do not readily elicit these same responses in neonates. Between six and nine months after birth, a child's immune system begins to respond more strongly to glycoproteins, but there is usually no marked improvement in their response to polysaccharides until they are at least one year old. This can be the reason for distinct time frames found in vaccination schedules.

During adolescence, the human body undergoes various physical, physiological and immunological changes triggered and mediated by hormones, of which the most significant in females is 17-β-oestradiol (an oestrogen) and, in males, is testosterone. Oestradiol usually begins to act around the age of 10 and testosterone some months later. There is evidence that these steroids act directly not only on the primary and secondary sexual characteristics but also have an effect on the development and regulation of the immune system, including an increased risk in developing pubescent and post-pubescent autoimmunity. There is also some evidence that cell surface receptors on B cells and macrophages may detect sex hormones in the system.

The female sex hormone 17-β-oestradiol has been shown to regulate the level of immunological response, while some male androgens such as testosterone seem to suppress the stress response to infection. Other androgens, however, such as DHEA, increase immune response. As in females, the male sex hormones seem to have more control of the immune system during puberty and post-puberty than during the rest of a male's adult life.

Physical changes during puberty such as thymic involution also affect immunological response.

Immunotherapy

The use of immune system components to treat a disease or disorder is known as immunotherapy. Immunotherapy is most commonly used in the context of the treatment of cancers together with chemotherapy (drugs) and radiotherapy (radiation). However, immunotherapy is also often used in the immunosuppressed (such as HIV patients) and people suffering from other immune deficiencies or autoimmune diseases. This includes regulating factors such as IL-2, IL-10, GM-CSF B, IFN-α.

Diagnostic Immunology

The specificity of the bond between antibody and antigen has made the antibody an excellent tool for the detection of substances by a variety of diagnostic techniques. Antibodies specific for a desired antigen can be conjugated with an isotopic (radio) or fluorescent label or with a color-forming enzyme in order to detect it. However, the similarity between some antigens can lead to false positives and other errors in such tests by antibodies cross-reacting with antigens that aren't exact matches.

Cancer Immunology

The study of the interaction of the immune system with cancer cells can lead to diagnostic tests and therapies with which to find and fight cancer.

Reproductive Immunology

This area of the immunology is devoted to the study of immunological aspects of the reproductive process including fetus acceptance. The term has also been used by fertility clinics to address fertility problems, recurrent miscarriages, premature deliveries and dangerous complications such as pre-eclampsia.

Theoretical Immunology

Immunology is strongly experimental in everyday practice but is also characterized by an ongoing theoretical attitude. Many theories have been suggested in immunology from the end of the nineteenth century up to the present time. The end of the 19th century and the beginning of the 20th century saw a battle between "cellular" and "humoral" theories of immunity. According to the cellular theory of immunity, represented in particular by Elie Metchnikoff, it was cells – more precisely, phagocytes – that were responsible for immune responses. In contrast, the humoral theory of immunity, held by Robert Koch and Emil von Behring, among others, stated that the active immune agents were soluble components (molecules) found in the organism's "humors" rather than its cells.

In the mid-1950s, Frank Burnet, inspired by a suggestion made by Niels Jerne, formulated the clonal selection theory (CST) of immunity. On the basis of CST, Burnet developed a theory of how an immune response is triggered according to the self/nonself distinction: "self" constituents (constituents of the body) do not trigger destructive immune responses, while "nonself" entities (e.g., pathogens, an allograft) trigger a destructive immune response. The theory was later modified to reflect new discoveries regarding histocompatibility or the complex "two-signal" activation of T cells. The self/nonself theory of immunity and the self/nonself vocabulary have been criticized, but remain very influential.

More recently, several theoretical frameworks have been suggested in immunology, including "autopoietic" views, "cognitive immune" views, the "danger model" (or "danger theory"), and the "discontinuity" theory. The danger model, suggested by Polly Matzinger and colleagues, has been very influential, arousing many comments and discussions.

Immunologist

According to the American Academy of Allergy, Asthma, and Immunology (AAAAI), "an immunologist is a research scientist who investigates the immune system of vertebrates (including the human immune system). Immunologists include research scientists (PhDs) who work in laboratories. Immunologists also include physicians who, for example, treat patients with immune system disorders. Some immunologists are physician-scientists who combine laboratory research with patient care."

Career in Immunology

Bioscience is the overall major in which undergraduate students who are interested in general well-being take in college. Immunology is a branch of bioscience for undergraduate programs but the major gets specified as students move on for graduate program in immunology. The aim of immunology is to study the health of humans and animals through effective yet consistent research, (AAAAI, 2013). The most important thing about being immunologists is the research because it is the biggest portion of their jobs.

Most graduate immunology schools follow the AAI courses immunology which are offered throughout numerous schools in the United States. For example, in New York State, there are several universities that offer the AAI courses immunology: Albany Medical College, Cornell University, Icahn School of Medicine at Mount Sinai, New York University Langone Medical Center, University at Albany (SUNY), University at Buffalo (SUNY), University of Rochester Medical Center and Upstate Medical University (SUNY). The AAI immunology courses include an Introductory Course and an Advance Course. The Introductory Course is a course that gives students an overview of the basics of immunology.

In addition, this Introductory Course gives students more information to complement general biology or science training. It also has two different parts: Part I is an introduction to the basic principles of immunology and Part II is a clinically-oriented lecture series. On the other hand, the Advanced Course is another course for those who are willing to expand or update their understanding of immunology. It is advised for students who want to attend the Advanced Course to have a background of the principles of immunology. Most schools require students to take electives in other to complete their degrees. A Master's degree requires two years of study following the attainment of a bachelor's degree. For a doctoral programme it is required to take two additional years of study.

The expectation of occupational growth in immunology is an increase of 36 percent from 2010 to 2020. The median annual wage was $76,700 in May 2010. However, the lowest 10 percent of immunologists earned less than $41,560, and the top 10 percent earned more than $142,800, (Bureau of Labor Statistics, 2013). The practice of immunology itself is not specified by the U.S. Department of Labor but it belongs to the practice of life science in general.

Virology

Virology is the study of viruses – submicroscopic, parasitic particles of genetic material contained in a protein coat – and virus-like agents. It focuses on the following aspects of viruses: their structure, classification and evolution, their ways to infect and exploit host cells for reproduction, their interaction with host organism physiology and immunity, the diseases they cause, the techniques to isolate and culture them, and their use in research and therapy. Virology is considered to be a subfield of microbiology or of medicine.

Virus Structure and Classification

A major branch of virology is virus classification. Viruses can be classified according to the host cell they infect: animal viruses, plant viruses, fungal viruses, and bacteriophages (viruses infecting bacterium, which include the most complex viruses). Another classification uses the geometrical shape of their capsid (often a helix or an icosahedron) or the virus's structure (e.g. presence or absence of a lipid envelope). Viruses range in size from about 30 nm to about 450 nm, which means that most of them cannot be seen with light microscopes. The shape and structure of viruses has been studied by electron microscopy, NMR spectroscopy, and X-ray crystallography.

The most useful and most widely used classification system distinguishes viruses according to the type of nucleic acid they use as genetic material and the viral replication method they employ to coax host cells into producing more viruses:

- DNA viruses (divided into double-stranded DNA viruses and single-stranded DNA viruses),

- RNA viruses (divided into positive-sense single-stranded RNA viruses, negative-sense single-stranded RNA viruses and the much less common double-stranded RNA viruses),

- reverse transcribing viruses (double-stranded reverse-transcribing DNA viruses and single-stranded reverse-transcribing RNA viruses including retroviruses).

The latest report by the International Committee on Taxonomy of Viruses (2005) lists 5450 viruses, organized in over 2,000 species, 287 genera, 73 families and 3 orders.

Virologists also study *subviral particles*, infectious entities notably smaller and simpler than viruses:

- viroids (naked circular RNA molecules infecting plants),

- satellites (nucleic acid molecules with or without a capsid that require a helper virus for infection and reproduction), and

- prions (proteins that can exist in a pathological conformation that induces other prion molecules to assume that same conformation).

Taxa in virology are not necessarily monophyletic, as the evolutionary relationships of the various virus groups remain unclear. Three hypotheses regarding their origin exist:

1. Viruses arose from non-living matter, separately from yet in parallel to cells, perhaps in the form of self-replicating RNA ribozymes similar to viroids.

2. Viruses arose by genome reduction from earlier, more competent cellular life forms that became parasites to host cells and subsequently lost most of their functionality; examples of such tiny parasitic prokaryotes are Mycoplasma and Nanoarchaea.

3. Viruses arose from mobile genetic elements of cells (such as transposons, retrotransposons or plasmids) that became encapsulated in protein capsids, acquired the ability to "break free" from the host cell and infect other cells.

Of particular interest here is mimivirus, a giant virus that infects amoebae and encodes much of the molecular machinery traditionally associated with bacteria. Is it a simplified version of a parasitic prokaryote, or did it originate as a simpler virus that acquired genes from its host?

The evolution of viruses, which often occurs in concert with the evolution of their hosts, is studied in the field of viral evolution.

While viruses reproduce and evolve, they do not engage in metabolism, do not move, and depend on a host cell for reproduction. The often-debated question of whether they are alive or not is a matter of definition that does not affect the biological reality of viruses.

Viral Diseases and Host Defenses

One main motivation for the study of viruses is the fact that they cause many important infectious diseases, among them the common cold, influenza, rabies, measles, many forms of diarrhea, hepatitis, Dengue fever, yellow fever, polio, smallpox and AIDS. Herpes simplex causes cold sores and genital herpes and is under investigation as a possible factor in Alzheimer's.

Some viruses, known as oncoviruses, contribute to the development of certain forms of cancer. The best studied example is the association between Human papillomavirus and cervical cancer: almost all cases of cervical cancer are caused by certain strains of this sexually transmitted virus. Another example is the association of infection with hepatitis B and hepatitis C viruses and liver cancer.

Some subviral particles also cause disease: the transmissible spongiform encephalopathies, which include Kuru, Creutzfeldt–Jakob disease and bovine spongiform encephalopathy ("mad cow disease"), are caused by prions, hepatitis D is due to a satellite virus.

The study of the manner in which viruses cause disease is viral pathogenesis. The degree to which a virus causes disease is its virulence.

When the immune system of a vertebrate encounters a virus, it may produce specific antibodies which bind to the virus and neutralize its infectivity or mark it for destruction. Antibody presence in blood serum is often used to determine whether a person has been exposed to a given virus in the past, with tests such as ELISA. Vaccinations protect against viral diseases, in part, by eliciting the production of antibodies. Monoclonal antibodies, specific to the virus, are also used for detection, as in fluorescence microscopy.

A second defense of vertebrates against viruses, cell-mediated immunity, involves immune cells known as T cells: the body's cells constantly display short fragments of their proteins on the cell's surface, and if a T cell recognizes a suspicious viral fragment there, the host cell is destroyed and the virus-specific T-cells proliferate. This mechanism is jump-started by certain vaccinations.

RNA interference, an important cellular mechanism found in plants, animals and many other eukaryotes, most likely evolved as a defense against viruses. An elaborate machinery of interacting enzymes detects double-stranded RNA molecules (which occur as part of the life cycle of many viruses) and then proceeds to destroy all single-stranded versions of those detected RNA molecules.

Every lethal viral disease presents a paradox: killing its host is obviously of no benefit to the virus, so how and why did it evolve to do so? Today it is believed that most viruses are relatively benign in their natural hosts; some viral infection might even be beneficial to the host. The lethal viral diseases are believed to have resulted from an "accidental" jump of the virus from a species in which it is benign to a new one that is not accustomed to it. For example, viruses that cause serious influenza in humans probably have pigs or birds as their natural host, and HIV is thought to derive from the benign non-human primate virus SIV.

While it has been possible to prevent (certain) viral diseases by vaccination for a long time, the development of antiviral drugs to *treat* viral diseases is a comparatively recent development. The first such drug was interferon, a substance that is naturally produced when an infection is detected and stimulates other parts of the immune system.

Molecular Biology Research and Viral Therapy

Bacteriophages, the viruses which infect bacteria, can be relatively easily grown as viral plaques on bacterial cultures. Bacteriophages occasionally move genetic material from one bacterial cell to another in a process known as transduction, and this horizontal gene transfer is one reason why they served as a major research tool in the early development of molecular biology. The genetic code, the function of ribozymes, the first recombinant DNA and early genetic libraries were all arrived at using bacteriophages. Certain genetic elements derived from viruses, such as highly effective promoters, are commonly used in molecular biology research today.

Growing animal viruses outside of the living host animal is more difficult. Classically, fertilized chicken eggs have often been used, but cell cultures are increasingly employed for this purpose today.

Since some viruses that infect eukaryotes need to transport their genetic material into the host cell's nucleus, they are attractive tools for introducing new genes into the host (known as transformation or transfection). Modified retroviruses are often used for this purpose, as they integrate their genes into the host's chromosomes.

This approach of using viruses as gene vectors is being pursued in the gene therapy of genetic diseases. An obvious problem to be overcome in viral gene therapy is the rejection of the transforming virus by the immune system.

Phage therapy, the use of bacteriophages to combat bacterial diseases, was a popular research topic before the advent of antibiotics and has recently seen renewed interest.

Oncolytic viruses are viruses that preferably infect cancer cells. While early efforts to employ these viruses in the therapy of cancer failed, there have been reports in 2005 and 2006 of encouraging preliminary results.

Other Uses of Viruses

A new application of genetically engineered viruses in nanotechnology was recently described; For a use in mapping neurons see the applications of pseudorabies in neuroscience.

History of Virology

The word *virus* appeared in 1599 and originally meant "venom".

A very early form of vaccination known as variolation was developed several thousand years ago in China. It involved the application of materials from smallpox sufferers in order to immunize others. In 1717 Lady Mary Wortley Montagu observed the practice in Istanbul and attempted to popularize it in Britain, but encountered considerable resistance. In 1796 Edward Jenner developed a much safer method, using cowpox to successfully immunize a young boy against smallpox, and this practice was widely adopted. Vaccinations against other viral diseases followed, including the successful rabies vaccination by Louis Pasteur in 1886. The nature of viruses however was not clear to these researchers.

Dmitri Ivanovsky

Martinus Beijerinck

In 1892, the Russian biologist Dmitry Ivanovsky used a Chamberland filter to try to isolate the bacteria that caused tobacco mosaic disease. His experiments showed that crushed leaf extracts from infected tobacco plants remained infectious after filtration. Ivanovsky reported a minuscule infectious agent or toxin, capable of passing the filter, may be being produced by a bacterium.

In 1898 Martinus Beijerinck repeated Ivanovski's work but went further and passed the "filterable agent" from plant to plant, found the action undiminished, and concluded it infectious—replicating in the host—and thus not a mere toxin. He called it *contagium vivum fluidum*. The question of whether the agent was a "living fluid" or a particle was however still open.

In 1903 it was suggested for the first time that transduction by viruses might cause cancer. In 1908 Bang and Ellerman showed that a filterable virus could transmit chicken leukemia, data largely ignored till the 1930s when leukemia became regarded as cancerous. In 1911 Peyton Rous reported the transmission of chicken sarcoma, a solid tumor, with a virus, and thus Rous became "father of tumor virology". The virus was later called Rous sarcoma virus 1 and understood to be a retrovirus. Several other cancer-causing retroviruses have since been described.

The existence of viruses that infect bacteria (bacteriophages) was first recognized by Frederick Twort in 1911, and, independently, by Félix d'Herelle in 1917. As bacteria could be grown easily in culture, this led to an explosion of virology research.

The cause of the devastating Spanish flu pandemic of 1918 was initially unclear. In late 1918, French scientists showed that a "filter-passing virus" could transmit the disease to people and animals, fulfilling Koch's postulates.

In 1926 it was shown that scarlet fever is caused by a bacterium that is infected by a certain bacteriophage.

While plant viruses and bacteriophages can be grown comparatively easily, animal viruses normally require a living host animal, which complicates their study immensely. In 1931 it was shown that influenza virus could be grown in fertilized chicken eggs, a method that is still used today to produce vaccines. In 1937, Max Theiler managed to grow the yellow fever virus in chicken eggs and produced a vaccine from an attenuated virus strain; this vaccine saved millions of lives and is still being used today.

Max Delbrück, an important investigator in the area of bacteriophages, described the basic "life cycle" of a virus in 1937: rather than "growing", a virus particle is assembled from its constituent pieces in one step; eventually it leaves the host cell to infect other cells. The Hershey–Chase experiment in 1952 showed that only DNA and not protein enters a bacterial cell upon infection with bacteriophage T2. Transduction of bacteria by bacteriophages was first described in the same year.

In 1949 John F. Enders, Thomas Weller and Frederick Robbins reported growth of poliovirus in cultured human embryonal cells, the first significant example of an animal virus grown outside of animals or chicken eggs. This work aided Jonas Salk in deriving a polio vaccine from deactivated polio viruses; this vaccine was shown to be effective in 1955.

The first virus that could be crystalized and whose structure could therefore be elucidated in detail was tobacco mosaic virus (TMV), the virus that had been studied earlier by Ivanovski and Beijerink. In 1935, Wendell Stanley achieved its crystallization for electron microscopy and showed that it remains active even after crystallization. Clear X-ray diffraction pictures of the crystallized virus were obtained by Bernal and Fankuchen in 1941. Based on such pictures, Rosalind Franklin proposed the full structure of the tobacco mosaic virus in 1955. Also in 1955, Heinz Fraenkel-Conrat and Robley Williams showed that purified TMV RNA and its capsid (coat) protein can self-assemble into functional virions, suggesting that this assembly mechanism is also used within the host cell, as Delbrück had proposed earlier.

From the 1950s-60's, Chester M. Southam, a leading virologist and cancer researcher, injected cancer patients, healthy individuals, and prison inmates from the Ohio Penitentiary with HeLa cancer cells in order to observe if cancer could be transmitted. Additionally, in hopes of creating a vaccine for cancer, he observed if the subjects could become immune to cancer by developing an acquired immune response. This experiment was highly controversial, as the cancer patient subjects were unaware that they were being injected with cancer cells.

In 1963, the Hepatitis B virus was discovered by Baruch Blumberg who went on to develop a hepatitis B vaccine.

In 1965, Howard Temin described the first retrovirus: a virus whose RNA genome was reverse transcribed into complementary DNA (cDNA), then integrated into the host's genome and expressed from that template. The viral enzyme reverse transcriptase, which along with integrase is a distinguishing trait of retroviruses, was first described in 1970, independently by Howard Temin and David Baltimore. The first retrovirus infecting humans was identified by Robert Gallo in 1974. Later it was found that reverse transcriptase is not specific to retroviruses; retrotransposons which code for reverse transcriptase are abundant in the genomes of all eukaryotes. About 10-40% of the human genome derives from such retrotransposons.

In 1975 the functioning of oncoviruses was clarified considerably. Until that time, it was thought that these viruses carried certain genes called oncogenes which, when inserted into the host's genome, would cause cancer. Michael Bishop and Harold Varmus showed that the oncogene of Rous sarcoma virus is in fact not specific to the virus but is contained in the genome of healthy animals of many species. The oncovirus can switch this pre-existing benign proto-oncogene on, turning it into a true oncogene that causes cancer.

1976 saw the first recorded outbreak of Ebola virus disease, a highly lethal virally transmitted disease.

In 1977, Frederick Sanger achieved the first complete sequencing of the genome of any organism, the bacteriophage Phi X 174. In the same year, Richard Roberts and Phillip Sharp independently showed that the genes of adenovirus contain introns and therefore require gene splicing. It was later realized that almost all genes of eukaryotes have introns as well.

A worldwide vaccination campaign led by the UN World Health Organization resulted in the eradication of smallpox in 1979.

In 1982, Stanley Prusiner discovered prions and showed that they cause scrapie.

The first cases of AIDS were reported in 1981, and HIV, the retrovirus causing it, was identified in 1983 by Luc Montagnier, Françoise Barré-Sinoussi and Robert Gallo. Tests detecting HIV infection by detecting the presence of HIV antibody were developed. Subsequent tremendous research efforts turned HIV into the best studied virus. Human Herpes Virus 8, the cause of Kaposi's sarcoma which is often seen in AIDS patients, was identified in 1994. Several antiretroviral drugs were developed in the late 1990s, decreasing AIDS mortality dramatically in developed countries.

The Hepatitis C virus was identified using novel molecular cloning techniques in 1987, leading to screening tests that dramatically reduced the incidence of post-transfusion hepatitis.

The first attempts at gene therapy involving viral vectors began in the early 1980s, when retroviruses were developed that could insert a foreign gene into the host's genome. They contained the foreign gene but did not contain the viral genome and therefore could not reproduce. Tests in mice were followed by tests in humans, beginning in 1989. The first human studies attempted to correct the genetic disease severe combined immunodeficiency (SCID), but clinical success was limited. In the period from 1990 to 1995, gene therapy was tried on several other diseases and with different viral vectors, but it became clear that the initially high expectations were overstated. In 1999 a further setback occurred when 18-year-old Jesse Gelsinger died in a gene therapy trial. He suffered a severe immune response after having received an adenovirus vector. Success in the gene therapy of two cases of X-linked SCID was reported in 2000.

In 2002 it was reported that poliovirus had been synthetically assembled in the laboratory, representing the first synthetic organism. Assembling the 7741-base genome from scratch, starting with the virus's published RNA sequence, took about two years. In 2003 a faster method was shown to assemble the 5386-base genome of the bacteriophage Phi X 174 in 2 weeks.

The giant mimivirus, in some sense an intermediate between tiny prokaryotes and ordinary viruses, was described in 2003 and sequenced in 2004.

The strain of Influenza A virus subtype H1N1 that killed up to 50 million people during the Spanish flu pandemic in 1918 was reconstructed in 2005. Sequence information was pieced together from preserved tissue samples of flu victims; viable virus was then synthesized from this sequence. The 2009 flu pandemic involved another strain of Influenza A H1N1, commonly known as "swine flu".

By 1985, Harald zur Hausen had shown that two strains of Human papillomavirus (HPV) cause most cases of cervical cancer. Two vaccines protecting against these strains were released in 2006.

In 2006 and 2007 it was reported that introducing a small number of specific transcription factor genes into normal skin cells of mice or humans can turn these cells into pluripotent stem cells, known as induced pluripotent stem cells. The technique uses modified retroviruses to transform the cells; this is a potential problem for human therapy since these viruses integrate their genes at a random location in the host's genome, which can interrupt other genes and potentially causes cancer.

In 2008, Sputnik virophage was described, the first known *virophage*: it uses the machinery of a helper virus to reproduce and inhibits reproduction of that helper virus. Sputnik reproduces in amoeba infected by mamavirus, a relative of the mimivirus mentioned above and the largest known virus to date.

An endogenous retrovirus (ERV) is a retrovirus whose genome has been permanently incorporated into the germ-line genome of some organism and that is therefore copied with each reproduction of that organism. It is estimated that about 9 percent of the human genome have their origin in ERVs. In 2015 it was shown that proteins from an ERV are actively expressed in 3-day-old human embryos and appear to play a role in embryonal development and protect embryos from infection by other viruses.

Microbial Ecology

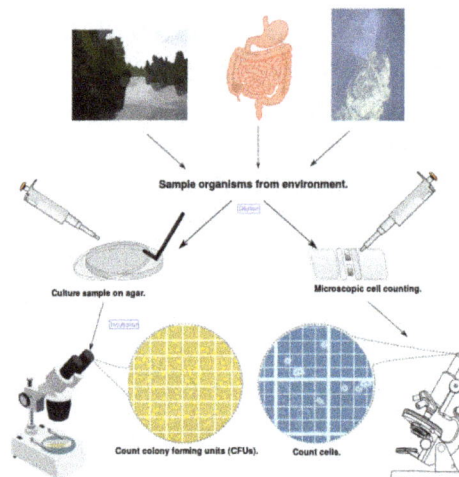

The great plate count anomaly. Counts of cells obtained via cultivation are orders of magnitude lower than those directly observed under the microscope. This is because microbiologists are able to cultivate only a minority of naturally occurring microbes using current laboratory techniques, depending on the environment.

Microbial ecology (or environmental microbiology) is the ecology of microorganisms: their relationship with one another and with their environment. It concerns the three major domains of life—Eukaryota, Archaea, and Bacteria—as well as viruses.

Microorganisms, by their omnipresence, impact the entire biosphere. Microbial life plays a primary role in regulating biogeochemical systems in virtually all of our planet's environments, including some of the most extreme, from frozen environments and acidic lakes, to hydrothermal vents at the bottom of deepest oceans, and some of the most familiar, such as the human small intestine. As a consequence of the quantitative magnitude of microbial life (Whitman and coworkers calculated 5.0×10^{30} cells, eight orders of magnitude greater than the number of stars in the observable universe) microbes, by virtue of their biomass alone, constitute a significant carbon sink. Aside from carbon fixation, microorganisms' key collective metabolic processes (including nitrogen fixation, methane metabolism, and sulfur metabolism) control global biogeochemical cycling. The immensity of microorganisms' production is such that, even in the total absence of eukaryotic life, these processes would likely continue unchanged.

History

While microbes have been studied since the seventeenth-century, this research was from a primarily physiological perspective rather than an ecological one. Martinus Beijerinck invented the enrichment culture, a fundamental method of studying microbes from the environment. He is often incorrectly credited with framing the microbial biogeographic idea that "everything is everywhere, but, the environment selects", which was stated by Lourens Baas Becking. Sergei Winogradsky was one of the first researchers to attempt to understand microorganisms outside of the medical context—making him among the first students of microbial ecology and environmental microbiology—discovering chemosynthesis, and developing the Winogradsky column in the process.

Beijerinck and Windogradsky, however, were focused on the physiology of microorganisms, not the microbial habitat or their ecological interactions. Modern microbial ecology was launched by Robert Hungate and coworkers, who investigated the rumen ecosystem. The study of the rumen required Hungate to develop techniques for culturing anaerobic microbes, and he also pioneered a quantitative approach to the study of microbes and their ecological activities that differentiated the relative contributions of species and catabolic pathways.

Symbiosis

Microbes, especially bacteria, often engage in symbiotic relationships (either positive or negative) with other organisms, and these relationships affect the ecosystem. One example of these fundamental symbioses are chloroplasts, which allow eukaryotes to conduct photosynthesis. Chloroplasts are considered to be endosymbiotic cyanobacteria, a group of bacteria that are thought to be the origins of aerobic photosynthesis. Some theories state that this invention coincides with a major shift in the early earth's atmosphere, from a reducing atmosphere to an oxygen-rich atmosphere. Some theories go as far as saying that this shift in the balance of gases might have triggered a global ice-age known as the Snowball Earth.

Roles

Microorganisms are the backbone of all ecosystems, but even more so in the zones where photosynthesis is unable to take place because of the absence of light. In such zones, chemosynthetic microbes provide energy and carbon to the other organisms.

Other microbes are decomposers, with the ability to recycle nutrients from other organisms' waste products. These microbes play a vital role in biogeochemical cycles. The nitrogen cycle, the phosphorus cycle, the sulphur cycle and the carbon cycle all depend on microorganisms in one way or another. For example, the nitrogen gas which makes up 78% of the earth's atmosphere is unavailable to most organisms, until it is converted to a biologically available form by the microbial process of nitrogen fixation.

Due to the high level of horizontal gene transfer among microbial communities, microbial ecology is also of importance to studies of evolution.

Microbial Resource Management

Biotechnology may be used alongside microbial ecology to address a number of environmental and economic challenges. For example, molecular techniques such as community fingerprinting can be used to track changes in microbial communities over time or assess their biodiversity. Managing the carbon cycle to sequester carbon dioxide and prevent excess methanogenesis is important in mitigating global warming, and the prospects of bioenergy are being expanded by the development of microbial fuel cells. Microbial resource management advocates a more progressive attitude towards disease, whereby biological control agents are favoured over attempts at eradication. Fluxes in microbial communities has to be better characterized for this field's potential to be realised. In addition, there are also clinical implications, as marine microbial symbioses are a valuable source of existing and novel antimicrobial agents, and thus offer another line of inquiry in the evolutionary arms race of antibiotic resistance, a pressing concern for researchers.

In Built Environment and Human Interaction

Microbes exist in all areas, including homes, offices, commercial centers, and hospitals. In 2016, the journal *Microbiome* published a collection of various works studying the microbial ecology of the built environment.

A 2006 study of pathogenic bacteria in hospitals found that their ability to survive varied by the type, with some surviving for only a few days while others survived for months.

The lifespan of microbes in the home varies similarly. Generally bacteria and viruses require a wet environment with a humidity of over 10 percent. *E. coli* can survive for a few hours to a day. Bacteria which form spores can survive longer, with *Staphylococcus aureus* surviving potentially for weeks or, in the case of *Bacillus anthracis*, years.

In the home, pets can be carriers of bacteria; for example, reptiles in particular are commonly carriers of salmonella.

S. aureus is particularly common, and asymptomatically colonizes about 30% of the human population; attempts to decolonize carriers have met with limited success and generally involve mupirocin nasally and chlorhexidine washing, potentially along with vancomycin and cotrimoxazole to address intestinal and urinary tract infections.

Antimicrobials

Some metals, particularly copper and silver, are relatively microbial. Using antimicrobial copper-alloy touch surfaces is a technique which has began to be used in the 21st century to prevent transmission of bacteria. Silver, particularly in nanoparticle form, has also began to be incorporated into building surfaces and fabrics, although concerns have been raised about the potential unhealthy side-effects of the tiny particles on the body.

Mycology

Mushrooms are a kind of fungal reproductive structure.

Mycology is the branch of biology concerned with the study of fungi, including their genetic and biochemical properties, their taxonomy and their use to humans as a source for tinder, medicine, food, and entheogens, as well as their dangers, such as poisoning or infection. A biologist specializing in mycology is called a mycologist.

From mycology arose the field of phytopathology, the study of plant diseases, and the two disciplines remain closely related because the vast majority of "plant" pathogens are fungi.

Historically, mycology was a branch of botany because, although fungi are evolutionarily more closely related to animals than to plants, this was not recognized until a few decades ago. Pioneer *mycologists* included Elias Magnus Fries, Christian Hendrik Persoon, Anton de Bary, and Lewis David von Schweinitz.

Many fungi produce toxins, antibiotics, and other secondary metabolites. For example, the cosmopolitan (worldwide) genus *Fusarium* and their toxins associated with fatal outbreaks of alimentary toxic aleukia in humans were extensively studied by Abraham Joffe.

Fungi are fundamental for life on earth in their roles as symbionts, e.g. in the form of mycorrhizae, insect symbionts, and lichens. Many fungi are able to break down complex organic biomolecules such as lignin, the more durable component of wood, and pollutants such as xenobiotics, petroleum, and polycyclic aromatic hydrocarbons. By decomposing these molecules, fungi play a critical role in the global carbon cycle.

Fungi and other organisms traditionally recognized as fungi, such as oomycetes and myxomycetes (slime molds), often are economically and socially important, as some cause diseases of animals (such as histoplasmosis) as well as plants (such as Dutch elm disease and Rice blast).

Field meetings to find interesting species of fungi are known as 'forays', after the first such meeting organized by the Woolhope Naturalists' Field Club in 1868 and entitled "A foray among the funguses"[sic].

Some fungi can cause disease in humans or other organisms. The study of pathogenic fungi is referred to as medical mycology.

History

It is presumed that humans started collecting mushrooms as food in Prehistoric times. Mushrooms were first written about in the works of Euripides (480-406 B.C.). The Greek philosopher Theophrastos of Eressos (371-288 B.C.) was perhaps the first to try to systematically classify plants; mushrooms were considered to be plants missing certain organs. It was later Pliny the elder (23–79 A.D.), who wrote about truffles in his encyclopedia Naturalis historia.

The Middle Ages saw little advancement in the body of knowledge about fungi. Rather, the invention of the printing press allowed some authors to disseminate superstitions and misconceptions about the fungi that had been perpetuated by the classical authors.

> *Fungi and truffles are neither herbs, nor roots, nor flowers, nor seeds, but merely the superfluous moisture or earth, of trees, or rotten wood, and of other rotting things. This is plain from the fact that all fungi and truffles, especially those that are used for eating, grow most commonly in thundery and wet weather.*
>
> — Jerome Bock (Hieronymus Tragus), 1552

The start of the modern age of mycology begins with Pier Antonio Micheli's 1737 publication of *Nova plantarum genera*. Published in Florence, this seminal work laid the foundations for the systematic classification of grasses, mosses and fungi. The term *mycology* and the complementary *mycologist* were first used in 1836 by M.J. Berkeley.

Medical Mycology

For centuries, certain mushrooms have been documented as a folk medicine in China, Japan, and Russia. Although the use of mushrooms in folk medicine is centered largely on the Asian continent, people in other parts of the world like the Middle East, Poland, and Belarus have been documented using mushrooms for medicinal purposes. Certain mushrooms, especially polypores like Reishi were thought to be able to benefit a wide variety of health ailments. Medicinal mushroom research in the United States is currently active, with studies taking place at City of Hope National

Medical Center, as well as the Memorial Sloan–Kettering Cancer Center.

Current research focuses on mushrooms that may have hypoglycemic activity, anti-cancer activity, anti-pathogenic activity, and immune system-enhancing activity. Recent research has found that the oyster mushroom naturally contains the cholesterol-lowering drug lovastatin, mushrooms produce large amounts of vitamin D when exposed to ultraviolet (UV) light, and that certain fungi may be a future source of taxol. To date, penicillin, lovastatin, ciclosporin, griseofulvin, cephalosporin, ergometrine, and statins are the most famous pharmaceuticals that have been isolated from the fungi kingdom.

Nematology

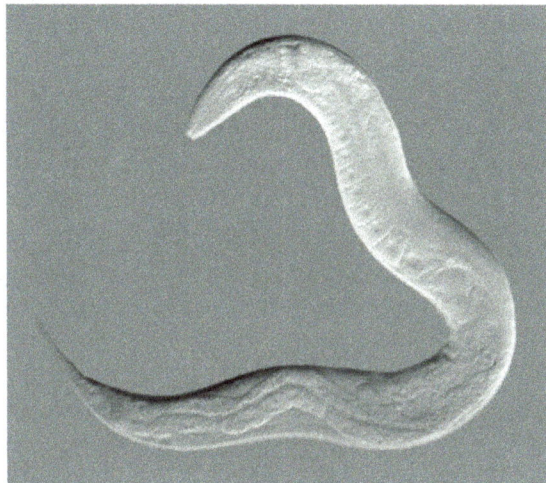

C. elegans

Nematology is the scientific discipline concerned with the study of nematodes, or roundworms. Although nematological investigation dates back to the days of Aristotle or even earlier, nematology as an independent discipline has its recognizable beginnings in the mid to late 19th century.

History: pre-1850

Nematology research, like most fields of science, has its foundations in observations and the recording of these observations. The earliest written account of a nematode "sighting," as it were, may be found in the Pentateuch of the Old Testament in the Bible, in the Fourth Book of Moses called Numbers: "And the Lord sent fiery serpents among the people, and they bit the people; and much people of Israel died". Although no empirical data exists to test the hypothesis, many nematologists assume and circumstantial evidence suggests the "fiery serpents" to be the Guinea worm, *Dracunculus medinensis*, as this nematode is known to inhabit the region near the Red Sea.

Before 1750, a large number of nematode observations were recorded, many by the notable great minds of ancient civilization. Hippocrates (ca. 420 B.C.), Aristotle (ca. 350 B.C.), Celsus (ca 10 B.C.), Galen (ca. 180 A.D.) and Redi (1684) all described nematodes parasitizing humans or other large animals and birds. Borellus (1653) was the first to observe and describe a free-living nematode, which he dubbed the "vinegar eel;" and Tyson (1683) used a crude microscope to describe the rough anatomy of the human intestinal roundworm, *Ascaris lumbricoides*.

Other well-known microscopists spent time observing and describing free-living and animal-parasitic nematodes: Hooke (1683), Leeuwenhoek (1722), Needham (1743), and Spallanzani (1769) are among these. Observations and descriptions of plant parasitic nematodes, which were less conspicuous to ancient scientists, didn't receive as much or as early attention as did animal parasites. The earliest allusion to a plant parasitic nematode is, however, preserved in famous writ. "Sowed cockle, reap'd no corn," a line by William Shakespeare penned in 1594 in *Love's Labour's Lost*, Act IV, Scene 3, most certainly has reference to blighted wheat caused by the plant parasite, *Anguina tritici*.

Needham (1743) solved the "riddle of cockle" when he crushed one of the diseased wheat grains and observed "Aquatic Animals...denominated Worms, Eels, or Serpents, which they very much resemble." It is likely that few or no other recorded observations of plant parasitic nematodes or their effects are to be found in ancient literature.

From 1750 to the early 1900s, nematology research continued to be descriptive and taxonomic, focusing primarily on free-living nematodes and plant and animal parasites. During this period a number of productive researchers contributed to the field of nematology in the United States and abroad. Beginning with Needham and continuing to Cobb, nematologists compiled and continuously revised a broad descriptive morphological taxonomy of nematodes.

History: 1850 to the Present

Kuhn (1874) is thought to be the first to use soil fumigation to control nematodes, applying carbon disulfide treatments in sugar beet fields in Germany. In Europe from 1870 to 1910, nematological research focused heavily on controlling the sugar beet nematode as sugar beet production became an important economy during this time in the Old World.

Although 18th and 19th century scientists yielded a considerable amount of important fundamental and applied knowledge about nematode biology, nematology research really began to advance in quality and quantity near the turn of the 20th century. In 1918, the first permanent nematology field station was constructed in the U.S. Post Office in Salt Lake City, Utah under the direction of Harry B. Shaw, after scientists observed the sugar beet nematode in a field south of the city. In this same year, Nathan Cobb (1918) published his Contributions to a Science of Nematology and his lab manual "Estimating the Nema Population of Soil." These two publications provide definitive resources for many methods and apparatus used in nematology even to this day.

Of Cobb's far-reaching influence on nematology research, Jenkins and Taylor write:

> *Although many workers have played important roles in development of plant nematology, none have had a greater impact, particularly in the United States, than N.A. Cobb. In 1913 Cobb published his first paper on nematology in the United States. From [1913] to 1932 he was the undisputed leader in [nematology] in this country. Through his efforts the widely renowned [USDA] nematology research program was initiated and developed. Many of his students and colleagues developed into the leaders in plant nematology in the 1930s, 1940's, and 1950's. In addition, during his productive career he contributed*

major discoveries in the areas of nematode taxonomy, morphology, and in methodology. Many of his techniques are still unsurpassed!

Perhaps no one person has had as favorable an impact on the field of nematology as has Nathan Augustus Cobb.

From 1900–1925 various state-run agricultural experimental stations investigated important problems relating to agro-economy, though few stations devoted much attention to plant-parasitic nematodes. Accounts of the history of nematology (the few that exist) mention three major events occurring between 1926 and 1950 that affected the relative importance of nematodes in the eyes of farmers, legislators and the U.S. public in general. These same events had profound worldwide effects on the course of nematology research over the next fifty to seventy-five years

First, the discovery of the golden nematode in the potato fields of Long Island led to a trip by U.S. quarantine officials to the potato fields of Europe, where the devastating effects of this parasite had been known for many years. This excursion allayed all skepticism about the seriousness of this agricultural pest. Second, the introduction of the soil fumigants, D-D and EDB made available for the first time nematicides that could be used effectively and practically on a field scale. Third, the development of nematode-resistant crop cultivars brought substantial government funding to applied nematology research.

These events contributed to a shift from broad taxonomy-based nematology research to deep, yet focused investigations of plant parasitic nematodes, especially the control of agricultural pests. From the early 1930s until recently, the bulk of researchers studying nematodes have been plant pathologists by training. Consequently, nematological research leaned heavily toward answering plant pathological and agro-economical questions for the last three-quarters of the 20th century.

Notable Nematologists

- Nathan Cobb

- Michel Luc

- Maynard Jack Ramsay

- Gregor W. Yeates

Contributions to Other Sciences

Nematologists in the 1800s also contributed to other scientific fields in important ways. Butschli (1875) first observed the formation of polar bodies by nuclear subdivision in a nematode, Beneden (1883) was studying *Ascaris megalocephala* when he discovered the separation of halves of each of the chromosomes from the two parents and the mechanism of Mendelian heredity, and Boveri (1893) showed evidence for continuity of the germ plasm and that the soma may be regarded as a by-product without influence upon heredity.

Caenorhabditis elegans is a widely used model species, initially for neural development, and then for genetics. WormBase collates research on the species.

Microbial Genetics

Microbial genetics is a subject area within microbiology and genetic engineering. It studies the genetics of very small (micro) organisms; bacteria, archaea, viruses and some protozoa and fungi. This involves the study of the genotype of microbial species and also the expression system in the form of phenotypes.

Since the discovery of microorganisms by two Fellows of The Royal Society, Robert Hooke and Antoni van Leeuwenhoek during the period 1665-1885 they have been used to study many processes and have had applications in various areas of study in genetics. For example: Microorganisms' rapid growth rates and short generation times are used by scientists to study evolution. Microbial genetics also has applications in being able to study processes and pathways that are similar to those found in humans such as drug metabolism.

Microorganisms Whose Study is Encompassed By Microbial Genetics

Bacteria

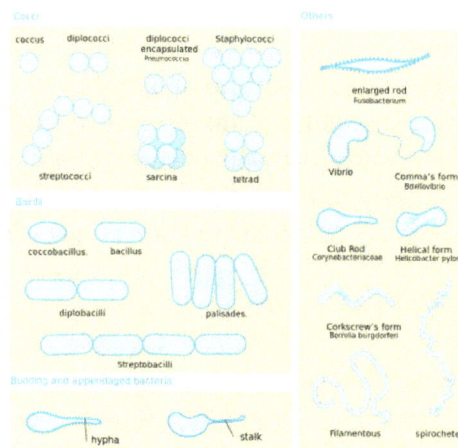

Bacteria are classified by their shape.

Bacteria have been on this planet for approximately 3.5 billion years, and are classified by their shape. Bacterial genetics studies the mechanisms of their heritable information, their chromosomes, plasmids, transposons, and phages.

Gene transfer systems that have been extensively studied in bacteria include genetic transformation, conjugation and transduction. Natural transformation is a bacterial adaptation for DNA transfer between two cells through the intervening medium. The uptake of donor DNA and its recombinational incorporation into the recipient chromosome depends on the expression of numerous bacterial genes whose products direct this process. In general, transformation is a complex, energy-requiring developmental process that appears to be an adaptation for repairing DNA damage.

Bacterial conjugation is the transfer of genetic material between bacterial cells by direct cell-to-cell contact or by a bridge-like connection between two cells. Bacterial conjugation has been extensively studied in *Escherichia coli*, but also occurs in other bacteria such as *Mycobacterium smegmatis*. Conjugation requires stable and extended contact between a donor and a recipient strain, is DNase resistant, and the transferred DNA is incorporated into the recipient chromosome by homologous recombination. *E. coli* conjugation is mediated by expression of plasmid genes, whereas mycobacterial conjugation is mediated by genes on the bacterial chromosome.

Transduction is the process by which foreign DNA is introduced into a cell by a virus or viral vector. Transduction is a common tool used by molecular biologists to stably introduce a foreign gene into a host cell's genome.

Archaea

Archaea is a domain of organisms that are prokaryotic, single-celled, and are thought to have developed 4 billion years ago. They share a common ancestor with bacteria, but are more closely related to eukaryotes in comparison to bacteria. Some Archaea are able to survive extreme environments, which leads to many applications in the field of genetics. One of such applications is the use of archaeal enzymes, which would be better able to survive harsh conditions *in vitro*.

Gene transfer and genetic exchange have been studied in the halophilic archaeon *Halobacterium volcanii* and the hyperthermophilic archaeons *Sulfolobus solfataricus* and *Sulfolobus acidocaldarius*. *H. volcani* forms cytoplasmic bridges between cells that appear to be used for transfer of DNA from one cell to another in either direction. When *S. solfataricus* and *S. acidocaldarius* are exposed to DNA damaging agents, species-specific cellular aggregation is induced. Cellular aggregation mediates chromosomal marker exchange and genetic recombination with high frequency. Cellular aggregation is thought to enhance species specific DNA transfer between *Sulfolobus* cells in order to provide increased repair of damaged DNA by means of homologous recombination.

Fungi

Fungi can be both multicellular and unicellular organisms, and are distinguished from other microbes by the way they obtain nutrients. Fungi secrete enzymes into their surroundings, to break down organic matter. Fungal genetics uses yeast, and filamentous fungi as model organisms for eukaryotic genetic research, including cell cycle regulation, chromatin structure and gene regulation.

Studies of the fungus *Neurospora crassa* have contributed substantially to understanding how genes work. *N. crassa* is a type of red bread mold of the phylum *Ascomycota*. It is used as a model organism because it is easy to grow and has a haploid life cycle that makes genetic analysis simple since recessive traits will show up in the offspring. Analysis of genetic recombination is facilitated by the ordered arrangement of the products of meiosis in ascospores. In its natural environment, *N. crassa* lives mainly in tropical and sub-tropical regions. It often can be found growing on dead plant matter after fires.

Neurospora was used by Edward Tatum and George Beadle in their experiments for which they won the Nobel Prize in Physiology or Medicine in 1958. The results of these experiments led directly to the one gene-one enzyme hypothesis that specific genes code for specific proteins. This concept proved to be the opening gun in what became molecular genetics and all the developments that have followed from that.

Saccharomyces cerevisiae is a yeast of the phylum *Ascomycota*. During vegetative growth that ordinarily occurs when nutrients are abundant, *S. cerevisiae* reproduces by mitosis as diploid cells. However, when starved, these cells undergo meiosis to form haploid spores. Mating occurs when haploid cells of opposite mating types MATa and MATα come into contact. Ruderfer et al. pointed out that, in nature, such contacts are frequent between closely related yeast cells for two reasons. The first is that cells of opposite mating type are present together in the same acus, the sac that contains the cells directly produced by a single meiosis, and these cells can mate with each other. The second reason is that haploid cells of one mating type, upon cell division, often produce cells of the opposite mating type. An analysis of the ancestry of natural *S. cerevisiae* strains concluded that outcrossing occurs very infrequently (only about once every 50,000 cell divisions). The relative rarity in nature of meiotic events that result from outcrossing suggests that the possible long-term benefits of outcrossing (e.g. generation of diversity) are unlikely to be sufficient for generally maintaining sex from one generation to the next. Rather, a short term benefit, such as meiotic recombinational repair of DNA damages caused by stressful conditions (such as starvation) may be the key to the maintenance of sex in *S. cerevisiae*.

Candida albicans is a diploid fungus that grows both as a yeast and as a filament. *C. albicans* is the most common fungal pathogen in humans. It causes both debilitating mucosal infections and potentially life-threatening systemic infections. *C. albicans* has maintained an elaborate, but largely hidden, mating apparatus. Johnson suggested that mating strategies may allow *C. albicans* to survive in the hostile environment of a mammalian host.

Among the 250 known species of aspergilli, about 33% have an identified sexual state. Among those *Aspergillus* species that exhibit a sexual cycle the overwhelming majority in nature are homothallic (self-fertilizing). Selfing in the homothallic fungus *Aspergillus nidulans* involves activation of the same mating pathways characteristic of sex in outcrossing species, i.e. self-fertilization does not bypass required pathways for outcrossing sex but instead requires activation of these pathways within a single individual. Fusion of haploid nuclei occurs within reproductive structures termed *cleistothecia*, in which the diploid zygote undergoes meiotic divisions to yield haploid ascospores.

Protozoa

Protozoa are unicellular organisms, which have nuclei, and ultramicroscopic cellular bodies within their cytoplasm. One particular aspect of protozoa that are of interest to human geneticists are their flagella, which are very similar to human sperm flagella.

Studies of *Paramecia* have contributed to our understanding of the function of meiosis. Like all ciliates, *Paramecia* have a polyploid macronucleus, and one or more diploid micronuclei. The macronucleus controls non-reproductive cell functions, expressing the genes needed for daily functioning. The micronucleus is the generative, or germline nucleus, containing the genetic material that is passed along from one generation to the next.

In the asexual fission phase of growth, during which cell divisions occur by mitosis rather than meiosis, clonal aging occurs leading to a gradual loss of vitality. In some species, such as the well studied *Paramecium tetraurelia*, the asexual line of clonally aging paramecia loses vitality and expires after about 200 fissions if the cells fail to undergo meiosis followed by either autogamy (self-fertilizaion) or conjugation (outcrossing). DNA damage increases dramatically during successive clonal cell divisions and is a likely cause of clonal aging in *P. tetraurelia*.

When clonally aged *P. tetraurelia* are stimulated to undergo meiosis in association with either autogamy or conjugation, the progeny are rejuvenated, and are able to have many more mitotic binary fission divisions. During either of these processes the micronuclei of the cell(s) undergo meiosis, the old macronucleus disintegrates and a new macronucleus is formed by replication of the micronuclear DNA that had recently undergone meiosis. There is apparently little, if any, DNA damage in the new macronucleus, suggesting that rejuvenation is associated with the repair of these damages in the micronucleus during meiosis.

Viruses

Viruses are capsid-encoding organisms composed of proteins and nucleic acids that can self-assemble after replication in a host cell using the host's replication machinery. There is a disagreement in science about whether viruses are living due to their lack of ribosomes. Comprehending the viral genome is important not only for studies in genetics but also for understanding their pathogenic properties.

Many types of virus are capable of genetic recombination. When two or more individual viruses of the same type infect a cell, their genomes may recombine with each other to produce recombinant virus progeny. Both DNA and RNA viruses can undergo recombination. When two or more viruses, each containing lethal genomic damage infect the same host cell, the virus genomes often can pair with each other and undergo homologous recombinational repair to produce viable progeny. This process is known as multiplicity reactivation. Enzymes employed in multiplicity reactivation are functionally homologous to enzymes employed in bacterial and eukaryotic recombinational repair. Multiplicity reactivation has been found to occur with pathogenic viruses including influenza virus, HIV-1, adenovirus simian virus 40, vaccinia virus, reovirus, poliovirus and herpes simplex virus as well as numerous bacteriophages.

Applications of Microbial Genetics

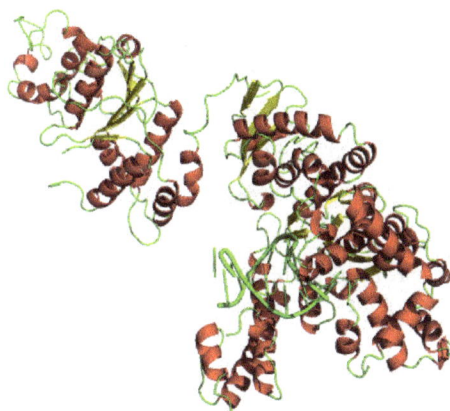

Taq polymerase which is used in Polymerase Chain Reaction(PCR)

Microbes are ideally suited for biochemical and genetics studies and have made huge contributions to these fields of science such as the demonstration that DNA is the genetic material, that the gene has a simple linear structure, that the genetic code is a triplet code, and that gene expression is regulated by specific genetic processes. Jacques Monod and François Jacob used *Escherichia coli*, a type of bacteria, in order to develop the operon model of gene expression, which lay down the basis of gene expression and regulation. Furthermore the hereditary processes of single-celled eukaryotic microorganisms are similar to those in multi-cellular organisms allowing researchers to gather information on this process as well. Another bacterium which has greatly contributed to the field of genetics is *Thermus aquaticus*, which is a bacterium that tolerates high temperatures. From this microbe scientists isolated the enzyme Taq polymerase, which is now used in the powerful experimental technique, Polymerase chain reaction(PCR). Additionally the development of recombinant DNA technology through the use of bacteria has led to the birth of modern genetic engineering and biotechnology.

Using microbes, protocols were developed to insert genes into bacterial plasmids, taking advantage of their fast reproduction, to make biofactories for the gene of interest. Such genetically engineered bacteria can produce pharmaceuticals such as insulin, human growth hormone, interferons and blood clotting factors. Microbes synthesize a variety of enzymes for industrial applications, such as fermented foods, laboratory test reagents, dairy products (such as renin), and even in clothing (such as *Trichoderma* fungus whose enzyme is used to give jeans a stone washed appearance).

References

- Goldsby RA; Kindt TK; Osborne BA & Kuby J (2003). Immunology (5th ed.). San Francisco: W.H. Freeman. ISBN 0-7167-4947-5.

- Crawford, Dorothy (2011). Viruses: A Very Short Introduction. New York, NY: Oxford University Press. p. 4. ISBN 0199574855.

- Evans, Alfred (1982). Viral Infections of Humans. New York, NY: Plenum Publishing Corporation. p. xxv-xxxi. ISBN 0306406764.

- Sussman, Max; Topley, W. W. C.; Wilson, Graham K.; Collier, L. H.; Balows, Albert (1998). Topley & Wilson's microbiology and microbial infections. London: Arnold. p. 3. ISBN 0-340-66316-2.

- Barton, Larry L.; Northup, Diana E. (9 September 2011). Microbial Ecology. Wiley-Blackwell. Oxford: John Wiley & Sons. p. 22. ISBN 978-1-118-01582-7. Retrieved 25 May 2013.

- Reddy, K. Ramesh; DeLaune, Ronald D. (15 July 2004). Biogeochemistry of Wetlands: Science and Applications. Boca Raton: Taylor & Francis. p. 116. ISBN 978-0-203-49145-4. Retrieved 25 May 2013.

- Konopka, A. (2009). "Ecology, Microbial in Encyclopedia of Microbiology": 91–106. doi:10.1016/B978-012373944-5.00002-X. ISBN 978-0-12-373944-5.

- Madigan, Michael T. (2012). Brock biology of microorganisms (13th ed.). San Francisco: Benjamin Cummings. ISBN 9780321649638.

- Fenchel, Tom; Blackburn, Henry; King, Gary M. (24 July 2012). Bacterial Biogeochemistry: The Ecophysiology of Mineral Cycling (3 ed.). Boston, Mass.: Academic Press/Elsevier. p. 3. ISBN 978-0-12-415974-7. Retrieved 25 May 2013.

- Ott, J. (2005). "Marine Microbial Thiotrophic Ectosymbioses". Oceanography and Marine Biology: An Annual Review. 42: 95–118. ISBN 9780203507810.

- "How long do microbes like bacteria and viruses live on surfaces in the home at normal room temperatures?". Retrieved 2016-09-18.

7

Applications of Microbiology

Humans use industrial fermentation to make products useful. Algae fuel is used as a substitute to liquid fossil fuels. A number of companies and governments have been trying to reduce the operating cost to make algae fuel production commercially feasible. The aspects elucidated in this chapter are of vital importance, and provides a better understanding of microbiology.

Industrial Fermentation

Industrial fermentation is the intentional use of fermentation by microorganisms such as bacteria and fungi to make products useful to humans. Fermented products have applications as food as well as in general industry. Some commodity chemicals, such as acetic acid, citric acid, and ethanol are made by fermentation. The rate of fermentation depends on the concentration of microorganisms, cells, cellular components, and enzymes as well as temperature, pH and for aerobic fermentation oxygen. Product recovery frequently involves the concentration of the dilute solution. Nearly all commercially produced enzymes, such as lipase, invertase and rennet, are made by fermentation with genetically modified microbes. In some cases, production of biomass itself is the objective, as in the case of baker's yeast and lactic acid bacteria starter cultures for cheesemaking. In general, fermentations can be divided into four types:

- Production of biomass (viable cellular material)
- Production of extracellular metabolites (chemical compounds)
- Production of intracellular components (enzymes and other proteins)
- Transformation of substrate (in which the transformed substrate is itself the product)

These types are not necessarily disjoint from each other, but provide a framework for understanding the differences in approach. The organisms used may be bacteria, yeasts, molds, algae, animal cells, or plant cells. Special considerations are required for the specific organisms used in the fermentation, such as the dissolved oxygen level, nutrient levels, and temperature.

General Process Overview

In most industrial fermentations, the organisms are submerged in a liquid medium; in others, such as the fermentation of cocoa beans, coffee cherries, and miso, fermentation takes place on the moist surface of the medium. There are also industrial considerations related to the fermentation process. For instance, to avoid biological process contamination, the fermentation medium, air, and equipment are sterilized. Foam control can be achieved by either mechanical foam destruction or chemical anti-foaming agents. Several other factors must be measured and controlled such as pressure, temperature, agitator shaft power, and viscosity. An important element for industrial fermentations is scale up. This is the conversion of a laboratory procedure to an industrial process. It is well established in the field of industrial microbiology that what works well at the laboratory scale may work poorly or not at all when first attempted at large scale. It is generally not possible to take fermentation conditions that have worked in the laboratory and blindly apply them to industrial-scale equipment. Although many parameters have been tested for use as scale up criteria, there is no general formula because of the variation in fermentation processes. The most important methods are the maintenance of constant power consumption per unit of broth and the maintenance of constant volumetric transfer rate.

Phases of Microbial Growth

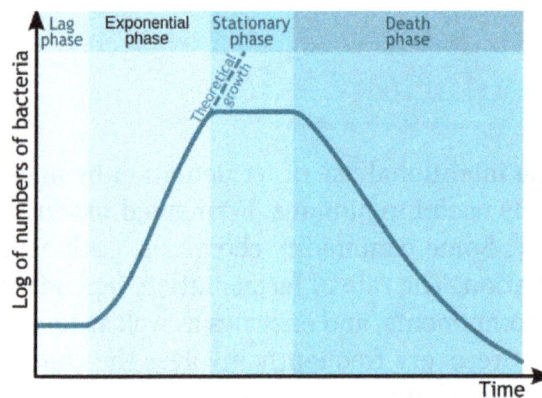

Bacterial growth curve\Kinetic Curve

When a particular organism is introduced into a selected growth medium, the medium is inoculated with the particular organism. Growth of the inoculum does not occur immediately, but takes a little while. This is the period of adaptation, called the lag phase. Following the lag phase, the rate of growth of the organism steadily increases, for a certain period—this period is the log or exponential phase. After a certain time of exponential phase, the rate of growth slows down, due to the continuously falling concentrations of nutrients and/or a continuously increasing (accumulating) concentrations of toxic substances. This phase, where the increase of the rate of growth is checked, is the deceleration phase. After the deceleration phase, growth ceases and the culture enters a stationary phase or a steady state. The biomass remains constant, except when certain accumulated chemicals in the culture lyse the cells (chemolysis). Unless other micro-organisms contaminate the culture, the chemical constitution remains unchanged. If all of the nutrients in the medium are consumed, or if the concentration of toxins is too great, the cells may become scenescent and begin to die off. The total amount of biomass may not decrease, but the number of viable organisms will decrease.

Fermentation Medium

The microbes used for fermentation grow in (or on) specially designed growth medium which supplies the nutrients required by the organisms. A variety of media exist, but invariably contain a carbon source, a nitrogen source, water, salts, and micronutrients. In the production of wine, the medium is grape must. In the production of bio-ethanol, the medium may consist mostly of whatever inexpensive carbon source is available.

Carbon sources are typically sugars or other carbohydrates, although in the case of substrate trans-formations (such as the production of vinegar) the carbon source may be an alcohol or something else altogether. For large scale fermentations, such as those used for the production of ethanol, in-expensive sources of carbohydrates, such as molasses, corn steep liquor, sugar cane juice, or sugar beet juice are used to minimize costs. More sensitive fermentations may instead use purified glu-cose, sucrose, glycerol or other sugars, which reduces variation and helps ensure the purity of the final product. Organisms meant to produce enzymes such as beta galactosidase, invertase or other amylases may be fed starch to select for organisms that express the enzymes in large quantity.

Fixed nitrogen sources are required for most organisms to synthesize proteins, nucleic acids and oth-er cellular components. Depending on the enzyme capabilities of the organism, nitrogen may be pro-vided as bulk protein, such as soy meal; as pre-digested polypeptides, such as peptone or tryptone; or as ammonia or nitrate salts. Cost is also an important factor in the choice of a nitrogen source. Phosphorus is needed for production of phospholipids in cellular membranes and for the production of nucleic acids. The amount of phosphate which must be added depends upon the composition of the broth and the needs of the organism, as well as the objective of the fermentation. For instance, some cultures will not produce secondary metabolites in the presence of phosphate.

Growth factors and trace nutrients are included in the fermentation broth for organisms incapable of producing all of the vitamins they require. Yeast extract is a common source of micronutrients and vita-mins for fermentation media. Inorganic nutrients, including trace elements such as iron, zinc, copper, manganese, molybdenum and cobalt are typically present in unrefined carbon and nitrogen sources, but may have to be added when purified carbon and nitrogen sources are used. Fermentations which produce large amounts of gas (or which require the addition of gas) will tend to form a layer of foam, since fermentation broth typically contains a variety of foam-reinforcing proteins, peptides or starch-es. To prevent this foam from occurring or accumulating, antifoaming agents may be added. Mineral buffering salts, such as carbonates and phosphates, may be used to stabilize pH near optimum. When metal ions are present in high concentrations, use of a chelating agent may be necessary.

Production of Biomass

Microbial cells or biomass is sometimes the intended product of fermentation. Examples include single cell protein, bakers yeast, lactobacillus, E. coli, and others. In the case of single-cell protein, algae is grown in large open ponds which allow photosynthesis to occur. If the biomass is to be used for inoculation of other fermentations, care must be taken to prevent mutations from occurring.

Production of Extracellular Metabolites

Microbial metabolites can be divided into two groups: those produced during the growth phase of the organism, called primary metabolites and those produced during the stationary phase, called secondary metabolites. Some examples of primary metabolites are ethanol, citric acid, glutamic acid, lysine, vitamins and polysaccharides. Some examples of secondary metabolites are penicillin, cyclosporin A, gibberellin, and lovastatin.

Primary Metabolites

Primary metabolites are compounds made during the ordinary metabolism of the organism during the growth phase. A common example is ethanol or lactic acid, produced during glycolysis. Citric acid is produced by some strains of *Aspergillus niger* as part of the citric acid cycle to acidify their environment and prevent competitors from taking over. Glutamate is produced by some *Micrococcus* species, and some *Corynebacterium* species produce lysine, threonine, tryptophan and other amino acids. All of these compounds are produced during the normal "business" of the cell and released into the environment. There is therefore no need to rupture the cells for product recovery.

Secondary Metabolites

Secondary metabolites are compounds made in the stationary phase; penicillin, for instance, prevents the growth of bacteria which could compete with *Penicillium* molds for resources. Some bacteria, such as *Lactobacillus* species, are able to produce bacteriocins which prevent the growth of bacterial competitors as well. These compounds are of obvious value to humans wishing to prevent the growth of bacteria, either as antibiotics or as antiseptics (such as gramicidin S). Fungicides, such as griseofulvin are also produced as secondary metabolites. Typically secondary metabolites are not produced in the presence of glucose or other carbon sources which would encourage growth, and like primary metabolites are released into the surrounding medium without rupture of the cell membrane.

Production of Intracellular Components

Of primary interest among the intracellular components are microbial enzymes: catalase, amylase, protease, pectinase, glucose isomerase, cellulase, hemicellulase, lipase, lactase, streptokinase and many others. Recombinant proteins, such as insulin, hepatitis B vaccine, interferon, granulocyte colony-stimulating factor, streptokinase and others are also made this way. The largest difference between this process and the others is that the cells must be ruptured (lysed) at the end of fermentation, and the environment must be manipulated to maximize the amount of the product. Furthermore, the product (typically a protein) must be separated from all of the other cellular proteins in the lysate to be purified.

Transformation of Substrate

Substrate transformation involves the transformation of a specific compound into another, such as in the case of phenylacetylcarbinol, and steroid biotransformation, or the transformation of a raw material into a finished product, in the case of food fermentations and sewage treatment.

Food Fermentation

Ancient fermented food processes, such as making bread, wine, cheese, curds, idli, dosa, etc., can be dated to more than seven thousand years ago. They were developed long before man had any knowledge of the existence of the microorganisms involved. Some foods such as Marmite are the byproduct of the fermentation process, in this case in the production of beer.

Ethanol Fuel

Fermentation is the main source of ethanol in the production of ethanol fuel. Common crops such as sugar cane, potato, cassava and corn are fermented by yeast to produce ethanol which is further processed to become fuel.

Sewage Treatment

In the process of sewage treatment, sewage is digested by enzymes secreted by bacteria. Solid organic matters are broken down into harmless, soluble substances and carbon dioxide. Liquids that result are disinfected to remove pathogens before being discharged into rivers or the sea or can be used as liquid fertilizers. Digested solids, known also as sludge, is dried and used as fertilizer. Gaseous byproducts such as methane can be utilized as biogas to fuel electrical generators. One advantage of bacterial digestion is that it reduces the bulk and odor of sewage, thus reducing space needed for dumping. The main disadvantage of bacterial digestion in sewage disposal is that it is a very slow process.

Agricultural Feed

A wide variety of agroindustrial waste products can be fermented to use as food for animals, especially ruminants. Fungi have been employed to break down cellulosic wastes to increase protein content and improve *in vitro* digestibility.

Microbial Biodegradation

Microbial biodegredation is the use of bioremediation and biotransformation methods to harness the naturally occurring ability of microbial xenobiotic metabolism to degrade, transform or accumulate environmental pollutants, including hydrocarbons (e.g. oil), polychlorinated biphenyls (PCBs), polyaromatic hydrocarbons (PAHs), heterocyclic compounds (such as pyridine or quinoline), pharmaceutical substances, radionuclides and metals.

Interest in the microbial biodegradation of pollutants has intensified in recent years, and recent major methodological breakthroughs have enabled detailed genomic, metagenomic, proteomic, bioinformatic and other high-throughput analyses of environmentally relevant microorganisms, providing new insights into biodegradative pathways and the ability of organisms to adapt to changing environmental conditions.

Biological processes play a major role in the removal of contaminants and take advantage of the catabolic versatility of microorganisms to degrade or convert such compounds. In environmental microbiology, genome-based global studies are increasing the understanding of metabolic and regulatory networks, as well as providing new information on the evolution of degradation pathways and molecular adaptation strategies to changing environmental conditions.

The increasing amount of bacterial genomic data provides new opportunities for understanding the genetic and molecular bases of the degradation of organic pollutants. Aromatic compounds are among the most persistent of these pollutants and lessons can be learned from the recent genomic studies of *Burkholderia xenovorans* LB400 and *Rhodococcus* sp. strain RHA1, two of the largest bacterial genomes completely sequenced to date. These studies have helped expand our understanding of bacterial catabolism, non-catabolic physiological adaptation to organic compounds, and the evolution of large bacterial genomes. First, the metabolic pathways from phylogenetically diverse isolates are very similar with respect to overall organization. Thus, as originally noted in pseudomonads, a large number of "peripheral aromatic" pathways funnel a range of natural and xenobiotic compounds into a restricted number of "central aromatic" pathways. Nevertheless, these pathways are genetically organized in genus-specific fashions, as exemplified by the b-ketoadipate and Paa pathways. Comparative genomic studies further reveal that some pathways are more widespread than initially thought. Thus, the Box and Paa pathways illustrate the prevalence of non-oxygenolytic ring-cleavage strategies in aerobic aromatic degradation processes. Functional genomic studies have been useful in establishing that even organisms harboring high numbers of homologous enzymes seem to contain few examples of true redundancy. For example, the multiplicity of ring-cleaving dioxygenases in certain rhodococcal isolates may be attributed to the cryptic aromatic catabolism of different terpenoids and steroids. Finally, analyses have indicated that recent genetic flux appears to have played a more significant role in the evolution of some large genomes, such as LB400's, than others. However, the emerging trend is that the large gene repertoires of potent pollutant degraders such as LB400 and RHA1 have evolved principally through more ancient processes. That this is true in such phylogenetically diverse species is remarkable and further suggests the ancient origin of this catabolic capacity.

Anaerobic Biodegradation of Pollutants

Anaerobic microbial mineralization of recalcitrant organic pollutants is of great environmental significance and involves intriguing novel biochemical reactions. In particular, hydrocarbons and halogenated compounds have long been doubted to be degradable in the absence of oxygen, but the isolation of hitherto unknown anaerobic hydrocarbon-degrading and reductively dehalogenating bacteria during the last decades provided ultimate proof for these processes in nature. While such research involved mostly chlorinated compounds initially, recent studies have revealed reductive dehalogenation of bromine and iodine moieties in aromatic pesticides. Other reactions, such as biologically induced abiotic reduction by soil minerals, has been shown to deactivate relatively persistent aniline-based herbicides far more rapidly than observed in aerobic environments. Many novel biochemical reactions were discovered enabling the respective metabolic pathways, but progress in the molecular understanding of these bacteria was rather slow, since genetic systems are not readily applicable for most of them. However, with the increasing application of genomics in the field of environmental microbiology, a new and promising perspective is now at hand to obtain molecular insights into these new metabolic properties. Several complete genome sequences were determined during the last few years from bacteria capable of anaerobic organic pollutant degradation. The ~4.7 Mb genome of the facultative denitrifying *Aromatoleum aromaticum* strain EbN1 was the first to be determined for an anaerobic hydrocarbon degrader (using toluene or ethylbenzene as substrates). The genome sequence revealed about two dozen gene clusters (including several paralogs) coding for a complex catabolic network for anaerobic and aerobic degradation of aromatic compounds. The genome sequence forms the basis for current detailed studies on regulation of pathways and enzyme structures. Further genomes of anaerobic hydrocarbon degrading bacteria were recently completed for the iron-reducing species *Geobacter metallireducens* (accession nr. NC_007517) and the perchlorate-reducing *Dechloromonas aromatica* (accession nr. NC_007298), but these are not yet evaluated in formal publications. Complete genomes were also determined for bacteria capable of anaerobic degradation of halogenated hydrocarbons by halorespiration: the ~1.4 Mb genomes of *Dehalococcoides ethenogenes* strain 195 and *Dehalococcoides* sp. strain CBDB1 and the ~5.7 Mb genome of *Desulfitobacterium hafniense* strain Y51. Characteristic for all these bacteria is the presence of multiple paralogous genes for reductive dehalogenases, implicating a wider dehalogenating spectrum of the organisms than previously known. Moreover, genome sequences provided unprecedented insights into the evolution of reductive dehalogenation and differing strategies for niche adaptation.

Recently, it has become apparent that some organisms, including *Desulfitobacterium chlororespirans*, originally evaluated for halorespiration on chlorophenols, can also use certain brominated compounds, such as the herbicide bromoxynil and its major metabolite as electron acceptors for growth. Iodinated compounds may be dehalogenated as well, though the process may not satisfy the need for an electron acceptor.

Bioavailability, Chemotaxis, and Transport of Pollutants

Bioavailability, or the amount of a substance that is physiochemically accessible to microorganisms is a key factor in the efficient biodegradation of pollutants. O'Loughlin *et al.* (2000) showed that, with the exception of kaolinite clay, most soil clays and cation exchange resins attenuated biodegradation of 2-picoline by *Arthrobacter* sp. strain R1, as a result of adsorption of the substrate to the clays. Chemotaxis, or the directed movement of motile organisms towards or away from chemicals in the environment is an important physiological response that may contribute to effective catabolism of molecules in the environment. In addition, mechanisms for the intracellular accumulation of aromatic molecules via various transport mechanisms are also important.

Oil Biodegradation

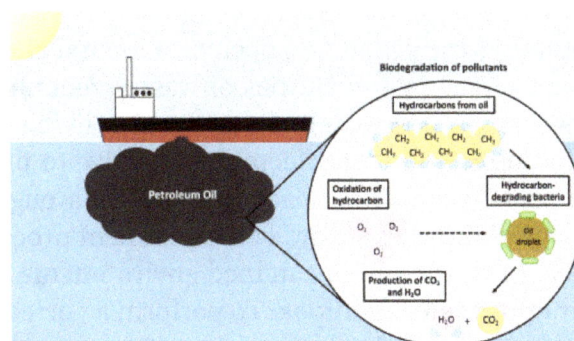

General overview of microbial biodegradation of petroleum oil by microbial communities. Some microorganisms, such as *A. borkumensis*, are able to use hydrocarbons as their source for carbon in metabolism. They are able to oxidize the environmentally harmful hydrocarbons while producing harmless products, following the general equation $C_nH_n + O_2 \rightarrow H_2O + CO_2$. In the figure, carbon is represented as yellow circles, oxygen as pink circles, and hydrogen as blue circles. This type of special metabolism allows these microbes to thrive in areas affected by oil spills and are important in the elimination of environmental pollutants.

Petroleum oil contains aromatic compounds that are toxic to most life forms. Episodic and chronic pollution of the environment by oil causes major disruption to the local ecological environment. Marine environments in particular are especially vulnerable, as oil spills near coastal regions and in the open sea are difficult to contain and make mitigation efforts more complicated. In addition to pollution through human activities, approximately 250 million litres of petroleum enter the marine environment every year from natural seepages. Despite its toxicity, a considerable fraction of petroleum oil entering marine systems is eliminated by the hydrocarbon-degrading activities of microbial communities, in particular by a recently discovered group of specialists, the hydrocarbonoclastic bacteria (HCB). *Alcanivorax borkumensis* was the first HCB to have its genome sequenced. In addition to hydrocarbons, crude oil often contains various heterocyclic compounds, such as pyridine, which appear to be degraded by similar mechanisms to hydrocarbons.

Cholesterol Biodegradation

Many synthetic steroidic compounds like some sexual hormones frequently appear in municipal and industrial wastewaters, acting as environmental pollutants with strong metabolic activities negatively affecting the ecosystems. Since these compounds are common carbon sources for many different microorganisms their aerobic and anaerobic mineralization has been extensively studied. The interest of these studies lies on the biotechnological applications of sterol transforming enzymes for the industrial synthesis of sexual hormones and corticoids. Very recently, the catabolism of cholesterol has acquired a high relevance because it is involved in the infectivity of the pathogen *Mycobacterium tuberculosis* (*Mtb*). *Mtb* causes tuberculosis disease, and it has been demonstrated that novel enzyme architectures have evolved to bind and modify steroid compounds like cholesterol in this organism and other steroid-utilizing bacteria as well. These new enzymes might be of interest for their potential in the chemical modification of steroid substrates.

Analysis of Waste Biotreatment

Sustainable development requires the promotion of environmental management and a constant search for new technologies to treat vast quantities of wastes generated by increasing anthropogenic activities. Biotreatment, the processing of wastes using living organisms, is an environmentally friendly, relatively simple and cost-effective alternative to physico-chemical clean-up options. Confined environments, such as bioreactors, have been engineered to overcome the physical, chemical and biological limiting factors of biotreatment processes in highly controlled systems. The great versatility in the design of confined environments allows the treatment of a wide range of wastes under optimized conditions. To perform a correct assessment, it is necessary to consider various microorganisms having a variety of genomes and expressed transcripts and proteins. A great number of analyses are often required. Using traditional genomic techniques, such assessments are limited and time-consuming. However, several high-throughput techniques originally developed for medical studies can be applied to assess biotreatment in confined environments.

Metabolic Engineering and Biocatalytic Applications

The study of the fate of persistent organic chemicals in the environment has revealed a large reservoir of enzymatic reactions with a large potential in preparative organic synthesis, which has already been exploited for a number of oxygenases on pilot and even on industrial scale. Novel catalysts can be obtained from metagenomic libraries and DNA sequence based approaches. Our increasing capabilities in adapting the catalysts to specific reactions and process requirements by rational and random mutagenesis broadens the scope for application in the fine chemical industry, but also in the field of biodegradation. In many cases, these catalysts need to be exploited in whole cell bioconversions or in fermentations, calling for system-wide approaches to understanding strain physiology and metabolism and rational approaches to the engineering of whole cells as they are increasingly put forward in the area of systems biotechnology and synthetic biology.

Fungal Biodegradation

In the ecosystem, different substrates are attacked at different rates by consortia of organisms from different kingdoms. *Aspergillus* and other moulds play an important role in these consortia because they are adept at recycling starches, hemicelluloses, celluloses, pectins and other sugar polymers. Some aspergilli are capable of degrading more refractory compounds such as fats, oils, chitin, and keratin. Maximum decomposition occurs when there is sufficient nitrogen, phosphorus and other essential inorganic nutrients. Fungi also provide food for many soil organisms.

For *Aspergillus* the process of degradation is the means of obtaining nutrients. When these moulds degrade human-made substrates, the process usually is called biodeterioration. Both paper and textiles (cotton, jute, and linen) are particularly vulnerable to *Aspergillus* degradation. Our artistic heritage is also subject to *Aspergillus* assault. To give but one example, after Florence in Italy flooded in 1969, 74% of the isolates from a damaged Ghirlandaio fresco in the Ognissanti church were *Aspergillus versicolor*.

Algae Fuel

A conical flask of "green" jet fuel made from algae

Algae fuel or algal biofuel is an alternative to liquid fossil fuels that uses algae as its source of energy-rich oils. Also, algae fuels are an alternative to common known biofuel sources, such as corn and sugarcane. Several companies and government agencies are funding efforts to reduce capital and operating costs and make algae fuel production commercially viable. Like fossil fuel, algae fuel releases CO_2 when burnt, but unlike fossil fuel, algae fuel and other biofuels only release CO_2 recently removed from the atmosphere via photosynthesis as the algae or plant grew. The energy crisis and the world food crisis have ignited interest in algaculture (farming algae) for making biodiesel and other biofuels using land unsuitable for agriculture. Among algal fuels' attractive characteristics are that they can be grown with minimal impact on fresh water resources, can be produced using saline and wastewater, have a high flash point, and are biodegradable and relative-ly harmless to the environment if spilled. Algae cost more per unit mass than other second-generation biofuel crops due to high capital and operating costs, but are claimed to yield between 10 and 100 times more fuel per unit area. The United States Department of Energy estimates that if algae fuel replaced all the petroleum fuel in the United States, it would require 15,000 square miles (39,000 km^2), which is only 0.42% of the U.S. map, or about half of the land area of Maine. This is less than $\frac{1}{7}$ the area of corn harvested in the United States in 2000.

According to the head of the Algal Biomass Organization, algae fuel can reach price parity with oil in 2018 if granted production tax credits. However, in 2013, Exxon Mobil Chairman and CEO Rex Tillerson said that after committing to spend up to $600 million over 10 years on development in a joint venture with J. Craig Venter's Synthetic Genomics in 2009, Exxon pulled back after four years (and $100 million) when it realized that algae fuel is "probably further" than 25 years away from commercial viability. On the other hand, Solazyme and Sapphire Energy already began commercial sales of algal biofuel in 2012 and 2013, respectively, and Algenol hopes to produce commercially in 2014.

History

In 1942 Harder and Von Witsch were the first to propose that microalgae be grown as a source of lipids for food or fuel. Following World War II, research began in the US, Germany, Japan, England, and Israel on culturing techniques and engineering systems for growing microalgae on larger scales, particularly species in the genus *Chlorella*. Meanwhile, H. G. Aach showed that *Chlorella pyrenoidosa* could be induced via nitrogen starvation to accumulate as much as 70% of its dry weight as lipids. Since the need for alternative transportation fuel had subsided after World War II, research at this time focused on culturing algae as a food source or, in some cases, for wastewater treatment.

Interest in the application of algae for biofuels was rekindled during the oil embargo and oil price surges of the 1970s, leading the US Department of Energy to initiate the Aquatic Species Program in 1978. The Aquatic Species Program spent $25 million over 18 years with the goal of developing liquid transportation fuel from algae that would be price competitive with petroleum-derived fuels. The research program focused on the cultivation of microalgae in open outdoor ponds, systems which are low in cost but vulnerable to environmental disturbances like temperature swings and biological invasions. 3,000 algal strains were collected from around the country and screened for desirable properties such as high productivity, lipid content, and thermal tolerance, and the most promising strains were included in the SERI microalgae collection at the Solar Energy Research Institute (SERI) in Golden, Colorado and used for further research. Among the program's most significant findings were that rapid growth and high lipid production were "mutually exclusive," since the former required high nutrients and the latter required low nutrients. The final report suggested that genetic engineering may be necessary to be able to overcome this and other natural limitations of algal strains, and that the ideal species might vary with place and season. Although it was successfully demonstrated that large-scale production of algae for fuel in outdoor ponds was feasible, the program failed to do so at a cost that would be competitive with petroleum, especially as oil prices sank in the 1990s. Even in the best case scenario, it was estimated that unextracted algal oil would cost $59–186 per barrel, while petroleum cost less than $20 per barrel in 1995. Therefore, under budget pressure in 1996, the Aquatic Species Program was abandoned.

Other contributions to algal biofuels research have come indirectly from projects focusing on different applications of algal cultures. For example, in the 1990s Japan's Research Institute of Innovative Technology for the Earth (RITE) implemented a research program with the goal of developing systems to fix CO2 using microalgae. Although the goal was not energy production, several studies produced by RITE demonstrated that algae could be grown using flue gas from power plants as a CO2 source, an important development for algal biofuel research. Other work focusing on harvesting hydrogen gas, methane, or ethanol from algae, as well as nutritional supplements and pharmaceutical compounds, has also helped inform research on biofuel production from algae.

Following the disbanding of the Aquatic Species Program in 1996, there was a relative lull in algal biofuel research. Still, various projects were funded in the US by the Department of Energy, Department of Defense, National Science Foundation, Department of Agriculture, National Laboratories, state funding, and private funding, as well as in other countries. More recently, rising oil prices in the 2000s spurred a revival of interest in algal biofuels and US federal funding has increased, numerous research projects are being funded in Australia, New Zealand, Europe, the Middle East, and other parts of the world, and a wave of private companies has entered the field. In November 2012, Solazyme and Propel Fuels made the first retail sales of algae-derived fuel, and in March 2013 Sapphire Energy began commercial sales of algal biofuel to Tesoro.

Fuels

Algae can be converted into various types of fuels, depending on the technique and the part of the cells used. The lipid, or oily part of the algae biomass can be extracted and converted into biodiesel through a process similar to that used for any other vegetable oil, or converted in a refinery into "drop-in" replacements for petroleum-based fuels. Alternatively or following lipid extraction, the carbohydrate content of algae can be fermented into bioethanol or butanol fuel.

Biodiesel

Biodiesel is a diesel fuel derived from animal or plant lipids (oils and fats). Studies have shown that some species of algae can produce 60% or more of their dry weight in the form of oil. Because the cells grow in aqueous suspension, where they have more efficient access to water, CO2 and dissolved nutrients, microalgae are capable of producing large amounts of biomass and usable oil in either high rate algal ponds or photobioreactors. This oil can then be turned into biodiesel which could be sold for use in automobiles. Regional production of microalgae and processing into biofuels will provide economic benefits to rural communities.

As they do not have to produce structural compounds such as cellulose for leaves, stems, or roots, and because they can be grown floating in a rich nutritional medium, microalgae can have faster growth rates than terrestrial crops. Also, they can convert a much higher fraction of their biomass to oil than conventional crops, e.g. 60% versus 2-3% for soybeans. The per unit area yield of oil from algae is estimated to be from 58,700 to 136,900 L/ha/year, depending on lipid content, which is 10 to 23 times as high as the next highest yielding crop, oil palm, at 5,950 L/ha/year.

The U.S. Department of Energy's Aquatic Species Program, 1978–1996, focused on biodiesel from microalgae. The final report suggested that biodiesel could be the only viable method by which to produce enough fuel to replace current world diesel usage. If algae-derived biodiesel were to replace the annual global production of 1.1bn tons of conventional diesel then a land mass of 57.3 million hectares would be required, which would be highly favorable compared to other biofuels.

Biobutanol

Butanol can be made from algae or diatoms using only a solar powered biorefinery. This fuel has an energy density 10% less than gasoline, and greater than that of either ethanol or methanol. In most gasoline engines, butanol can be used in place of gasoline with no modifications. In several tests, butanol consumption is similar to that of gasoline, and when blended with gasoline, provides better performance and corrosion resistance than that of ethanol or E85.

The green waste left over from the algae oil extraction can be used to produce butanol. In addition, it has been shown that macroalgae (seaweeds) can be fermented by *Clostridia* genus bacteria to butanol and other solvents.

Biogasoline

Biogasoline is gasoline produced from biomass. Like traditionally produced gasoline, it contains between 6 (hexane) and 12 (dodecane) carbon atoms per molecule and can be used in internal-combustion engines.

Methane

Methane, the main constituent of natural gas can be produced from algae in various methods, namely Gasification, Pyrolysis and Anaerobic Digestion. In Gasification and Pyrolysis methods methane is extracted under high temperature and pressure. Anaerobic Digestion is a straightforward method involved in decomposition of algae into simple components then transforming it into fatty acids using microbes like acidific bacteria followed by removing any solid particles and finally adding methanogenic bacteria to release a gas mixture containing methane. A number of studies have successfully shown that biomass from microalgae can be converted into biogas via anaerobic digestion. Therefore, in order to improve the overall energy balance of microalgae cultivation operations, it has been proposed to recover the energy contained in waste biomass via anaerobic digestion to methane for generating electricity.

Ethanol

The Algenol system which is being commercialized by BioFields in Puerto Libertad, Sonora, Mexico utilizes seawater and industrial exhaust to produce ethanol. Porphyridium cruentum also have shown to be potentially suitable for ethanol production due to its capacity for accumulating large amount of carbohydrates.

Green Diesel

Algae can be used to produce 'green diesel' (also known as renewable diesel, hydrotreating vegetable oil or hydrogen-derived renewable diesel) through a hydrotreating refinery process that breaks molecules down into shorter hydrocarbon chains used in diesel engines. It has the same chemical properties as petroleum-based diesel meaning that it does not require new engines, pipelines or infrastructure to distribute and use. It has yet to be produced at a cost that is competitive with petroleum. While hydrotreating is currently the most common pathway to produce fuel-like hydrocarbons via decarboxylation/decarbonylation, an alternative process offering a number of important advantages over hydrotreating. In this regard, the work of Crocker et al. and Lercher et al. is particularly noteworthy. for of oil refining, research is underway for catalytic conversion of renewable fuels by decarboxylation. As the oxygen is present in crude oil at rather low levels, of the order of 0.5%, deoxygenation in petroleum refining is not of

much concern, and no catalysts are specifically formulated for oxygenates hydrotreating. Hence, one of the critical technical challenges to make the hydrodeoxygenation of algae oil process economically feasible is related to the research and development of effective catalysts.

Jet Fuel

Rising jet fuel prices are putting severe pressure on airline companies, creating an incentive for algal jet fuel research. The International Air Transport Association, for example, supports research, development and deployment of algal fuels. IATA's goal is for its members to be using 10% alternative fuels by 2017.

Trials have been carried with aviation biofuel by Air New Zealand, Lufthansa, and Virgin Airlines.

In February 2010, the Defense Advanced Research Projects Agency announced that the U.S. military was about to begin large-scale oil production from algal ponds into jet fuel. After extraction at a cost of $2 per gallon, the oil will be refined at less than $3 a gallon. A larger-scale refining operation, producing 50 million gallons a year, is expected to go into production in 2013, with the possibility of lower per gallon costs so that algae-based fuel would be competitive with fossil fuels. The projects, run by the companies SAIC and General Atomics, are expected to produce 1,000 gallons of oil per acre per year from algal ponds.

Species

Research into algae for the mass-production of oil focuses mainly on microalgae (organisms capable of photosynthesis that are less than 0.4 mm in diameter, including the diatoms and cyanobacteria) as opposed to macroalgae, such as seaweed. The preference for microalgae has come about due largely to their less complex structure, fast growth rates, and high oil-content (for some species). However, some research is being done into using seaweeds for biofuels, probably due to the high availability of this resource.

As of 2012 researchers across various locations worldwide have started investigating the following species for their suitability as a mass oil-producers:

- *Botryococcus braunii*

- *Chlorella*

- *Dunaliella tertiolecta*

- *Gracilaria*

- *Pleurochrysis carterae* (also called CCMP647).

- *Sargassum*, with 10 times the output volume of *Gracilaria*.

The amount of oil each strain of algae produces varies widely. Note the following microalgae and their various oil yields:

- *Ankistrodesmus* TR-87: 28–40% dry weight

- *Botryococcus braunii*: 29–75% dw

- *Chlorella* sp.: 29%dw

- *Chlorella protothecoides*(autotrophic/ heterothrophic): 15–55% dw

- *Crypthecodinium cohnii*: 20%dw

- *Cyclotella* DI- 35: 42%dw

- *Dunaliella tertiolecta* : 36–42%dw

- *Hantzschia* DI-160: 66%dw

- *Nannochloris*: 31(6–63)%dw

- *Nannochloropsis* : 46(31–68)%dw

- *Neochloris oleoabundans*: 35–54%dw

- *Nitzschia* TR-114: 28–50%dw

- *Phaeodactylum tricornutum*: 31%dw

- *Scenedesmus* TR-84: 45%dw

- *Schizochytrium* 50–77%dw

- *Stichococcus*: 33(9–59)%dw

- *Tetraselmis suecica*: 15–32%dw

- *Thalassiosira pseudonana*: (21–31)%dw

In addition, due to its high growth-rate, *Ulva* has been investigated as a fuel for use in the *SOFT cycle*, (SOFT stands for Solar Oxygen Fuel Turbine), a closed-cycle power-generation system suitable for use in arid, subtropical regions.

Algae Cultivation

Photobioreactor from glass tubes

Open Pond

Design of a race-way open pond commonly used for algal culture

Algae grow much faster than food crops, and can produce hundreds of times more oil per unit area than conventional crops such as rapeseed, palms, soybeans, or jatropha. As algae have a harvesting cycle of 1–10 days, their cultivation permits several harvests in a very short time-frame, a strategy differing from that associated with annual crops. In addition, algae can be grown on land unsuitable for terrestrial crops, including arid land and land with excessively saline soil, minimizing competition with agriculture. Most research on algae cultivation has focused on growing algae in clean but expensive photobioreactors, or in open ponds, which are cheap to maintain but prone to contamination.

Closed-loop System

The lack of equipment and structures needed to begin growing algae in large quantities has inhibited widespread mass-production of algae for biofuel production. Maximum use of existing agriculture processes and hardware is the goal.

Closed systems (not exposed to open air) avoid the problem of contamination by other organisms blown in by the air. The problem for a closed system is finding a cheap source of sterile CO_2. Several experimenters have found the CO_2 from a smokestack works well for growing algae. For reasons of economy, some experts think that algae farming for biofuels will have to be done as part of cogeneration, where it can make use of waste heat and help soak up pollution.

Photobioreactors

Most companies pursuing algae as a source of biofuels pump nutrient-rich water through plastic or borosilicate glass tubes (called "bioreactors") that are exposed to sunlight (and so-called photobioreactors or PBR).

Running a PBR is more difficult than using an open pond, and costlier, but may provide a higher level of control and productivity. In addition, a photobioreactor can be integrated into a closed loop cogeneration system much more easily than ponds or other methods.

Open Pond

Raceway pond used for the cultivation of microalgae

Open-pond systems for the most part have been given up for the cultivation of algae with especially high oil content. Many believe that a major flaw of the Aquatic Species Program was the decision to focus their efforts exclusively on open-ponds; this makes the entire effort dependent upon the hardiness of the strain chosen, requiring it to be unnecessarily resilient in order to withstand wide swings in temperature and pH, and competition from invasive algae and bacteria. Open systems using a monoculture are also vulnerable to viral infection. The energy that a high-oil strain invests into the production of oil is energy that is not invested into the production of proteins or carbohydrates, usually resulting in the species being less hardy, or having a slower growth rate. Algal species with a lower oil content, not having to divert their energies away from growth, can be grown more effectively in the harsher conditions of an open system.

Some open sewage-ponds trial production has taken place in Marlborough, New Zealand.

Algal Turf Scrubber

The algal turf scrubber (ATS) is a system designed primarily for cleaning nutrients and pollutants out of water using algal turfs. ATS mimics the algal turfs of a natural coral reef by taking in nutrient rich water from waste streams or natural water sources, and pulsing it over a sloped surface. This surface is coated with a rough plastic membrane or a screen, which allows naturally occurring algal spores to settle and colonize the surface. Once the algae has been established, it can be harvested every 5–15 days, and can produce 18 metric tons of algal biomass per hectare per year. In contrast to other methods, which focus primarily on a single high yielding species of algae, this method focuses on naturally occurring polycultures of algae. As such, the lipid content of the algae in an ATS system is usually lower, which makes it more suitable for a fermented fuel product, such as ethanol, methane, or butanol. Conversely, the harvested algae could be treated with a hydrothermal liquefaction process, which would make possible biodiesel, gasoline, and jet fuel production.

2.5 acre ATS system, installed by Hydromentia on a farm creek in Florida

There are three major advantages of ATS over other systems. The first advantage is documented higher productivity over open pond systems. The second is lower operating and fuel production costs. The third is the elimination of contamination issues due to the reliance on naturally occurring algae species. The projected costs for energy production in an ATS system are $0.75/kg, compared to a photobioreactor which would cost $3.50/kg. Furthermore, due to the fact that the primary purpose of ATS is removing nutrients and pollutants out of water, and these costs have been shown to be lower than other methods of nutrient removal, this may incentivize the use of this technology for nutrient removal as the primary function, with biofuel production as an added benefit.

Algae being harvested and dried from an ATS system

Fuel Production

After harvesting the algae, the biomass is typically processed in a series of steps, which can differ based on the species and desired product; this is an active area of research. and also is the bottleneck of this technology: the cost of extraction is higher than those obtained. One of the solutions is to use filter feeders to "eat" them. Improved animals can provide both foods and fuels.

Dehydration

Often, the algae is dehydrated, and then a solvent such as hexane is used to extract energy-rich compounds like triglycerides from the dried material. Then, the extracted compounds can be processed into fuel using standard industrial procedures. For example, the extracted triglycerides are reacted with methanol to create biodiesel via transesterification. The unique composition of fatty acids of each species influences the quality of the resulting biodiesel and thus must be taken into account when selecting algal species for feedstock.

Hydrothermal Liquefaction

An alternative approach called Hydrothermal liquefaction employs a continuous process that subjects harvested wet algae to high temperatures and pressures—350 °C (662 °F) and 3,000 pounds per square inch (21,000 kPa).

Products include crude oil, which can be further refined into aviation fuel, gasoline, or diesel fuel using one or many upgrading processes. The test process converted between 50 and 70 percent of the algae's carbon into fuel. Other outputs include clean water, fuel gas and nutrients such as nitrogen, phosphorus, and potassium.

Nutrients

Nutrients like nitrogen (N), phosphorus (P), and potassium (K), are important for plant growth and are essential parts of fertilizer. Silica and iron, as well as several trace elements, may also be considered important marine nutrients as the lack of one can limit the growth of, or productivity in, an area.

Carbon Dioxide

Bubbling CO_2 through algal cultivation systems can greatly increase productivity and yield (up to a saturation point). Typically, about 1.8 tonnes of CO_2 will be utilised per tonne of algal biomass (dry) produced, though this varies with algae species. The Glenturret Distillery in Perthshire, UK – home to The Famous Grouse Whisky – percolate CO_2 made during the whisky distillation through a microalgae bioreactor. Each tonne of microalgae absorbs two tonnes of CO_2. Scottish Bioenergy, who run the project, sell the microalgae as high value, protein-rich food for fisheries. In the future, they will use the algae residues to produce renewable energy through anaerobic digestion.

Nitrogen

Nitrogen is a valuable substrate that can be utilized in algal growth. Various sources of nitrogen can be used as a nutrient for algae, with varying capacities. Nitrate was found to be the preferred source of nitrogen, in regards to amount of biomass grown. Urea is a readily available source that shows comparable results, making it an economical substitute for nitrogen source in large scale culturing of algae. Despite the clear increase in growth in comparison to a nitrogen-less medium, it has been shown that alterations in nitrogen levels affect lipid content within the algal cells. In one study nitrogen deprivation for 72 hours caused the total fatty acid content (on a per cell basis) to increase by 2.4-fold. 65% of the total fatty acids were esterified to triacylglycerides in oil bodies, when compared to the initial culture, indicating that the algal cells utilized de novo synthesis of fatty acids. It is vital for the lipid content in algal cells to be of high enough quantity, while maintaining adequate cell division times, so parameters that can maximize both are under investigation.

Wastewater

A possible nutrient source is waste water from the treatment of sewage, agricultural, or flood plain run-off, all currently major pollutants and health risks. However, this waste water cannot feed algae directly and must first be processed by bacteria, through anaerobic digestion. If waste water is not processed before it reaches the algae, it will contaminate the algae in the reactor, and at the very least, kill much of the desired algae strain. In biogas facilities, organic waste is often converted to a mixture of carbon dioxide, methane, and organic fertilizer. Organic fertilizer that comes out of the digester is liquid, and nearly suitable for algae growth, but it must first be cleaned and sterilized.

The utilization of wastewater and ocean water instead of freshwater is strongly advocated due to the continuing depletion of freshwater resources. However, heavy metals, trace metals, and other contaminants in wastewater can decrease the ability of cells to produce lipids biosynthetically and also impact various other workings in the machinery of cells. The same is true for ocean water, but the contaminants are found in different concentrations. Thus, agricultural-grade fertilizer is the preferred source of nutrients, but heavy metals are again a problem, especially for strains of algae that are susceptible to these metals. In open pond systems the use of strains of algae that can deal with high concentrations of heavy metals could prevent other organisms from infesting these systems. In some instances it has even been shown that strains of algae can remove over 90% of nickel and zinc from industrial wastewater in relatively short periods of time.

Environmental Impact

In comparison with terrestrial-based biofuel crops such as corn or soybeans, microalgal production results in a much less significant land footprint due to the higher oil productivity from the microalgae than all other oil crops. Algae can also be grown on marginal lands useless for ordinary crops and with low conservation value, and can use water from salt aquifers that is not useful for agriculture or drinking. Algae can also grow on the surface of the ocean in bags or floating screens. Thus microalgae could provide a source of clean energy with little impact on the provisioning of adequate food and water or the conservation of biodiversity. Algae cultivation also requires no external subsidies of insecticides or herbicides, removing any risk of generating associated pesticide waste streams. In addition, algal biofuels are much less toxic, and degrade far more readily than petroleum-based fuels. However, due to the flammable nature of any combustible fuel, there is potential for some environmental hazards if ignited or spilled, as may occur in a train derailment or a pipeline leak. This hazard is reduced compared to fossil fuels, due to the ability for algal biofuels to be produced in a much more localized manner, and due to the lower toxicity overall, but the hazard is still there nonetheless. Therefore, algal biofuels should be treated in a similar manner to petroleum fuels in transportation and use, with sufficient safety measures in place at all times.

Studies have determined that replacing fossil fuels with renewable energy sources, such as biofuels, have the capability of reducing CO_2 emissions by up to 80%. An algae-based system could capture approximately 80% of the CO_2 emitted from a power plant when sunlight is available. Although this CO_2 will later be released into the atmosphere when the fuel is burned, this CO_2 would have entered the atmosphere regardless. The possibility of reducing total CO_2 emissions therefore lies in the prevention of the release of CO_2 from fossil fuels. Furthermore, compared to fuels like diesel and petroleum, and even compared to other sources of biofuels, the production and combustion of algal biofuel does not produce any sulfur oxides or nitrous oxides, and produces a reduced amount of carbon monoxide, unburned hydrocarbons, and reduced emission of other harmful pollutants. Since terrestrial plant sources of biofuel production simply do not have the production capacity to meet current energy requirements, microalgae may be one of the only options to approach complete replacement of fossil fuels.

Microalgae production also includes the ability to use saline waste or waste CO_2 streams as an energy source. This opens a new strategy to produce biofuel in conjunction with waste water treatment, while being able to produce clean water as a byproduct. When used in a microalgal bioreactor, harvested microalgae will capture significant quantities of organic compounds as well as heavy metal contaminants absorbed from wastewater streams that would otherwise be directly discharged into surface and ground-water. Moreover, this process also allows the recovery of phosphorus from waste, which is an essential but scarce element in nature – the reserves of which are estimated to have depleted in the last 50 years. Another possibility is the use of algae production systems to clean up non-point source pollution, in a system known as an algal turf scrubber (ATS). This has been demonstrated to reduce nitrogen and phosphorus levels in rivers and other large bodies of water affected by eutrophication, and systems are being built that will be capable of processing up to 110 million liters of water per day. ATS can also be used for treating point source pollution, such as the waste water mentioned above, or in treating livestock effluent.

Polycultures

Nearly all research in algal biofuels has focused on culturing single species, or monocultures, of microalgae. However, ecological theory and empirical studies have demonstrated that plant and algae polycultures, i.e. groups of multiple species, tend to produce larger yields than monocultures. Experiments have also shown that more diverse aquatic microbial communities tend to be more stable through time than less diverse communities. Recent studies found that polycultures of microalgae produced significantly higher lipid yields than monocultures. Polycultures also tend to be more resistant to pest and disease outbreaks, as well as invasion by other plants or algae. Thus culturing microalgae in polyculture may not only increase yields and stability of yields of biofuel, but also reduce the environmental impact of an algal biofuel industry.

Economic Viability

There is clearly a demand for sustainable biofuel production, but whether a particular biofuel will be used ultimately depends not on sustainability but cost efficiency. Therefore, research is focusing on cutting the cost of algal biofuel production to the point where it can compete with conventional petroleum. The production of several products from algae has been mentioned as the most important factor for making algae production economically viable. Other factors are the improving of the solar energy to biomass conversion efficiency (currently 3%, but 5 to 7% is theoretically attainable) and making the oil extraction from the algae easier.

In a 2007 report a formula was derived estimating the cost of algal oil in order for it to be a viable substitute to petroleum diesel:

$$C_{(algal\ oil)} = 25.9 \times 10^{-3}\ C_{(petroleum)}$$

where: $C_{(algal\ oil)}$ is the price of microalgal oil in dollars per gallon and $C_{(petroleum)}$ is the price of crude oil in dollars per barrel. This equation assumes that algal oil has roughly 80% of the caloric energy value of crude petroleum.

With current technology available, it is estimated that the cost of producing microalgal biomass is $2.95/kg for photobioreactors and $3.80/kg for open-ponds. These estimates assume that carbon dioxide is available at no cost. If the annual biomass production capacity is increased to 10,000 tonnes, the cost of production per kilogram reduces to roughly $0.47 and $0.60, respectively. Assuming that the biomass contains 30% oil by weight, the cost of biomass for providing a liter of oil would be approximately $1.40 ($5.30/gal) and $1.81 ($6.85/gal) for photobioreactors and raceways, respectively. Oil recovered from the lower cost biomass produced in photobioreactors is estimated to cost $2.80/L, assuming the recovery process contributes 50% to the cost of the final recovered oil. If existing algae projects can achieve biodiesel production price targets of less than $1 per gallon, the United States may realize its goal of replacing up to 20% of transport fuels by 2020 by using environmentally and economically sustainable fuels from algae production.

Whereas technical problems, such as harvesting, are being addressed successfully by the industry, the high up-front investment of algae-to-biofuels facilities is seen by many as a major obstacle to the success of this technology. Only few studies on the economic viability are publicly available, and must often rely on the little data (often only engineering estimates) available in the public domain. Dmitrov examined the GreenFuel's photobioreactor and estimated that algae oil would only be competitive at an oil price of $800 per barrel. A study by Alabi et al. examined raceways, photobioreactors and anaerobic fermenters to make biofuels from algae and found that photobioreactors are too expensive to make biofuels. Raceways might be cost-effective in warm climates with very low labor costs, and fermenters may become cost-effective subsequent to significant process improvements. The group found that capital cost, labor cost and operational costs (fertilizer, electricity, etc.) by themselves are too high for algae biofuels to be cost-competitive with conventional fuels. Similar results were found by others, suggesting that unless new, cheaper ways of harnessing algae for biofuels production are found, their great technical potential may never become economically accessible. Recently, Rodrigo E. Teixeira demonstrated a new reaction and proposed a process for harvesting and extracting raw materials for biofuel and chemical production that requires a fraction of the energy of current methods, while extracting all cell constituents.

Use of Byproducts

Many of the byproducts produced in the processing of microalgae can be used in various applications, many of which have a longer history of production than algal biofuel. Some of the products not used in the production of biofuel include natural dyes and pigments, antioxidants, and other high-value bio-active compounds. These chemicals and excess biomass have found numerous use in other industries. For example, the dyes and oils have found a place in cosmetics, commonly as thickening and water-binding agents. Discoveries within the pharmaceutical industry include antibiotics and antifungals derived from microalgae, as well as natural health products, which have been growing in popularity over the past few decades. For instance *Spirulina* contains numerous polyunsaturated fats (Omega 3 and 6), amino acids, and vitamins, as well as pigments that may be beneficial, such as beta-carotene and chlorophyll.

Advantages

Ease of Growth

One of the main advantages that using microalgae as the feedstock when compared to more traditional crops is that it can be grown much more easily. Algae can be grown in land that would not be considered suitable for the growth of the regularly used crops. In addition to this, wastewater that would normally hinder plant growth has been shown to be very effective in growing algae. Because of this, algae can be grown without taking up arable land that would otherwise be used for producing food crops, and the better resources can be reserved for normal crop production. Microalgae also require fewer resources to grow and little attention is needed, allowing the growth and cultivation of algae to be a very passive process.

Impact on Food

Many traditional feedstocks for biodiesel, such as corn and palm, are also used as feed for livestock on farms, as well as a valuable source of food for humans. Because of this, using them as biofuel reduces the amount of food available for both, resulting in an increased cost for both the food and the fuel produced. Using algae as a source of biodiesel can alleviate this problem in a number of ways. First, algae is not used as a primary food source for humans, meaning that it can be used solely for fuel and there would be little impact in the food industry. Second, many of the waste-product extracts produced during the processing of algae for biofuel can be used as a sufficient animal feed. This is an effective way to minimize waste and a much cheaper alternative to the more traditional corn- or grain-based feeds.

Minimization of Waste

Growing algae as a source of biofuel has also been shown to have numerous environmental benefits, and has presented itself as a much more environmentally friendly alternative to current biofuels. For one, it is able to utilize run-off, water contaminated with fertilizers and other nutrients that are a by-product of farming, as its primary source of water and nutrients. Because of this, it prevents this contaminated water from mixing with the lakes and rivers that currently supply our drinking water. In addition to this, the ammonia, nitrates, and phosphates that would normally render the water unsafe actually serve as excellent nutrients for the algae, meaning that fewer resources are needed to grow the algae. Many algae species used in biodiesel production are excellent bio-fixers, meaning they are able to remove carbon dioxide from the atmosphere to use as a form of energy for themselves. Because of this, they have found use in industry as a way to treat flue gases and reduce GHG emissions.

Disadvantages

Commercial Viability

Algae biodiesel is still a fairly new technology. Despite the fact that research began over 30 years ago, it was put on hold during the mid-1990s, mainly due to a lack of funding and a relatively low petroleum cost. For the next few years algae biofuels saw little attention; it was not until the gas peak of the early 2000s that it eventually had a revitalization in the search for alternative fuel sources. While the technology exists to harvest and convert algae into a usable source of biodiesel, it still hasn't been implemented into a large enough scale to support the current energy needs. Further research will be required to make the production of algae biofuels more efficient, and at this point it is currently being held back by lobbyists in support of alternative biofuels, like those produced from corn and grain. In 2013, Exxon Mobil Chairman and CEO Rex Tillerson said that after originally committing to spending up to $600 million on development in a joint venture with J. Craig Venter's Synthetic Genomics, algae is "probably further" than "25 years away" from commercial viability, although Solazyme and Sapphire Energy already began small-scale commercial sales in 2012 and 2013, respectively.

Stability

The biodiesel produced from the processing of microalgae differs from other forms of biodiesel in the content of polyunsaturated fats. Polyunsaturated fats are known for their ability to retain fluidity at lower temperatures. While this may seem like an advantage in production during the colder temperatures of the winter, the polyunsaturated fats result in lower stability during regular seasonal temperatures.

Current Projects

United States

The National Renewable Energy Laboratory (NREL) is the U.S. Department of Energy's primary national laboratory for renewable energy and energy efficiency research and development. This program is involved in the production of renewable energies and energy efficiency. One of its most current divisions is the biomass program which is involved in biomass characterization, biochemical and thermochemical conversion technologies in conjunction with biomass process engineering and analysis. The program aims at producing energy efficient, cost-effective and environmentally friendly technologies that support rural economies, reduce the nations dependency in oil and improve air quality.

At the Woods Hole Oceanographic Institution and the Harbor Branch Oceanographic Institution the wastewater from domestic and industrial sources contain rich organic compounds that are being used to accelerate the growth of algae. The Department of Biological and Agricultural Engineering at University of Georgia is exploring microalgal biomass production using industrial wastewater. Algaewheel, based in Indianapolis, Indiana, presented a proposal to build a facility in Cedar Lake, Indiana that uses algae to treat municipal wastewater, using the sludge byproduct to produce biofuel. A similar approach is being followed by Algae Systems, a company based in Daphne, Alabama.

Sapphire Energy (San Diego) has produced green crude from algae.

Solazyme (South San Francisco, California) has produced a fuel suitable for powering jet aircraft from algae.

Europe

Universities in the United Kingdom which are working on producing oil from algae include: University of Manchester, University of Sheffield, University of Glasgow, University of Brighton, University of Cambridge, University College London, Imperial College London, Cranfield University and Newcastle University. In Spain, it is also relevant the research carried out by the CSIC´s Instituto de Bioquímica Vegetal y Fotosíntesis (Microalgae Biotechnology Group, Seville).

The Marine Research station in Ketch Harbour, Nova Scotia, has been involved in growing algae for 50 years. The National Research Council (Canada) (NRC) and National Byproducts Program have provided $5 million to fund this project. The aim of the program has been to build a 50 000 litre cultivation pilot plant at the Ketch harbor facility. The station has been involved in assessing how best to grow algae for biofuel and is involved in investigating the utilization of numerous algae species in regions of North America. NRC has joined forces with the United States Department of Energy, the National Renewable Energy Laboratory in Colorado and Sandia National Laboratories in New Mexico.

The European Algae Biomass Association (EABA) is the European association representing both research and industry in the field of algae technologies, currently with 79 members. The association is headquartered in Florence, Italy. The general objective of the EABA is to promote mutual interchange and cooperation in the field of biomass production and use, including biofuels uses and all other utilisations. It aims at creating, developing and maintaining solidarity and links between its Members and at defending their interests at European and international level. Its main target is to act as a catalyst for fostering synergies among scientists, industrialists and decision makers to promote the development of research, technology and industrial capacities in the field of Algae.

CMCL innovations and the University of Cambridge are carrying out a detailed design study of a C-FAST (Carbon negative Fuels derived from Algal and Solar Technologies) plant. The main objective is to design a pilot plant which can demonstrate production of hydrocarbon fuels (including diesel and gasoline) as sustainable carbon-negative energy carriers and raw materials for the chemical commodity industry. This project will report in June 2013.

Ukraine plans to produce biofuel using a special type of algae.

The European Commission's Algae Cluster Project, funded through the Seventh Framework Programme, is made up of three algae biofuel projects, each looking to design and build a different algae biofuel facility covering 10ha of land. The projects are BIOFAT, All-Gas and InteSusAl.

Since various fuels and chemicals can be produced from algae, it has been suggested to investigate the feasibility of various production processes(conventional extraction/separation, hydrothermal liquefaction, gasification and pyrolysis) for application in an integrated algal biorefinery.

India

Reliance industries in collaboration with Algenol, USA commissioned a pilot project to produce algal bio-oil in the year 2014. Spirulina which is an alga rich in proteins content has been commercially cultivated in India. Algae is used in India for treating the sewage in open/natural oxidation ponds This reduces the Biological Oxygen Demand (BOD) of the sewage and also provides algal biomass which can be converted to fuel.

Other

The Algae Biomass Organization (ABO) is a non-profit organization whose mission is "to promote the development of viable commercial markets for renewable and sustainable commodities derived from algae".

The National Algae Association (NAA) is a non-profit organization of algae researchers, algae production companies and the investment community who share the goal of commercializing algae oil as an alternative feedstock for the biofuels markets. The NAA gives its members a forum to efficiently evaluate various algae technologies for potential early stage company opportunities.

Pond Biofuels Inc. in Ontario, Canada has a functioning pilot plant where algae is grown directly off of smokestack emissions from a cement plant, and dried using waste heat. In May 2013, Pond Biofuels announced a partnership with the National Research Council of Canada and Canadian Natural Resources Limited to construct a demonstration-scale algal biorefinery at an oil sands site near Bonnyville, Alberta.

Ocean Nutrition Canada in Halifax, Nova Scotia, Canada has found a new strain of algae that appears capable of producing oil at a rate 60 times greater than other types of algae being used for the generation of biofuels.

VG Energy, a subsidiary of Viral Genetics Incorporated, claims to have discovered a new method of increasing algal lipid production by disrupting the metabolic pathways that would otherwise divert photosynthetic energy towards carbohydrate production. Using these techniques, the company states that lipid production could be increased several-fold, potentially making algal biofuels cost-competitive with existing fossil fuels.

Algae production from the warm water discharge of a nuclear power plant has been piloted by Patrick C. Kangas at Peach Bottom Nuclear Power Station, owned by Exelon Corporation. This process takes advantage of the relatively high temperature water to sustain algae growth even during winter months.

Companies such as Sapphire Energy and Bio Solar Cells are using genetic engineering to make algae fuel production more efficient. According to Klein Lankhorst of Bio Solar Cells, genetic engineering could vastly improve algae fuel efficiency as algae can be modified to only build short carbon chains instead of long chains of carbohydrates. Sapphire Energy also uses chemically induced mutations to produce algae suitable for use as a crop.

Some commercial interests into large-scale algal-cultivation systems are looking to tie in to existing infrastructures, such as cement factories, coal power plants, or sewage treatment facilities.

This approach changes wastes into resources to provide the raw materials, CO2 and nutrients, for the system.

A feasibility study using marine microalgae in a photobioreactor is being done by The International Research Consortium on Continental Margins at the Jacobs University Bremen.

The Department of Environmental Science at Ateneo de Manila University in the Philippines, is working on producing biofuel from a local species of algae.

Genetic Engineering

Genetic engineering algae has been used to increase lipid production or growth rates. Current research in genetic engineering includes either the introduction or removal of enzymes. In 2007 Oswald et al. introduced a monoterpene synthase from sweet basil into Saccharomyces cerevisiae, a strain of yeast. This particular monoterpene synthase causes the de novo synthesis of large amounts of geraniol, while also secreting it into the medium. Geraniol is a primary component in rose oil, palmarosa oil, and citronella oil as well as essential oils, making it a viable source of triacylglycerides for biodiesel production.

The enzyme ADP-glucose pyrophosphorylase is vital in starch production, but has no connection to lipid synthesis. Removal of this enzyme resulted in the sta6 mutant, which showed increased lipid content. After 18 hours of growth in nitrogen deficient medium the sta6 mutants had on average 17 ng triacylglycerides/1000 cells, compared to 10 ng/1000 cells in WT cells. This increase in lipid production was attributed to reallocation of intracellular resources, as the algae diverted energy from starch production.

In 2013 researchers used a "knock-down" of fat-reducing enzymes (multifunctional lipase/phospholipase/acyltransferase) to increase lipids (oils) without compromising growth. The study also introduced an efficient screening process. Antisense-expressing knockdown strains 1A6 and 1B1 contained 2.4- and 3.3-fold higher lipid content during exponential growth, and 4.1- and 3.2-fold higher lipid content after 40 h of silicon starvation.

Funding Programs

Numerous Funding programs have been created with aims of promoting the use of Renewable Energy. In Canada, the ecoAgriculture biofuels capital initiative (ecoABC) provides $25 million per project to assist farmers in constructing and expanding a renewable fuel production facility. The program has $186 million set aside for these projects. The sustainable development (SDTC) program has also applied $500 millions over 8 years to assist with the construction of next-generation renewable fuels. In addition, over the last 2 years $10 million has been made available for renewable fuel research and analysis

In Europe, the Seventh Framework Programme (FP7) is the main instrument for funding research. Similarly, the NER 300 is an unofficial, independent portal dedicated to renewable energy and grid integration projects. Another program includes the horizon 2020 program which will start 1 January, and will bring together the framework program and other EC innovation and research funding into a new integrated funding system

The American NBB's Feedstock Development program is addressing production of algae on the horizon to expand available material for biodiesel in a sustainable manner.

International Policies

Canada

Numerous policies have been put in place since the 1975 oil crisis in order to promote the use of Renewable Fuels in the United States, Canada and Europe. In Canada, these included the implementation of excise taxes exempting propane and natural gas which was extended to ethanol made from biomass and methanol in 1992. The federal government also announced their renewable fuels strategy in 2006 which proposed four components: increasing availability of renewable fuels through regulation, supporting the expansion of Canadian production of renewable fuels, assisting farmers to seize new opportunities in this sector and accelerating the commercialization of new technologies. These mandates were quickly followed by the Canadian provinces:

BC introduced a 5% ethanol and 5% renewable diesel requirement which was effective by January 2010. It also introduced a low carbon fuel requirement for 2012 to 2020.

Alberta introduced a 5% ethanol and 2% renewable diesel requirement implemented April 2011. The province also introduced a minimum 25% GHG emission reduction requirement for qualifying renewable fuels.

Saskatchewan implemented a 2% renewable diesel requirement in 2009.

Additionally, in 2006, the Canadian Federal Government announced its commitment to using its purchasing power to encourage the biofuel industry. Section three of the 2006 alternative fuels act stated that when it is economically feasible to do so-75% per cent of all federal bodies and crown corporation will be motor vehicles.

The National Research Council of Canada has established research on Algal Carbon Conversion as one of its flagship programs. As part of this program, the NRC made an announcement in May 2013 that they are partnering with Canadian Natural Resources Limited and Pond Biofuels to construct a demonstration-scale algal biorefinery near Bonnyville, Alberta.

United States

Policies in the United States have included a decrease in the subsidies provided by the federal and state governments to the oil industry which have usually included $2.84 billion. This is more than what is actually set aside for the biofuel industry. The measure was discussed at the G20 in Pittsburgh where leaders agreed that "inefficient fossil fuel subsidies encourage wasteful consumption, reduce our energy security, impede investment in clean sources and undermine efforts to deal with the threat of climate change". If this commitment is followed through and subsidies are removed, a fairer market in which algae biofuels can compete will be created. In 2010, the U.S. House of Representatives passed a legislation seeking to give algae-based biofuels parity with cellulose biofuels in federal tax credit programs. The algae-based renewable fuel promotion act (HR 4168) was implemented to give biofuel projects access to a $1.01 per gal production tax credit and 50% bonus depreciation for biofuel plant property. The U.S Government also introduced the domestic Fuel for

Enhancing National Security Act implemented in 2011. This policy constitutes an amendment to the Federal property and administrative services act of 1949 and federal defense provisions in order to extend to 15 the number of years that the Department of Defense (DOD) multiyear contract may be entered into the case of the purchase of advanced biofuel. Federal and DOD programs are usually limited to a 5-year period

Other

The European Union (EU) has also responded by quadrupling the credits for second-generation algae biofuels which was established as an amendment to the Biofuels and Fuel Quality Directives

Companies

With algal biofuel being a relatively new alternative to conventional petroleum products, it leaves numerous opportunities for drastic advances in all aspects of the technology. Producing algae biofuel is not yet a cost-effective replacement for gasoline, but alterations to current methodologies can change this. The two most common targets for advancements are the growth medium (open pond vs. photobioreactor) and methods to remove the intracellular components of the algae. Below are companies that are currently innovating algal biofuel technologies.

Algenol Biofuels

Founded in 2006, Algenol Biofuels is a global, industrial biotechnology company that is commercializing its patented algae technology for production of ethanol and other fuels. Based in Southwest Florida, Algenol's patented technology enables the production of the four most important fuels (ethanol, gasoline, jet, and diesel fuel) using proprietary algae, sunlight, carbon dioxide and saltwater for around $1.27 per gallon and at production levels of 8,000 total gallons of liquid fuel per acre per year. Algenol's technology produces high yields and relies on patented photobioreactors and proprietary downstream techniques for low-cost fuel production using carbon dioxide from industrial sources.

Blue Marble Production

Blue Marble Production is a Seattle-based company that is dedicated to removing algae from algae-infested water. This in turn cleans up the environment and allows this company to produce biofuel. Rather than just focusing on the mass production of algae, this company focuses on what to do with the byproducts. This company recycles almost 100% of its water via reverse osmosis, saving about 26,000 gallons of water every month. This water is then pumped back into their system. The gas produced as a byproduct of algae will also be recycled by being placed into a photobioreactor system that holds multiple strains of algae. Whatever gas remains is then made into pyrolysis oil by thermochemical processes. Not only does this company seek to produce biofuel, but it also wishes to use algae for a variety of other purposes such as fertilizer, food flavoring, anti-inflammatory, and anti-cancer drugs.

Solazyme

Solazyme is one of a handful of companies which is supported by oil companies such as Chevron. Additionally, this company is also backed by Imperium Renewables, Blue Crest Capital Finance, and The Roda Group. Solazyme has developed a way to use up to 80% percent of dry algae as oil. This process requires the algae to grow in a dark fermentation vessel and be fed by carbon substrates within their growth media. The effect is the production of triglycerides that are almost identical to vegetable oil. Solazyme's production method is said to produce more oil than those algae cultivated photosynthetically or made to produce ethanol. Oil refineries can then take this algal oil and turn it into biodiesel, renewable diesel or jet fuels.

Part of Solazyme's testing, in collaboration with Maersk Line and the US Navy, placed 30 tons of Soladiesel(RD) algae fuel into the 98,000-tonne, 300-meter container ship Maersk Kalmar. This fuel was used at blends from 7% to 100% in an auxiliary engine on a month-long trip from Bremerhaven, Germany to Pipavav, India in Dec 2011. In Jul 2012, The US Navy used 700,000 gallons of HRD76 biodiesel in three ships of the USS Nimitz "Green Strike Group" during the 2012 RIMPAC exercise in Hawaii. The Nimitz also used 200,000 gallons of HRJ5 jet biofuel. The 50/50 biofuel blends were provided by Solazyme and Dynamic Fuels.

Sapphire Energy

Sapphire Energy is a leader in the algal biofuel industry backed by the Wellcome Trust, Bill Gates' Cascade Investment, Monsanto, and other large donors. After experimenting with production of various algae fuels beginning in 2007, the company now focuses on producing what it calls "green crude" from algae in open raceway ponds. After receiving more than $100 million in federal funds in 2012, Sapphire built the first commercial demonstration algae fuel facility in New Mexico and has continuously produced biofuel since completion of the facility in that year. In 2013, Sapphire began commercial sales of algal biofuel to Tesoro, making it one of the first companies, along with Solazyme, to sell algae fuel on the market.

Diversified Technologies Inc.

Diversified Technologies Inc. has created a patent pending pre-treatment option to reduce costs of oil extraction from algae. This technology, called Pulsed Electric Field (PEF) technology, is a low cost, low energy process that applies high voltage electric pulses to a slurry of algae. The electric pulses enable the algal cell walls to be ruptured easily, increasing the availability of all cell contents (Lipids, proteins and carbohydrates), allowing the separation into specific components downstream. This alternative method to intracellular extraction has shown the capability to be both integrated in-line as well as scalable into high yield assemblies. The Pulse Electric Field subjects the algae to short, intense bursts of electromagnetic radiation in a treatment chamber, electroporating the cell walls. The formation of holes in the cell wall allows the contents within to flow into the surrounding solution for further separation. PEF technology only requires 1-10 microsecond pulses, enabling a high-throughput approach to algal extraction.

Preliminary calculations have shown that utilization of PEF technology would only account for $0.10 per gallon of algae derived biofuel produced. In comparison, conventional drying and solvent-based extractions account for $1.75 per gallon. This inconsistency between costs can be attributed to the fact that algal drying generally accounts for 75% of the extraction process. Although a relatively new technology, PEF has been successfully used in both food decomtamination processes as well as waste water treatments.

Origin Oils Inc.

Origin Oils Inc. has been researching a revolutionary method called the Helix Bioreactor, altering the common closed-loop growth system. This system utilizes low energy lights in a helical pattern, enabling each algal cell to obtain the required amount of light. Sunlight can only penetrate a few inches through algal cells, making light a limiting reagent in open-pond algae farms. Each lighting element in the bioreactor is specially altered to emit specific wavelengths of light, as a full spectrum of light is not beneficial to algae growth. In fact, ultraviolet irradiation is actually detrimental as it inhibits photosynthesis, photoreduction, and the 520 nm light-dark absorbance change of algae.

This bioreactor also addresses another key issue in algal cell growth; introducing CO_2 and nutrients to the algae without disrupting or over-aerating the algae. Origin Oils Inc. combats this issues through the creation of their Quantum Fracturing technology. This process takes the CO_2 and other nutrients, fractures them at extremely high pressures and then deliver the micron sized bubbles to the algae. This allows the nutrients to be delivered at a much lower pressure, maintaining the integrity of the cells.

Proviron

Proviron has been working on a new type of reactor (using flat plates) which reduces the cost of algae cultivation. At AlgaePARC similar research is being conducted using 4 grow systems (1 open pond system and 3 types of closed systems). According to René Wijffels the current systems do not yet allow algae fuel to be produced competitively. However using new (closed) systems, and by scaling up the production it would be possible to reduce costs by 10X, up to a price of 0,4 € per kg of algae.

Genifuels

Genifuel Corporation has licensed the high temperature/pressure fuel extraction process and has been working with the team at the lab since 2008. The company intends to team with some industrial partners to create a pilot plant using this process to make biofuel in industrial quantities. Genifuel process combines hydrothermal liquefaction with catalytic hydrothermal gasification in reactor running at 350 Celsius (662 Fahrenheit) and pressure of 3000 PSI.

Baya® Biofuel

The QMAB started to mobilize its 100 hectares micro algae cultivation farm in Qeshm Island free zone. Development of the farm mainly focuses on 2 phases, production of nutraceutical products and green crude oil to produce biofuel. main product of microalgae culture is crude oil, which can be fractioned into the same kinds of fuels and chemical compounds.

Fungiculture

Fungiculture is the process of producing food, medicine, and other products by the cultivation of mushrooms and other fungi.

The word is also commonly used to refer to the practice of cultivating fungi by leafcutter ants, termites, ambrosia beetles, and marsh periwinkles.

Introduction

Mushrooms are not plants, and require different conditions for optimal growth. Plants develop through photosynthesis, a process that converts atmospheric carbon dioxide into carbohydrates, especially cellulose. While sunlight provides an energy source for plants, mushrooms derive all of their energy and growth materials from their growth medium, through biochemical decomposition processes. This does not mean that light is an irrelevant requirement, since some fungi use light as a signal for fruiting. However, all the materials for growth must already be present in the growth medium. Mushrooms grow well at relative humidity levels of around 95–100%, and substrate moisture levels of 50 to 75%.

Instead of seeds, mushrooms reproduce asexually through spores. Spores can be contaminated with airborne microorganisms, which will interfere with mushroom growth and prevent a healthy crop.

Mycelium, or actively growing mushroom culture, is placed on a substrate—usually sterilized grains such as rye or millet—and induced to grow into those grains. This is called inoculation. Inoculated grains are referred to as spawn. Spores are another inoculation option, but are less developed than established mycelium. Since they are also contaminated easily, they are only manipulated in laboratory conditions with a laminar flow cabinet.

Techniques

All mushroom growing techniques require the correct combination of humidity, temperature, substrate (growth medium) and inoculum (spawn or starter culture). Wild harvests, outdoor log inoculation and indoor trays all provide these elements.

Wild Harvesting

Due to its climate, the Pacific Northwest of the USA produces commercially valuable mushrooms. Valued species include:

- American matsutake or pine mushroom (*Tricholoma magnivelare*)

- Chanterelles (*Cantharellus formosus*, *Cantharellus subalbidus*, and *Cantharellus cibarius*)

- Horn of plenty (*Craterellus cornucopioides*)

- Boletes (*Boletus edulis* and others)

- Truffles (*Tuber gibbosum* and *Leucangium carthusiana*)

- Hedgehogs / "spreading-hedgehog mushroom" (*Hydnum repandum*)

- Edible morel (*Morchella esculenta*)

- Coral tooth mushroom (*Hericium abietis*)

- Shaggy parasol (*Lepiota rhacodes*)

- Black picoa (*Picoa carthusiana*)

- Cauliflower mushroom (*Sparassis crispa*)

Mushroom gatherers have few requirements to begin business. Gatherers only need to supply funds for possible park fees, knowledge for identifying mushrooms and gathering time.

There are significant disadvantages to relying on natural mushroom production. These sales may be unregulated, placing buyers at risk for buying toxic or inedible mushrooms. By honest error, harvests may include toxic or inedible species. No controls exist to regulate the quality or frequency of harvests, since gatherers rely on favorable natural conditions and weather to produce fruiting. Conflicts may arise between competing gatherers trying to harvest from the same location.

State parks in the Pacific Northwest or elsewhere may charge fees for mushroom gathering permits. Appalachia also produces edible wild mushrooms, including chanterelles and morels. Pickers may sell directly to distributors, restaurants, or sell their harvest through roadside stands wherever a natural supply of mushrooms is plentiful.

While there may be concern that harvesting wild mushrooms may exploit or damage a natural environment, harvesting wild mushrooms is different from harvesting wild plants, fishing or hunting animals. In these last three cases, removing individuals decreases the ability of a wild population to reproduce, since fewer adults remain. Removing adults leaves fewer individuals capable of reproducing and reduces genetic diversity.

Harvesting wild mushrooms removes only fruiting bodies and their attached spores. However, the fruiting bodies (mushrooms) have likely dropped spores before the harvest time, or will likely drop them en route to the harvester's destination, further expanding the fungi's habitat. Arguably, the practice of mushroom harvesting may actually help the species being harvested. While truffles also represent the fruiting body of a larger underground network, they are an exception, since they rely on animal spore dispersal.

Additionally, reproduction and propagation can still occur by propagation of the parent mycelium. Harvesting removes none of the parent mycelium, which remains intact underground.

Outdoor Logs

Mushrooms can be grown on logs placed outdoors in stacks or piles, as has been done for hundreds of years. Sterilization is not performed in this method. Since production may be unpredictable and seasonal, less than 5% of commercially sold mushrooms are produced this way. Here, tree logs are inoculated with spawn, then allowed to grow as they would in wild conditions. Fruiting, or pinning, is triggered by seasonal changes, or by briefly soaking the logs in cool water. Shiitake and oyster mushrooms have traditionally been produced using the outdoor log technique, although controlled techniques such as indoor tray growing or artificial logs made of compressed substrate have been substituted.

Shiitake mushrooms grown under a forested canopy are considered non-timber forest products In the Northeast shiitake mushrooms can be cultivated on a variety of hardwood logs including oak, American beech, sugar maple and hophornbeam. Softwood should not be used to cultivate shiitake mushrooms. The resin of softwoods will oftentimes inhibit the growth of the shiitake mushroom making it impractical as a growing substrate.

In order to produce shiitake mushrooms, 3 foot hardwood logs with a diameter ranging between 4 and 6 inches are inoculated with the mycelium of the shiitake fungus. Inoculation is completed by drilling holes in hardwood logs, filling the holes with cultured shiitake mycelium or inoculum, and then sealing the filled holes with hot wax. After inoculation, the logs are placed under the closed canopy of a coniferous stand and are left to incubate for 12 to 15 months. Once incubation is complete, the logs are soaked in water for 24 hours. 7 to 10 days after soaking, shiitake mushrooms will begin to fruit and can be harvested once fully ripe.

Indoor Trays

Indoor growing provides the ability to tightly regulate light, temperature and humidity while excluding contaminants and pests. This allows consistent production, regulated by spawning cycles. This is typically accomplished in windowless, purpose-built buildings, for large scale commercial production.

Indoor tray growing is the most common commercial technique, followed by containerized growing. The tray technique provides the advantages of scalability and easier harvesting. Unlike wild harvests, indoor techniques provide tight control over growing substrate composition and growing conditions. Indoor harvests are much more predictable.

According to Daniel Royse and Robert Beelman, "[Indoor] Mushroom farming consists of six steps, and although the divisions are somewhat arbitrary, these steps identify what is needed to form a production system. The six steps are phase I composting, phase II fertilizing, spawning, casing, pinning, and cropping."

Six Phases of Mushroom Cultivation

Phase	Time span	Temperature	Key points	Process(procedure)
1. Phase I composting	6–14 days		Regulate water and NH_3 content through microbial action. Add fertilizer / additives	
2. Phase II composting or pasteurization	7–18 days via composting method, ~2 hours for pasteurization (heat sterilization)		Reduce number of potentially harmful microbes through further composting, or apply heat sterilization. Remove unwanted NH_3.	

3. Spawning and growth	14–21 days	75 °F; to 80 °F; must be above 74 °F; for rapid growth. Must be below 80 °F; to 85 °F to avoid damaging mycelia	Add starter culture. Allow mycelium to grow through substrate and form a colony. Depends on substrate dimensions and composition. Finished when mycelium has propagated through entire substrate layer	
4. Casing	13–20 days		Promote the formation of primordia, or mushroom pins. Add a top covering or dressing to the colonized substrate. Fertilizing with nitrogen increases yields. Induces pinning	
5. Pinning	18–21 days		Earliest formation of recognizable mushrooms from mycelium. Adjusting temperature, humidity and CO_2 will also affect the number of pins, and mushroom size	
6. Cropping	Repeated over 7- to 10-day cycles		Harvest	

Complete sterilization is not always required or performed during composting. In some cases, a pasteurization step is not included to allow some beneficial microorganisms to remain in the growth substrate.

Specific time spans and temperatures required during stages 3–6 will vary respective to species and variety. Substrate composition and the geometry of growth substrate will also affect the ideal times and temperatures.

Pinning is the trickiest part for a mushroom grower, since a combination of carbon dioxide (CO_2) concentration, temperature, light, and humidity triggers mushrooms towards fruiting. Up until the point when rhizomorphs or mushroom "pins" appear, the mycelium is an amorphous mass spread throughout the growth substrate, unrecognizable as a mushroom.

Carbon dioxide concentration becomes elevated during the vegetative growth phase, when mycelium is sealed in a gas-resistant plastic barrier or bag which traps gases produced by the growing mycelium. To induce pinning, this barrier is opened or ruptured. CO_2 concentration then decreases from about 0.08% to 0.04%, the ambient atmospheric level.

Substrates

Mushroom production converts the raw natural ingredients into mushroom tissue, most notably the carbohydrate chitin.

An ideal substrate will contain enough nitrogen and carbohydrate for rapid mushroom growth. Common bulk substrates include several of the following ingredients:

- Wood chips or sawdust
- Mulched straw (usually wheat, but also rice and other straws)
- Strawbedded horse or poultry manure
- Corncobs
- Waste or recycled paper
- coffee pulp or grounds
- Nut and seed hulls
- Cottonseed hulls
- Cocoa bean hulls
- Cottonseed meal
- Soybean meal
- Brewer's grain
- Ammonium nitrate
- Urea

Mushrooms metabolize complex carbohydrates in their substrate into glucose, which is then transported through the mycelium as needed for growth and energy. While it is used as a main energy source, its concentration in the growth medium should not exceed 2%. For ideal fruiting, closer to 1% is ideal.

Pests and Diseases

Parasitic insects, bacteria and other fungi all pose risks to indoor production. The sciarid fly or phorid fly may lay eggs in the growth medium, which hatch into maggots and damage developing mushrooms during all growth stages. Bacterial blotch caused by *Pseudomonas* bacteria or patches of *Trichoderma* green mold also pose a risks during the fruiting stage. Pesticides and sanitizing agents are available to use against these infestations. Biological controls for insect sciarid and phorid flies have also been proposed.

A recent epidemic of Trichoderma green mold has significantly affected mushroom production: "From 1994–96, crop losses in Pennsylvania ranged from 30 to 100%".

Commercially Cultivated Fungi

Home cultivated *shiitake* developing over approximately 24 hours.

- *Agaricus bisporus*, also known as champignon and the button mushroom. This species also includes the portobello and crimini mushrooms.

- *Auricularia polytricha* or *Auricularia auricula-judae* (Tree ear fungus), two closely related species of jelly fungi that are commonly used in Chinese cuisine.

- *Flammulina velutipes*, the "winter mushroom", also known as *enokitake* in Japan

- *Hypsizygus tessulatus* (also *Hypsizygus marmoreus*), called *shimeji* in Japanese, it is a common variety of mushroom available in most markets in Japan. Known as "Beech mushroom" in Europe.

- *Lentinus edodes*, also known as shiitake, oak mushroom. *Lentinus edodes* is largely produced in Japan, China and South Korea. *Lentinus edodes* accounts for 10% of world production of cultivated mushrooms. Common in Japan, China, Australia and North America.

- *Pleurotus* species are the second most important mushrooms in production in the world, accounting for 25% of total world production. *Pleurotus* mushrooms are cultivated worldwide; China is the major producer. Several species can be grown on carbonaceous matter such as straw or newspaper. In the wild they are usually found growing on wood.

  o *Pleurotus cornucopiae*

  o *Pleurotus eryngii* (king trumpet mushroom)

  o *Pleurotus ostreatus* (oyster mushroom)

Harvesting *Pleurotus ostreatus* cultivated using spawns embedded in sawdust mixture placed in plastic containers

Details of the gill structure of the edible oyster mushroom Pleurotus ostreatus.

- *Rhizopus oligosporus* - the fungal starter culture used in the production of tempeh. In tempeh the mycelia of *R. oligosporus* are consumed.

- *Sparassis crispa* - recent developments have led to this being cultivated in California. It is cultivated on large scale in Korea and Japan.

- *Tremella fuciformis* (Snow fungus), another type of jelly fungus that is commonly used in Chinese cuisine.

- *Tuber* species, (the truffle), Truffles belong to the ascomycete grouping of fungi. The truffle fruitbodies develop underground in mycorrhizal association with certain trees e.g. oak, poplar, beech, and hazel. Being difficult to find, trained pigs or dogs are often used to sniff them out for easy harvesting.

    o   *Tuber aestivum* (Summer or St. Jean truffle)

    o   *Tuber magnatum* (Piemont white truffle)

    o   *Tuber melanosporum* (Périgord truffle)

    o   *T.melanosporum* x *T.magnatum* (Khanaqa truffle)

    o   *Terfezia sp.* (Desert truffle)

- *Ustilago maydis* (Corn smut), a fungal pathogen of the maize plants. Also called the Mexican truffle, although not a true truffle.

- *Volvariella volvacea* (the "Paddy straw mushroom.") *Volvariella* mushrooms account for 16% of total production of cultivated mushrooms in the world.

- *Fusarium venenatum* - the source for mycoprotein which is used in Quorn, a meat analogue.

Production Regions in North America

Pennsylvania is the top-producing mushroom state in the United States, and celebrates September as "Mushroom Month".

The borough of Kennett Square is a historical and present leader in mushroom production. It currently leads production of Agaricus-type mushrooms, followed by California, Florida and Michigan.

Other mushroom-producing states:

- East: Connecticut, Delaware, Florida, Maryland, New York, Pennsylvania, Tennessee, and Vermont

- Central: Illinois, Oklahoma, Texas, and Wisconsin

- West: California, Colorado, Montana, Oregon, Utah and Washington

Vancouver, British Columbia, also has a significant number of producers — about 60 as of 1998 — mostly located in the lower Fraser Valley.

References

- Yusuf Chisti (1999). Robinson, Richard K., ed. Encyclopedia of Food Microbiology (PDF). London: Academic Press. pp. 663–674. ISBN 978-0-12-227070-3.

- Stanbury, Peter F.; Whiitaker, Allan; Hall, Stephen J. (1999). Principles of Fermentation Technology (Second ed.). Butterworth-Heinemann. ISBN 978-0750645010.

- Koukkou, Anna-Irini, ed. (2011). Microbial Bioremediation of Non-metals: Current Research. Caister Academic Press. ISBN 978-1-904455-83-7.

- Díaz, Eduardo, ed. (2008). Microbial Biodegradation: Genomics and Molecular Biology (1st ed.). Caister Academic Press. ISBN 978-1-904455-17-2.

- McLeod MP & Eltis LD (2008). "Genomic Insights Into the Aerobic Pathways for Degradation of Organic Pollutants". Microbial Biodegradation: Genomics and Molecular Biology. Caister Academic Press. ISBN 978-1-904455-17-2.

- Heider J & Rabus R (2008). "Genomic Insights in the Anaerobic Biodegradation of Organic Pollutants". Microbial Biodegradation: Genomics and Molecular Biology. Caister Academic Press. ISBN 978-1-904455-17-2.

- Parales RE, et al. (2008). "Bioavailability, Chemotaxis, and Transport of Organic Pollutants". Microbial Biodegradation: Genomics and Molecular Biology. Caister Academic Press. ISBN 978-1-904455-17-2.

- Martins dos Santos VA, et al. (2008). "Genomic Insights into Oil Biodegradation in Marine Systems". In Díaz E. Microbial Biodegradation: Genomics and Molecular Biology. Caister Academic Press. ISBN 978-1-904455-17-2.

- Watanabe K & Kasai Y (2008). "Emerging Technologies to Analyze Natural Attenuation and Bioremediation". Microbial Biodegradation: Genomics and Molecular Biology. Caister Academic Press. ISBN 978-1-904455-17-2.

- Meyer A & Panke S (2008). "Genomics in Metabolic Engineering and Biocatalytic Applications of the Pollutant Degradation Machinery". Microbial Biodegradation: Genomics and Molecular Biology. Caister Academic Press. ISBN 978-1-904455-17-2.

- Bennett JW (2010). "An Overview of the Genus Aspergillus" (PDF). Aspergillus: Molecular Biology and Genomics. Caister Academic Press. ISBN 978-1-904455-53-0.

- Machida, Masayuki; Gomi, Katsuya, eds. (2010). Aspergillus: Molecular Biology and Genomics. Caister Academic Press. ISBN 978-1-904455-53-0.

Permissions

Index